D1281136

Ergonomics in Sport and Physical Activity

Enhancing Performance and Improving Safety

THOMAS REILLY

Director, Research Institute for Sport and Exercise Sciences
Liverpool John Moores University

Human Kinetics

Library of Congress Cataloging-in-Publication Data

Reilly, Thomas, 1941-2009
Ergonomics in sport and physical activity : enhancing performance and improving safety / Thomas Reilly.
p. cm.
Includes bibliographical references and index.
ISBN-13: 978-0-7360-6932-8 (hard cover)
ISBN-10: 0-7360-6932-1 (hard cover)
1. Sports--Physiological aspects. 2. Human engineering. I. Title.
RC1235.R45 2010
613.7'1--dc22

2009033693

ISBN-10 (print): 0-7360-6932-1
ISBN-13 (print) : 978-0-7360-6932-8
ISBN-10 (Adobe PDF): 0-7360-8555-6
ISBN-13 (Adobe PDF): 978-0-7360-8555-6

The Web addresses cited in this text were current as of xxx, unless otherwise noted.

Acquisitions Editor: Loarn D. Robertson, PhD; **Developmental Editor:** Judy Park; **Assistant Editor:** Lee Alexander; **Copyeditor:** Julie Anderson; **Indexer:** Bobbi Swanson; **Permission Manager:** Dalene Reeder; **Graphic Designer:** Nancy Rasmus; **Graphic Artist:** Kathleen Boudreau-Fuoss; **Cover Designer:** Keith Blomberg; **Photographer (cover):** Adam Pretty/Getty Images Sport; **Photographer (Interior):** Thomas Reilly; **Photo Asset Manager:** Jason Allen; **Art Manager:** Kelly Hendren; **Associate Art Manager:** Alan L. Wilborn; **Illustrator:** Mike Meyer; **Printer:** Sheridan Books

Printed in the United States of America 10 9 8 7 6 5 4 3 2 1

The paper in this book is certified under a sustainable forestry program.

Human Kinetics
Web site: www.HumanKinetics.com

United States: Human Kinetics
P.O. Box 5076, Champaign, IL 61825-5076
800-747-4457
e-mail: humank@hkusa.com

Canada: Human Kinetics
475 Devonshire Road Unit 100,
Windsor, ON N8Y 2L5
800-465-7301 (in Canada only)
e-mail: info@hkcanada.com

Europe: Human Kinetics
107 Bradford Road, Stanningley,
Leeds LS28 6AT, United Kingdom
+44 (0) 113 255 5665
e-mail: hk@hkeurope.com

Australia: Human Kinetics
57A Price Avenue, Lower Mitcham,
South Australia 5062
08 8372 0999
e-mail: info@hkaustralia.com

New Zealand: Human Kinetics
Division of Sports Distributors NZ Ltd.
P.O. Box 300 226 Albany, North Shore City, Auckland
0064 9 448 1207
e-mail: info@humankinetics.co.nz

E4126

preface

THE idea for this book has experienced a long gestation period. During a research career that straddled occupational ergonomics and sport science, a text that bridged the gap between these domains was a project that appeared logical to me. The notion of the book remained dormant until the arrival of a proposal from Loarn Robertson at the publisher's office. My challenge was to see the proposal through into a concrete reality.

Given the overlap between the sport sciences and ergonomics, the absence hitherto of a textbook on ergonomics in sports is surprising. The reason may lie in common misunderstandings of both these areas, not only as academic topics of study but also as professional avenues of work. Indeed, specialists in other areas frequently make connections with these human sciences, including public health and physical training. The motivation in writing this book is to take the opportunity to bring together approaches common in ergonomics and sport science so that the mutual benefits for the separate areas are evident.

The purpose of this book is to outline the scope of ergonomics in the field of sport and recreational activity and describe how ergonomics is applied in these domains to solve problems in human factors. The textbook bridges a gap in the current literature and has no direct competitor or prototype to follow.

The content of the book is broadly based and ranges from principles and concepts in classical ergonomics to contemporary problems in special populations. It includes issues relating to design, assessment, and training in competitive sport and in leisure and other physical activities.

The intention is to provide the reader with research-based information on how an ergonomics approach can be used to improve physical performance and enhance safety. The physical properties of the body and the factors limiting performance are considered. The essential concepts, terms, and principles of ergonomics are presented and related to physical activity. Injury risk factors are identified in relation to body mechanics within different physical activities, and the interactions between the individual, the task, and the environment are analyzed. There is an emphasis on injury prevention and individual protection in the review of sports equipment and sports environments. Comfort, efficiency, and safety are reviewed along with systems criteria where design issues are relevant. The overall theme is the quest for improved performance, optimal efficiency, enhanced comfort, and fewer injuries in sport and exercise.

The common features of ergonomics and sport science are their central concerns with the individual and the scientific disciplines they employ. Ergonomics is not merely an altruistic endeavor to make things better for people at work, play, or leisure; it is a technology with practical approaches to solving problems that arise in these environments. There are many approaches to arriving at solutions and many theoretical models and analytical techniques to draw on. The potential for these approaches to be extended across work, recreation, and sport is emphasized throughout the book.

Ergonomics focuses on the individual in a variety of scenarios to understand his or her relationship with the task at hand, the equipment or devices used, and the environment in which this interface takes place. Novel insights are provided by the integration of the human sciences into an ergonomics approach. This book explains what ergonomics is, how ergonomists solve practical problems in the workplace, and how principles of ergonomics are applied in the context of sport and other physical activities. The classical concepts of ergonomics, or human factors, are related in the text to sport, leisure, and other physical activities.

This text explains how the practical tools of ergonomics can be applied at various levels from a broad base of systems analysis to a matching of specific tasks and individual characteristics. Examples are given of activities and events that are potentially harmful to human health and well-being. The reader is provided with an outline of methods for assessing risk in such situations and procedures for dealing with stress, eliminating hazards, and evaluating the challenges posed by particular working environments.

The content provides insight into the mindset of the competitive athlete, prepared to operate in extremis, train at the edge of human capabilities, and set new levels of performance. The role of ergonomics in human enhancement technologies is explained, as are the ways in which participants can benefit. The downside to a continuous striving for optimal performance is that overloading may occur, hindering performance or causing an injury. Predispositions to injury and acceleration of the recovery process are placed in a clinical context. Strategies for offsetting fatigue are presented, whether transient or attributable to a depletion of energy resources.

The book shows readers how ergonomics principles can be applied to solve practical problems related to human characteristics and capabilities. There is a wealth of information about the tools that can be used to increase safety and promote efficiency in performance of a variety of activities. With an improved awareness of how human capabilities are best matched to physical activities, the reader concerned with human factors should become a better professional as a result.

The book is divided into three main parts. These are preceded by an outline of the history and a description of the professional world of ergonomics. The concepts and principles of ergonomics are explained. In part I, the emphasis is on risk and safety, methods of assessment, and the adoption of preventive measures. Part II focuses on sport ergonomics, detailing the appropriate design of equipment with the user in mind and various sources of environmental stress. Ergonomics in physical activities and the needs of special populations, including a clinical context, are addressed in part III. The culmination of the text is an afterword integrating the content as a whole and condensing the recommendations into tools to be used in practical settings.

Rather than addressing a single disciplinary group or a specific set of practitioners, the book has a wide appeal. It is relevant to students and postgraduates in ergonomics and human factors educational programs as well as ergonomists in professional practice. It is especially useful at these levels in sport and exercise sciences and to students in hybrid programs such as sport technology and sport engineering. Physical trainers, physiotherapists, personal trainers, exercise leaders, sport science support workers, and sport development personnel are also likely to benefit from the subject matter. The multidisciplinary content will be attractive to biomechanists, physiologists, and behavioral scientists in special topics courses. Although the book is broad in scope, each of the topics addressed is covered in depth to provide readers with the rationale for ergonomics solutions to the questions raised.

introduction to ergonomics

DEFINITIONS

aerobic power—The highest rate of muscular work that can be produced from the combustion of oxygen within the body.

delayed-onset muscle soreness—Discomfort felt 24 to 72 hours following exercise caused by microtrauma to muscle on resisting a forced stretch.

ergometers—Device for measuring mechanical work accomplished.

ergonomics—The application of the human sciences to individuals or groups in the working environment.

fatigue—Inability to maintain the required or desired level of force or work.

manual materials handling—Lifting, lowering, pushing, pulling, or carrying physical loads without the use of assistive devices.

proteomics—The study of proteins.

repetitive strain injury—Harm due to multiple repetitions of actions, usually in the soft tissues of the limbs

simulation—Presentation of a model of the real world, task, or machine; representation in some form of the situation or artifact being assessed. Machines or environments that represent such models are known as *simulators*.

stress—An individual response syndrome denoting internal load; unwanted stimulation that calls for attention to be removed, reduced, or faced.

tendinitis—Inflammation of connective tissue known as tendons.

intro

ERGONOMICS refers to the application of the human sciences to individuals or groups in the working environment (Reilly and Lees, 2009). This environment can extend from professional occupational domains to domestic, sport, and leisure contexts. The human sciences embrace a number of disciplines that include applied anatomy, biomechanics, physiology, psychology, and social sciences. The broad field encompassed by ergonomics is also informed by other technologies and subdisciplines that range from engineering and industrial design to information science and neuroscience. Irrespective of the discipline being applied, a central theme of ergonomics is the focus on the individual and the activity involved (Reilly, 1991a).

Sport and leisure traditionally were viewed as entirely separate from occupational work in the domain of people's lives. The advent of mass participation in sport, exercise, and recreational activities over recent decades necessitated a reappraisal of human factors in these areas. Over this period also there was a growth in professional sport that is now readily accepted as a form of mass entertainment. Parallel with these developments has been an increasingly systematic approach to preparing athletes for competitive engagements and analyzing the stresses that sport places on participants. Furthermore, innovative engineering technologies have led to changes in design of sports equipment with the aim of enhancing performance.

The International Ergonomics Association at its Council meeting in August 2000 agreed on the following definition: "ergonomics (or human factors) is the scientific discipline concerned with the understanding of the interactions among humans and other elements of a system, and the profession that applies theoretical principles, data and methods to design in order to optimize human well-being and overall system performance." This broad definition applies to sport as well as to industry. An interpretation restricted to occupational work would apply only to professional sport, where talented individuals earn their livelihood by virtue of their specialized competitive competencies. Nevertheless, sport in general presents many of the issues tackled in the pioneering years of ergonomics: high levels of energy expenditure, thermoregulatory strain, precompetition emotional stress, unique postural loadings, severe information processing demands, fatigue in sustained activities, and a myriad of other problems familiar to ergonomists. It has been suggested that with the possible exception of military contexts, human limits are seldom so systematically explored and so ruthlessly exposed as in high-level sport (Reilly, 1984). Indeed, the margin between success and failure is most often less in sport than in warfare.

The human operator (the athlete) forms the centerpiece of a sport ergonomics model, the task or interface with machine or equipment being immediate connections. Then the environmental conditions can be considered, including workspace, temperature, pollution, and ambient pressure (see figure I.1). More global parameters involve travel, social aspects, and organizational factors. Relationships with coaches, mentors, and trainers may apply at this outer level, forming an aspect of team dynamics; alternatively, these relationships may have a more critical central role in the individual's well-being and motivation for performance.

It is not surprising, therefore, that ergonomists find employment and challenging problems in many areas. This richness is reflected in the academic programs for studying ergonomics. Whereas the original programs were termed *ergonomics* or *human factors*, there is now opportunity to specialize in transport ergonomics; health

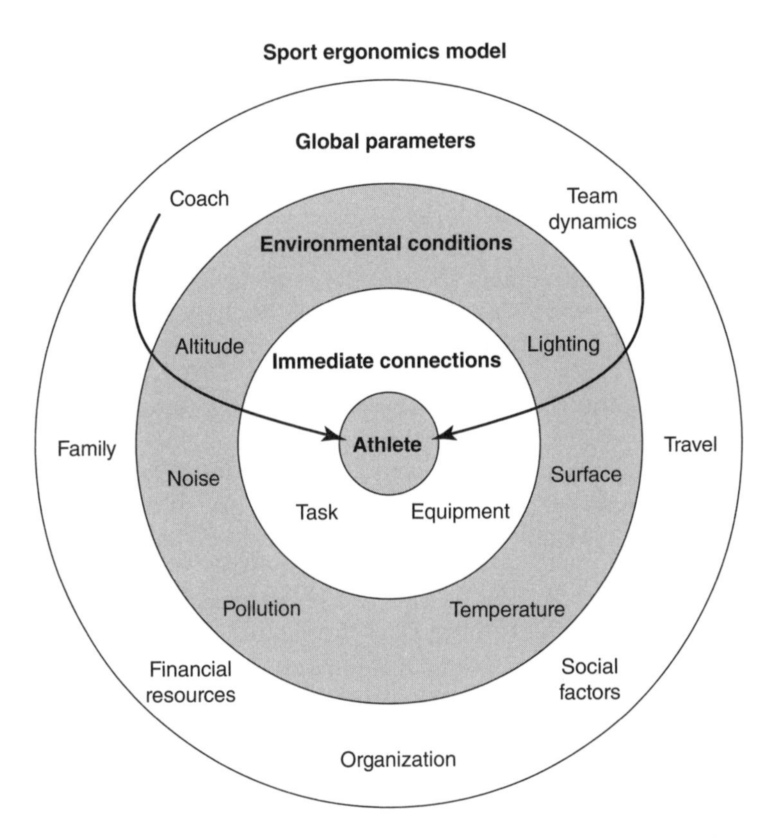

Figure I.1 The interface between the individual and the sport environment in the ergonomics framework.

care ergonomics; and ergonomics, health, and safety, among others. In many cases ergonomics modules constitute a major component of academic programs in safety engineering, occupational health, and related areas. Ergonomics principles and material are incorporated into academic courses, research training, and applied projects within the sport and exercise sciences. Tracing the subject area back to its development and the roles of its professional bodies provides insight into the nature and scope of ergonomics.

Historical Development

Ergonomics emerged as a technology from the realization during the Second World War that performances of workers in munitions factories were variable, affected by environmental conditions, workplace design, hours of work, and the subjective state of the individual. Working longer hours became counterproductive rather than helpful as productivity declined and accidents increased. No single scientific discipline was capable of explaining the increases in errors, accidents, and fluctuations in output that were occurring. Similar problems related to the war effort were noted away from the factories. Sailors on watch duty were noticed to miss radar signals during long periods on vigilance tasks, despite being focused on their display screens. Mental fatigue posed a similar enigma as did physical fatigue during hard and long periods of work. It was evident that these problems called for an interdisciplinary approach, because no single science could provide a satisfactory answer.

Following the wartime experiences in Europe and its allied nations, the Ergonomics Research Society was formed in 1949. Its title was later changed to the Ergonomics Society as the applied focus became more clearly established. This title became the Institute of Ergonomics and Human Factors in 2009 after members approved a further name change. The parallel organization in North America is the Human Factors Society. These two bodies, together with a host of national and regional professional societies, are affiliated with the International Ergonomics Association.

Both the Ergonomics Society and the Human Factors Society hold an annual conference as well as regular events and workshops during each year. Their agendas include public engagement to inform others about the nature of ergonomics and the contributions it can make to the well-being of workers. The International Ergonomics Association holds a triennial congress, the 17th such event being scheduled for Beijing, China, during 2009. Since 1987, the Ergonomics Society has supported the International Conference on Sport, Leisure and Ergonomics. Its sixth event was held during 2007 in Burton Manor College, Cheshire, United Kingdom, cosponsored by the International Society for Advancement of Kinanthropometry and the World Commission for Science and Sports.

Publications

A range of scientific periodicals relating to ergonomics are published. The journal *Ergonomics* is the official publication of the Institute of Ergonomics and Human Factors and is endorsed by the International Ergonomics Association. The imprint on its current cover indicates that it is "an international journal of research and practice in human factors and ergonomics." It reached its 50th year of publication in 2007 and is the primary outlet for work in the field.

Applied Ergonomics is expressly aimed at ergonomists and all those interested in applying ergonomics or human factors in the design, planning, and management of technical and social systems at work or leisure. Among the professionals for whom the journal is deemed by the publishers to be of interest are ergonomists, designers, industrial engineers, health and safety specialists, system engineers, design engineers, organizational psychologists, occupational health specialists, and specialists in human–computer interaction. Areas of application include office, industry, consumer products, information technology, and military design. This journal—espoused as the journal of people's relationships with equipment, environment, and work systems—first appeared in 1969 and is published by Elsevier Ltd in cooperation with the Institute of Ergonomics and Human Factors.

The counterpart of these journals in the United States is *Human Factors,* the official publication of the Human Factors Society. It was first published in 1958 and is the primary outlet for ergonomics and human factors research in North America. Complementary journals commenced publication later (see table I.1) such as the *Journal of Industrial Ergonomics, Human Computer Interaction Journal, Journal of Human Stress,* and *Theoretical Issues in Ergonomics Science* (first issue 2000). The *Journal of Human Ergology* is published in Japan, having first appeared in 1971, and the *Inter-*

Table I.1 Major Ergonomics Journals and Dates of First Publication

Journal	Year
Ergonomics	1957
International Journal of Man-Machine Studies	1957
Human Factors	1958
Safety Science	1967
Applied Ergonomics	1970
Journal of Human Ergology	1971
Journal of Human Stress	1975
Scandinavian Journal of Work, Environment & Health	1975
Human Movement Science	1982
Journal of Sports Sciences	1983
Journal of Applied Biomechanics	1985
International Journal of Human-Computer Interaction	1986
Work and Stress	1987
Human Performance	1988
Ergonomics in Design	1993
International Journal of Cognitive Ergonomics	1997
Transportation Human Factors	1999
Cognition, Technology, and Work	1999

national Journal of Man-Machine Studies is older still, having first appeared in 1957. Their content is testimony to the enormous variety of projects in which ergonomists are engaged across many countries.

Many conference proceedings contribute to the ergonomics literature. These may appear as special issues of the mainstream journals or as international conference proceedings published as books. The *Proceedings of the Annual Conference of the Ergonomics Society* have been published in book form, commencing in 1983 (Coombes, 1983), and a similar arrangement applies to the Human Factors Society and the International Ergonomics Association. By the time of the XVth Triennial Congress of the International Ergonomics Association (incorporating the 7th Joint Conference of the Ergonomics Society of Korea and the Japan Ergonomics Society), the congress proceedings ran to four volumes, each with approximately 700 pages.

The field of ergonomics has a rich literature that includes a range of classical textbooks (e.g., Grandjean, 1969; McCormick, 1976; Pheasant, 1991) and academic manuals (e.g., Pheasant, 1986). Despite the large number of texts devoted to mainstream ergonomics, this literature has been insufficient for the more specialized areas of application. Topics that have required their own dedicated texts include transport ergonomics, systems engineering, **manual materials handling,** and hospital-based ergonomics. The recent literature on sport ergonomics includes three special issues of *Applied Ergonomics;* five special issues of *Ergonomics* devoted to the quadrennial International Conference on Sport, Leisure and Ergonomics; and three books forming the proceedings of two of these events (Atkinson and Reilly, 1995; Reilly and Atkinson, 2009; Reilly and Greeves, 2002). The purpose of this book is to outline the role and scope of ergonomics in sport, leisure, and recreational activity and describe how ergonomics is applied in these domains to solve problems in human factors.

Scope of Ergonomics

The scope of ergonomics is evident from the publication of special issues devoted to sport ergonomics, first in the journal *Human Factors* in 1976 and later in the journal *Applied Ergonomics* (Reilly, 1984; Reilly, 1991b). The topics included, for example, novel techniques for measurement of motion (Atha, 1984), the emerging uses of computers in sport (Lees, 1985), the applications of hydrodynamics and electromyography to water-based sports (Clarys, 1985), and controlling system uncertainty in sport and work (Davids et al., 1991). A breakdown of the material published in the proceedings of the first five International Conferences on Sport, Leisure and Ergonomics shows the main areas of application of ergonomics to sport (table I.2). The material reviewed has been published in texts (Atkinson and Reilly, 1995; Reilly and Greeves, 2002) or in special issues of the journal *Ergonomics.* The areas of application range from health-related exercise to combinations of environmental conditions that pose challenges for elite performers.

The scope of sport science and its overlap with the field of ergonomics were outlined some years ago (Reilly, 1984). The overlap in methods includes physiological and psychological techniques for monitoring work stress, motion analysis, computer modeling, and simulations, whereas areas of mutual concerns embrace equipment design, footwear and clothing, and systems performance. In both domains, particular strengths are the insights and flexibility offered by an interdisciplinary approach to problem solving.

Primary Concepts

The classical approach in ergonomics is to match task demands and the capacity of the individual. The main focus is on the human (i.e., the individual), and the principle is that the task and associated equipment or machinery should be designed around human capabilities. A recurrent theme is that of limited human capacity, implying a ceiling to current functional ability. This ceiling sets an upper limit to both quantitative and qualitative aspects of performance. The expectation is that the individual will be unsuccessful, or even will incur injury, if task demands outstrip the person's capacity to meet or tolerate those demands. Assessing these demands may entail an interdisciplinary perspective: The National Institute for Occupational Safety and Health (NIOSH) lift equation has been an influential ergonomics force in the United States, integrating physiological, biomechanical, and psychophysical variables along with population demographics into a predictive model (NIOSH, 1977). Major contribu-

Table I.2 Subject Areas Contained Within the Five Published Proceedings of the International Conference on Sport, Leisure and Ergonomics

Subject area	Reports published
Aging	3
Body composition and functional anatomy	10
Circadian factors	6
Computers in sport	3
Corporate health and fitness	17
Disability	13
Environmental stress	10
Ergogenic aids and drugs	6
Fitness assessment	5
Injury	2
Measurement methods	14
Musculoskeletal loading	12
Pediatric ergonomics	4
Psychological stress	2
Sports coaching	4
Technique analysis	19
Training responses	7
Equipment design	
Clothing	2
Machines	5
Protective devices	4
Shoes and orthotics	5

tions of original work in this area have come from the efforts of Chaffin (1975) and of Snook's group (Snook et al., 1970; Snook and Ciriello, 1974).

Physiological Capacity

Extensive data are available on typical maximal values of such functional aspects as aerobic power and physical working capacity, anaerobic power and muscular strength, working memory, information processing, and stress tolerance. In recent years attention has been paid to the capacities of female athletes and of young and veteran sport participants. These sources provide a background for scientifically guiding participants toward realistic sporting aspirations and selecting capable playing personnel.

Because the task of achieving victory in top-flight sport is irrevocably set by the physical, physiological, and mental demands of the activity, the possibility of success at the elite level is limited to a select group. An abundance of proven paradigms are available that allow coaches to use physiological variables in selecting individuals and teams for success in international sport. Such precision is not limited to individual sports such as running, cycling, and swimming, where competitive performance is easily defined. Assessment of the demands of rowing, for example, at Olympic performance level permits sport science personnel to set down the values for maximal aerobic level and anaerobic power required of members of the team to achieve this standard. This type of calculation is relatively straightforward for competent sport scientists and forms the basis for what has become known as sport science support. Use of rowing **ergometers,** swimming flumes, and **simulators** for canoeing and Nordic skiing has considerably facilitated functional testing and made predictions of field performance from laboratory measures more accurate and reliable. The early investigations of Costill (1972) served as a model for identification of top marathon runners on physiological criteria. Since that time, laboratory assessments have become more refined and sport specific. Using simple dynamics, Keller (1976) showed that the physiological attributes of runners can be correlated with track records and the optimal race strategy determined. More recently, Atkinson and colleagues (2003) developed a comprehensive model for accommodating the factors that determine performance in cycling. The predictive model included not only anthropometric and physiological variables but also posture, bicycle characteristics, pacing strategy, biomechanics, distance, terrain, and environmental features. The model as further defined for cycling time trials is shown in figure I.2 (Davison et al., 2008).

Psychological Factors

The contribution of psychological factors to performance was evidenced by Morgan and Pollock's (1977) finding that elite performers cope with high metabolic demands by associating with the accompanying physical discomfort to negotiate pain zones. Dissociation, on the other hand, was found to carry the attendant risk of tissue, organ, and systems trauma. Other authorities claim to have identified personality dimensions essential to competitive success, although attitude and commitment to training may be more influential. Integration of behavioral, physiological, and subjective methods has helped to quantify the emotional load imposed on performers immediately prior to the event and to develop methods of coping following the event. Although a scientific approach can help to single out individuals ill-equipped to cope with competition stress, athletes tend to gravitate toward the event and the competitive standard they are psychologically suited to and for which they have the appropriate physical attributes. Unique individual traits can, however, be used to advantage in team sports by astute allocation of tactical roles, thereby molding the task to suit the individual.

The notion of a limited capacity is not as easily accepted in sports as in an occupational environment. The mindset of competitive athletes is to extend the boundaries of performance, challenge existing records, and train at extremes. By adopting optimal training programs, athletes overcome their limitations and improve their performance capabilities. Elite performers display a persistence and mental toughness that help them overcome barriers to success and learn lessons when defeated that turn them into winners. Even so, there is a limit to which physiological characteristics can be enhanced by training, and training effects are constrained by genetic factors.

Those with the potential to become world-class athletes form a very small part of the population. For example, the ability to produce a large power output from aerobic energy sources is important for endurance exercise, and the best overall physiological measure of **aerobic power** is generally taken to be maximal oxygen uptake ($\dot{V}O_2max$). This function indicates the maximal rate at which oxygen can be consumed in strenuous exercise. It is determined by monitoring physiological responses to a progressive exercise protocol continued to voluntary exhaustion (see figure I.3). The variance in this physiological characteristic between individuals is considerably greater than that

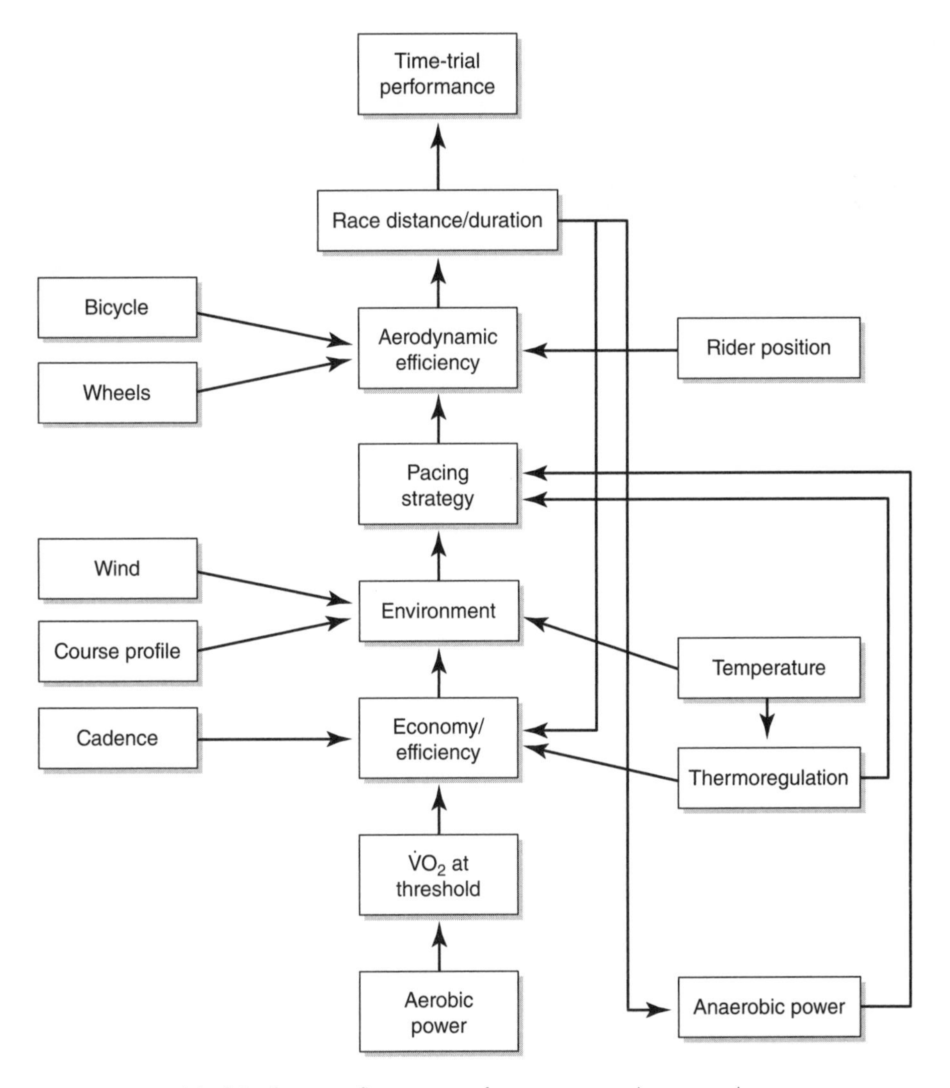

Figure I.2. Model of the factors influencing performance in cycle time trials.
From Davison and colleagues 2008.

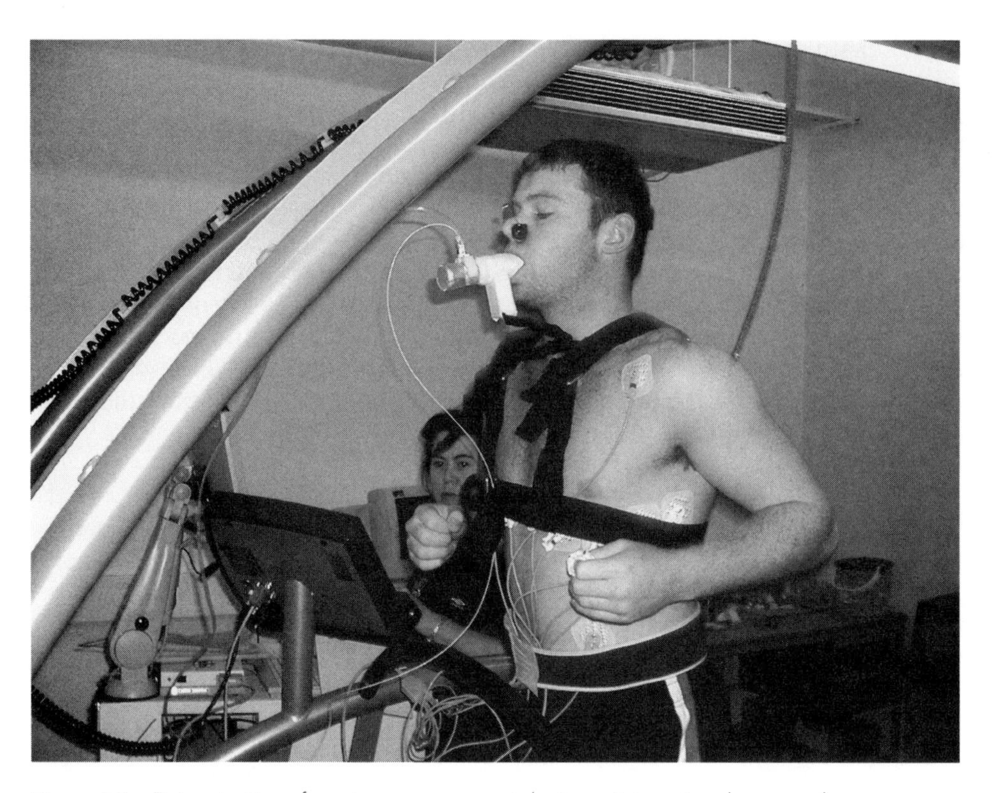

Figure I.3 Determination of maximum oxygen uptake in an international soccer player.
Reprinted, by permission, from T. Reilly, 2007, *The Science of Training: Soccer* (London: Routledge), 155 © Tom Reilly.

associated with the effect of different training regimens, a fact that underlines the importance of heredity. For this reason, casting the net wide to recruit individuals with the potential and high aerobic power would seem more productive than attempting to train individuals of low aerobic power, however highly motivated they are. This observation raises fundamental questions about the essential nature of sport and the value of its intrinsic satisfaction to the individual participant irrespective of achievement level. These questions, in turn, indicate that a national philosophy of sport must be developed prior to any systemization of sport science services. Establishment of national centers such as the Australian Institute of Sport in Canberra, the Japan Institute of Sports Science in Tokyo, the United States Olympic Committee center in Colorado Springs, and the Swedish center in Boson near Stockholm was the outcome of political decisions to support elite athletes as part of a national policy.

Limited Capacity

The predominance of endowment over environment in determining $\dot{V}O_2max$ was first clearly demonstrated by Klissouras (1971) following investigations of intrapair differences in identical and nonidentical twins: The variance in maximal aerobic power accounted for by heredity was 93%. Although this figure has been revised downward on the basis of molecular studies, the influence of genetics on $\dot{V}O_2max$ is still regarded to be at least 50%. Such figures might be misleading when considering top athletes, because nature and nurture are intricately intertwined so that an organic attribute cannot develop without a genetic basis and an appropriate environment. With many biophysical attributes, heredity sets limits on the magnitude of change in the expression of genetic potential even in the presence of optimal training stimuli (Klissouras, 1976). More recently molecular biologists have explored the human genome in search of a "performance gene," but a single candidate gene has proven elusive.

The development of individual capacity by physical training embodies a philosophy of fitting the man or woman for the job at hand. The degree of possible improvement in capacity varies with the individual, his or her initial fitness status, and the biological system being trained. However, the maximal possible predicted change in capacity will always be tempered by the genetic contribution. Besides elevating the ceiling of functional ability, training may contribute to a greater fractional utilization of capacity, allowing the athlete to exercise at a higher percent of the maximal value, throughout any sport contest. Top marathon runners, for example, are not necessarily those with the highest $\dot{V}O_2$max but rather those who can work close to a high maximum capacity for the complete race. They also can tolerate high internal body temperatures, and their muscles use the oxygen delivered to the muscle cells more effectively and remove metabolites associated with energy produced from anaerobic sources.

An improvement of 20% to 25% in $\dot{V}O_2$max has traditionally been regarded by physiologists as a good training effect. However, improvements approaching 40% have been found when intensive training programs are conducted for longer than the customary experimental periods. The $\dot{V}O_2$max presents an overall picture of the functional integration of lungs, heart, blood, and active muscles in aerobic work. Noninvasive cardiac assessment, the muscle biopsy technique, **proteomics,** and more sophisticated biochemical analyses and functional imaging techniques have broadened our understanding of the relative roles of central and peripheral factors in the chronic responses to training that are evidenced by changes in $\dot{V}O_2$max. We also know that there are important adaptations that increase the relative exercise intensity that can be sustained for a prolonged period.

Whereas the $\dot{V}O_2$max and the proportion of $\dot{V}O_2$max that can be used in continuous exercise are good indexes of capability for endurance sports, many activities place priority on muscle strength rather than stamina. Improvements in strength are first attributed to neuromuscular factors and subsequently to muscle hypertrophy. Adaptation to high-intensity exercise entails functional changes in the skeletal muscles and other tissues that have been engaged in exercise. At the molecular level, the exercise stimulus switches on signal transduction processes that activate responses within the muscle fibers. Genes carry the genetic information encoded in DNA to build proteins, and messenger RNA (mRNA) levels for several metabolic genes are acutely elevated after a single bout of exercise (Hawley et al., 2006). Alterations in muscle ultrastructure occur while muscles are recovering from the exercise session that induced skeletal muscle overload. Testosterone and insulin-like growth factor-1 play important roles in both recovering from these sessions and promoting protein synthesis. These anabolic processes gradually cause the muscle cross-sectional area to increase, which explains the accompanying improvement in capability to generate force.

The search for optimum training programs based on sound scientific principles has attracted numerous scientific investigators over the years. The main theme has been the attempt to identify training stimulus thresholds and relate them to the intensity, frequency, and duration of training and to fitness status. Habitually sedentary individuals undertaking a basic recreational program may be satisfied with attaining this threshold to acquire a modicum of fitness, whereas highly driven elite athletes seek to optimize their training stimulus. Evaluation of physical training programs is often based on the degree to which assumed training stimulus thresholds are exceeded or indicators of fitness are improved over time. Monitoring fitness throughout preseason and competitive periods is now advocated, because specific fitness measures tend to fluctuate seasonally. Regular assessment helps trainers evaluate the athlete's preparation for competition and can assist in team selection as well as indicate specific individual weaknesses that require training.

At the other end of the performance spectrum is the recreation participant attracted to physical training to improve health. Exercise is advocated as prevention for a host of diseases and as therapy for stress. Among the diseases are metabolic syndrome, cardiovascular events, and some forms of cancer. Whereas the competitive athlete seeks the optimal dose-response associated with strenuous training, the recreational participant is primarily interested in exercise as a vehicle for maintaining health and enhancing well-being.

Empirically devised training programs tend to precede those that have been scientifically validated. The scientific principles underlying strength, speed, stamina, and flexibility training have slowly evolved to support the rich fund of coaching acumen generated by insightful experiences. Experimental work has, for example, helped to explain why, at the cellular level, many repetitions at low intensity promote endurance whereas short-duration, high-intensity efforts promote power and in certain conditions muscle hypertrophy. Different elements may interact in complex fashions in a combined training program, such as when speed and endurance or intermittent and continuous work regimens are used. Each type of training has its own specific hazards as well as effects, the shortcut to fitness provided by intermittent work regimens being offset by the increased risk of soft-tissue injury. It is known also that aerobic training alters muscle glycolytic enzymes and that some intermediate fiber types between the slow-twitch and fast-twitch varieties are affected by the quality of training. All possible interactions suggested by the preceding discussion may be too complex for researchers to represent adequately in any single experimental model. Competitive performance in real life continually provides feedback to modify ongoing training; recognizing and reacting to these subtle adjustments are as much an art as science. The scientist, coach, and athlete should operate as a team, with the scientist providing guidance and physiological monitoring of athletes to ensure that blind alleys in advancing the training regimen are avoided and implications of experimental research findings are understood by the coach and athlete. The foreseeable future is likely to produce more systems modeling of the effects of training on physical performance, such as applied by Impellizzeri and colleagues (2006), and further transfers of knowledge from theory to practice.

Ergonomic Criteria

A few principles guide the ergonomist in the conduct of practical work. These may be prioritized or combined to establish a framework for making decisions about problems that arise in a practical setting. The ergonomist may, for example, be prepared to accept a transient subjective discomfort or periodic inefficiency provided safety is not compromised. The guide may be in the form of consequences for a key criterion or dependent variable.

The emphasis in ergonomics is placed on the human operator, irrespective of whether the individual acts within a complex system or a relatively simple one. The focus is on the immediate interface with the task but may shift gradually to the interface with any equipment, tools, or machines used in the task. The layout of the entire workstation, environmental aspects, and broader features can also be relevant. These factors can include sociodynamics, group interactions, and, in a sports context, relationships between team members and between participants and coaching staff, as illustrated in figure I.1.

Safety

A tenet of ergonomics is that the work environment will not harm the individual. Hence, safety is an overriding criterion. This aim may have been based initially on humanitarian values but is now subject to legislative penalties. Regulations for health and safety at work are taken seriously by management. In some institutions, a culture of safety has been promoted by participatory ergonomics, empowering workers to have more involvement in working practices that prevent accidents. The injured or absentee worker is unable to contribute to the productivity of the organization. In a similar manner, the injured athlete cannot contribute to team performance and loses fitness when not able to train. Injury prevention in both cases will be more effective than a reliance on treatment and rehabilitation.

Elite sport involves a constant drive to gain a competitive edge over opponents to secure victory in major competitions. At lower levels of participation, the objectives may be participation for personal enjoyment, social reasons, or health-related benefits. Regardless of the aspirations of participants, safety is a fundamental criterion. Risk of injury varies among sports, as it does in occupational contexts, but in both domains the use of preventive strategies is a feature of ergonomics.

For occupational contexts, the NIOSH (1977) lift equation has been a very important ergonomic force in the United States and elsewhere. This model integrates physiological, biomechanical, and psychophysical measures along with population demographics. The object is to predict at what stage musculoskeletal forces are at such a magnitude to endanger the operator, whether male (Snook et al., 1970) or female (Snook and Ciriello, 1974).

Fatigue

Fatigue is an elusive concept, although the term has been used in the scientific nomenclature for some time. The concept likely originated in occupational and military settings rather than in sports, although it transferred readily to the latter. In munitions factories, for example, it became evident during World War II that performance of workers was not invariant but deteriorated with time on the task, directly attributable to attempts to continue or maintain performance. The work of Sir Frederick Bartlett at the Psychological Laboratory in Cambridge helped to pioneer human factors research into the problems of fatigue (Bartlett, 1943). The decrement in performance, despite rewards and strong motivation, was to become one of the early markers of fatigue, and its study was reflected in the establishment of the Ergonomics Research Society in 1949. The first major symposium on fatigue followed soon after and is recorded in the classical proceedings of Floyd and Welford (1953).

In the United States, the phenomenon of fatigue was formally recognized much earlier with the establishment of the Harvard Fatigue Laboratory at Harvard Business School in 1927. Many of the century's outstanding exercise physiologists were associated with activities at this laboratory. Its research programs for the next 3 decades or so provided the cornerstone for developments in contemporary exercise physiology and for establishing its respectability as a scientific discipline. This laboratory's achievements have been described in detail by Horvath and Horvath (1973) and Buskirk and Tipton (1997).

It was relatively recently that fatigue, defined as a decrease in force production, took on a specific meaning in research related to human performance. The physiological concomitants and potential causes of fatigue are legion in competitive sports, largely because of the heterogeneity in physiological demands of sports contests. Fatigue may be manifest as an early marker in a continuum of phenomena that ultimately lead to exhaustion. Fatigue may occur transiently or repeatedly during high-intensity intermittent exercise. It is known that fatigue involves central factors whose failure impairs function before peripheral factors begin to fail. This richness of context stimulates research into the causes and mechanisms of fatigue and means of alleviating or countering it.

Stress

Stress has many different meanings but is usually associated with inducing adverse reactions in individuals experiencing it. The causes of stress are varied and range from specific environmental factors to more abstract social phenomena such as harassment from managers or mentors. In engineering terms, stress is an agent inducing unwelcome physiological or psychological responses in individuals. These reactions are expressed as strain and may entail behavioral and physiological measures. Measures that can be used to indicate psychological or emotional strain in the absence of high physiological loads are listed in the highlight box on this page.

Emotional Strain Measures in Human Factors Research

Behavioral

- Subjective scales
- Behavior observation
- Hand tremor
- Errors
- Nonverbal communication

Physiological

- Muscle tone
- Skin conductance
- Blood content
- Urine content
- Blood pressure
- Heart rate
- Electroencephalography

It is increasingly recognized that the physical environment in which a person works, competes, or relaxes is a possible source of stressors such as heat, cold, pressure, noise, vibration, and pollution. Although considerable information has been gathered on the effects of discrete environmental stressors, there is much to be learned about their interactions or synergistic effects. Similarly, there is incomplete understanding of how or why tensions arise among team members when placed under stresses of extreme environments.

Efficiency

The concept of efficiency implies a relationship between the output achieved and the energy expended in doing so. In engineering terms, mechanical efficiency is represented by the work output expressed as a percentage of energy expended. This model applies equally to the operations of motorized vehicles and of human operators. In the latter case, the human is relatively inefficient because about 80% of the energy expended in physical activity is lost as heat.

In exercise and occupational work, it may not be easy to determine mechanical efficiency. It is a straightforward task to measure the energy expended and the power output on a cycle ergometer: Measurements required include $\dot{V}O_2$, carbon dioxide ($\dot{V}CO_2$), ventilation rate ($\dot{V}E$), pedal rate, and load or resistance. Calculating the mechanical work done is more difficult in running and swimming, although $\dot{V}O_2$ can be measured for these activities. In these instances, the oxygen cost of exercising at a fixed workload can be used as an index of the economy of motion: The lower the oxygen cost, the more economical the action. On this basis, the concept of running economy tends to be preferred over mechanical efficiency when runners are evaluated.

A reduction in energy expended for a given power production (or work rate) should constitute a decreased physiological strain. A similar design applies to mental exertion, so redesign of tasks to reduce cognitive loading should decrease mental strain . Although this principle applies readily to occupational contexts, it does not necessarily apply to competitive sport. The athlete can accept a reduction in efficiency and an increased energy cost provided the overall energy or power output is commensurate with victory.

Comfort

Comfort is a subjective state in which the individual accepts the severity of the load imposed and the environment in which it is experienced. The concept may be applied to posture, temperature, work surroundings, and both task-specific and general features of the work environment. Comfort is a condition in which a person feels neutral with respect to the external environment.

The comfort zone lies within a relatively narrow range of responses to external stimuli that the individual can accommodate physiologically. Discomfort arises with excursions outside of this range and suggests that performance may be impaired. Prolonged discomfort may reflect poor posture or undue environmental stress, and it carries the potential for adverse reactions if the discomfort is sustained. For these reasons, comfort is an important criterion in occupational environments.

Some element of discomfort is implicit in exercise training and sport participation. Athletes must operate outside of their comfort zones to allow a training overload that improves physical fitness and promotes physiological adaptations. The competitive nature of sport also implies the imposition of discomfort on the opponent to gain a winning edge.

The concept of comfort also applies to the fit between the athlete and any equipment, apparatus, or apparel used. These include athletic shoes and clothing; rackets, skis, or other implements; and machines such as cycles, rowing shells, boats, and bobsleds. The notion of "fit" applies also to the match between individual, shoe, and surface, where *surface* can refer to synthetic pitches for game play or to conditions for outdoor winter sports.

Ergonomics Applications

Much of the early emphasis in ergonomics was placed on measuring the physical load on the worker and formulating a means of alleviating it. Jobs and equipment were redesigned according to criteria for safety, efficiency, comfort, and fit. The application of such regenerative ergonomics helped to improve the operator's well-being, and creative design helped engineers to enhance the workplace with respect to its quality and safety.

Work Applications

At the other end of the spectrum of energy expenditure, the attention of ergonomists was directed to low-level activities that required repetitive work cycles or tasks involving vigilance. In these activities, errors were sometimes attributable to a low level of arousal, and thus providing a more stimulating environment and restructuring the tasks were called for. Methods of task analysis evolved for examining these work-related issues, and it was found that advanced techniques were needed when dealing with complex organizations. A systems approach was developed to allocate tasks to humans and to machinery in work processes that became more automated with modernization.

The evolution of computer technology introduced a host of new ways in which humans and machines would interact in controlling work processes and the working environment. There was a gradual shift in emphasis to studying the interface between human and computer, an area informed by the growth of cognitive ergonomics. This trend reflected the change from heavy manual activities that were highly stressful physically—agriculture, building and construction, forestry, mining, fishing—to relatively sedentary office-based activities. Novel human-factors methods were generated to address decision making, error identification, situation awareness, and teamwork assessment in these contexts, some of the approaches having potential also for use in sport and leisure.

In office work, novel problems arose with repetitive activities that were sustained with use of small-muscle groups at low levels of energy expenditure. **Repetitive strain injury** (RSI) became a recognized clinical entity that was work related, despite the absence of high physical load. The design of chairs, displays, and controls for computerized activities has continued to occupy mainstream ergonomists, who work on establishment of optimal viewing angles, avoidance of glare, use of compatible and appropriate seating, and design of user-friendly products.

Despite the general shift in ergonomics toward computer-based work, safety legislation has highlighted the incidence and causes of musculoskeletal disorders incurred by workers. Attention initially was directed toward periodic manual materials handling, lifting, and carrying loads. Occupations that entail lifting of heavy loads include delivery jobs, health care professions, and furniture removals, to name but a few. The emphasis changed to adopting correct techniques for lifting and carrying, using hoists or similar assistive devices, reducing the load according to national guidelines, and allowing adequate recovery between bouts of activity.

Exercise and Sport Applications

The developments in occupational ergonomics are mirrored in exercise and sport contexts. The intensity of exercise can be monitored so that the athlete is not unduly overloaded. The stimulus must be varied regularly to avoid habituation and boredom. Safety is a priority if training is to be effective, and injuries are often caused by use of faulty techniques. For example, a majority of weight training injuries occur when athletes handle loads that are too heavy or adopt poor lifting techniques.

It may not always be practical to study ergonomics interventions in a real-life situation, and simulation of an activity may be appropriate. Such simulations vary in sophistication from simple prototype designs or anthropometric dummies for testing equipment or clothing to complex dynamic setups such as virtual reality scenarios (see figure I.4). The former may be appropriate where design principles relate

Figure I.4. A virtual reality scenario using a computer-controlled platform and a visual display. The participant negotiates the passage of the track by using body movements.

to sizes or percentile bandwidths. The latter are suitable where there is undue risk in conducting experiments in the real-life event and where immediate feedback on posture and balance can be presented visually (Lees et al., 2007). In such instances it may be possible also to apply tests to robots with inbuilt mechanical characteristics close to those of humans. **Simulations** help to reduce the cost and risk associated with putative ergonomics interventions, but the question remains about how well findings apply in the real world.

Postural Stress

Identification of postural stress is a continuing task for ergonomists. Posture refers to the orientation of the various segments or anatomical links associated with an activity. Parts of the body may be static during activity, forming a platform for dynamic movements in other segments. The posture adopted may induce strain in the body's soft tissues or sensory organs and, if uncorrected, can cause harm. Various methods are available to quantify and isolate postural stress and recommend appropriate remedial action

OWAS Method

The OWAS method (Oslo Working Postural Analyzing System) of Karhu and colleagues (1981) was one of the early methods of analyzing working postures. Its theoretical basis lies in observation analysis and risk assessment rather than physiological principles. The assumption is that by observing postures at certain intervals, originally 30 s intervals, the scientist can form a reliable description of the postural stress of the job.

The OWAS method can be used to determine the most common postures for the back, arms, and legs and to estimate the load handled by each. Each major body part is assessed for its displacement, and a number is assigned according to the increase in distance of that body segment from a defined neutral posture. The values for the various body parts are combined on reference to a set of tables. The result is a score indicating the severity of the posture and the degree of urgency to effect a change.

The OWAS method has gained wide acceptance in occupational ergonomics because of its simplicity and low cost. It is a survey technique that can be executed easily; it can be implemented by means of video analysis, which may cause delay in arriving at solutions but increase reliability. Because OWAS allows the user to identify remedial measures, it has value among physical therapists when treating athletic injuries associated with postural stress.

The coding and analyzing procedures of OWAS were originally conducted using manual recording methods. The computer system designed for its application includes an element for postural coding in the field, a second element for transfer of data to PC, and a third element for analysis and presentation of data. An action category is associated with the aggregate score (see table I.3).

Rapid Upper-Limb Assessment (RULA)

Many work-related musculoskeletal disorders involve the upper limbs and are defined as alterations of the musculotendon unit, peripheral nerves, and the vascular system. These disorders can be triggered or exacerbated by repetitive movements, physical

Table I.3 Categories for the OWAS Method of Evaluating Working Postures

Action category	Action needed
I: No particular harmful effect on the musculoskeletal system	No action is needed to change work postures.
II: Some harmful effect on the musculoskeletal system	No immediate action is needed; changes to be made during future planning.
III: Distinctly harmful effect on the musculoskeletal system	Working methods should be changed as soon as possible.
IV: Extremely harmful effect on the musculoskeletal system	Immediate changes should be made to working postures.

Modified from Mattila 2001.

strain, or postural discomfort and can affect neck, shoulder, elbow, wrist, and finger joints. Examples include (1) tension neck syndrome, thought to be caused by static postures and isometric loads; (2) shoulder **tendinitis,** linked to work or exercise with a highly repetitive rhythm (e.g., swimming); (3) lateral epicondylitis, caused by exposure to a combination of force and posture (as in tennis and squash play); (4) carpal tunnel syndrome, in which awkward postures are implicated in the compression of the median l nerve; and (5) hand–wrist tendinitis, for example, De Quervain's tendinitis, caused by entrapment of the tendons of extensor pollicis brevis and abductor pollicis longus.

McAtamney and Corlett (1993) designed a system for rapid upper-limb assessment (RULA), which has gained wide recognition within occupational ergonomics. This system provides an easily calculated rating of musculoskeletal loads in tasks with an inherent risk of high loading of the neck and upper limbs. The procedure leads to a single score as a snapshot of the task involved, computed from a rating of the posture, force, and movement. The risk is calculated according to a scale from 1 to 7, where 1 is low risk and 7 is high risk. The scores are arranged into four action levels, including the time during which initiation of risk control is expected. This approach has potential for use in active recreational contexts.

Originally a paper-based procedure, RULA is now applied with its own detailed software. Validity and reliability have been established in both industrial and office-based locations with little application as yet to sport contexts. It may be necessary to assess a variety of working postures during a work cycle to provide a complete picture of overall musculoskeletal loading. In some cases video analysis is helpful. When tasks involving manual materials handling entail whole-body movements or engage the back and legs as well as upper limbs and neck in musculoskeletal risk, alternative methods should be adopted.

Postural Discomfort

A discomfort diagram was described by Corlett and Bishop (1976) whereby individuals could rate the severity of discomfort felt over different body parts. The diagrams can be used to determine the onset of postural discomfort over a working period and highlight sources of the more serious manual loadings on the body.

Although designed for occupational settings, this method may be applied to many sport contexts. The discomfort diagram could be used in design evaluations in human–machine sports, after long-duration training activities, and in whole-body activities likely to induce **delayed-onset muscle soreness.**

Professional Organizations and Accreditation Systems

Ergonomics is an interdisciplinary subject and so it is not surprising that ergonomics specialists may enter the profession from different backgrounds. Engineers, physiologists, psychologists, physiotherapists, and sport scientists may operate in this field after securing an ergonomics qualification. This award is more frequently at the level of master's in science or MSc. The academic content of the training varies among institutions depending on the local flavor and expertise of teaching personnel. Undergraduate programs also provide a route into the profession, the most established being the BSc (ergonomics) at the University of Loughborough in the United Kingdom.

Programs must be approved by the Institute of Ergonomics and Human Factors for graduates to become members of that body. A list of currently approved programs at both BSc and MSc levels is available from the Institute's office. Some programs offer specialization in the area of application, for example, transport ergonomics or hospital ergonomics. Alternatively, programs with substantial ergonomics material in the syllabus focus on safety, enabling graduates to gain employment in industry as safety specialists. This route is emphasized in the master's programs in Ireland, notably at the University of Limerick and the National University of Ireland, Galway. In the latter case there is a firm link between the study of safety, ergonomics, and occupational health. Programs from the United States and Canada have met the requirements of the Institute of Ergonomics and Human Factors, formed from an engineering base (e.g., Cincinnati University). The syllabus satisfying the ergonomics imprimatur from the European community is more specific, allowing limited scope for local niches.

A formal academic qualification, or appropriate practical experience, allows individuals access to membership in the professional community of ergonomists. There is the further option of gaining recognition as an ergonomics practitioner. Fellowship is awarded after some years of experience and track record, enabling the ergonomist to use the letters FErgS.

Although sport scientists and ergonomists follow different academic and professional paths, they acquire techniques and areas of knowledge that are common. Ergonomics may form modules or parts of modules in postgraduate programs within the sport and exercise sciences. Nevertheless, it is more likely that students are exposed to ergonomics principles indirectly through topics on applied human anatomy, environmental physiology, or cognitive psychology, for example.

In recent years the Ergonomics Society instituted continuing professional development courses as a central element in its application for chartered status. These short courses are available at different levels and contain options in five knowledge areas:

- Anatomy, anthropometry, and physiology in human activities
- Environmental stressors (performance shaping factors) and psychophysiology
- Sociotechnical systems

Table I.4 Venues for the Annual Congress of the European College of Sport Science

Year	Venue	Country
1996	Nice	France
1997	Copenhagen	Denmark
1998	Manchester	United Kingdom
1999	Rome	Italy
2000	Jyvaskyla	Finland
2001	Koln	Germany
2002	Athens	Greece
2003	Salzburg	Austria
2004	Clermont-Ferrand	France
2005	Belgrade	Serbia
2006	Lausanne	Switzerland
2007	Jyvaskyla	Finland
2008	Lisbon	Portugal
2009	Oslo	Norway
2010	Antalya	Turkey
2011	Liverpool	United Kingdom

- General and organizational psychology
- Survey and research methods

There are nationally and regionally organized associations that represent the sport science and sports medicine communities. Typically these are membership bodies that tailor their professional services to accredited subscribers. The strongest organizations are in North America and Europe. The American College of Sports Medicine (ACSM) offers a range of services, including certification for clinical exercise testing; its courses have professional credibility worldwide. Attendance at the annual ACSM conference can be translated into educational credits in designated medical, paramedical, and exercise science programs.

In Europe separate societies represent sport science and sports medicine. The European College of Sport Science was established in 1995 and held its inaugural annual congress the following year (see table I.4). A major feature of its annual congress is the Young Investigators Award. The competition is open to postgraduate and postdoctoral researchers under 32 years of age, and there are 10 prizes in both oral and poster categories. The top four candidates present their communication a second time in plenary session, and an award represents a significant achievement in the career progression of young sport scientists. The college has its office in Köln, Germany, and its own publication, the *European Journal of Sport Science*. Its position statements are designed to guide practitioners and scientists and are published in this outlet. Examples include position statements on stretching (Magnusson and Renstrom, 2006), overtraining syndrome (Meeusen et al., 2006), and coping with jet lag (Reilly et al., 2007).

The British Association of Sport and Exercise Sciences, like the Institute of Ergonomics and Human Factors, has its own codes of practice, despite lacking chartered status. The association allows members to apply for accreditation for either research or support status. The system was set up in the United Kingdom in the late 1980s to ensure that individuals funded to work on sport science support programs had the expertise to deliver the scientific services required to do so (Reilly, 1992). Many of these services embrace skills and competencies that overlap with those of the ergonomist. The national sport science support program first implemented in the late 1980s achieved several goals:

- Sport science support was directed toward athletes' needs as perceived by coaches.
- Quality control was ensured by allocating projects only to accredited sport scientists and laboratories.
- Mutual trust was developed among sport scientists, coaches, athletes, and lay and professional administrators.
- A coordinated structure enabled national sport governing bodies to access sport science expertise.

Physiotherapists (physical therapists) and athletic trainers have their own professional bodies and schemes of vocational qualification. The training of physiotherapists is in the hands of a chartered body in the United Kingdom, and a parallel status applies to physical trainers in North America. In European countries, notably Belgium and the Netherlands, there has been a strong historical link between manual therapy, kinesetherapie (movement therapy or physical therapy in Flemish), physical education, and sport science. A similar link is evident in the academic programs in East European countries: In the National Sports Academy in Sofia, Bulgaria, the three main programs of education are physical education, sports coaching, and kinesetherapie. Specialists from this background tend to have a sound appreciation of postural stress, preventive procedures, and rehabilitation. It is therefore a short step from physiotherapy to ergonomics in these countries.

The discipline of athletic training and therapy is considered to bring together theoretical and practical perspectives from several interrelated bodies of knowledge (Ortega and Ferrara, 2008). The training of practitioners is directed toward developing competencies in specific areas, such as risk management and injury, pathology of injury and illnesses, therapeutic modalities, and health care administration. Its research tools include basic sciences, clinical studies, educational research methods, sport epidemiology, and observational studies. Common ground between different countries is represented in the World Federation of Athletic Training and Therapy, with which a number of national professional organizations are affiliated. These include the National Athletic Trainers' Association (United States), the Japan Athletic

Trainers' Organization, the Biokinetics Association of South Africa, and Association of Chartered Physiotherapists in Sports Medicine (United Kingdom). Several of the associations maintain their own databases on injury statistics and on related health care materials for athletes.

Although anthropometric measures are commonly used by practicing ergonomists and physiotherapists, there is no formal requirement for specialist training in anthropometry. Quality control procedures are applied when research reports are refereed for publication, so local professional training in the relevant techniques is normally assumed. The International Society for Advancement of Kinanthropometry (ISAK) is the membership organization that implements accreditation for surface anthropometry. Its standards entail certification at level 1 (3-day course) and level 2 (5-day course) and progression to levels 3 and 4 (to become a criterion anthropometrist) after more rigorous examinations. Candidates must satisfy thresholds in accuracy and precision before achieving certification (see figure I.5).

It is evident that there are many entry points into the ergonomics profession. There is also richness in the diversity of applications of ergonomics techniques and principles. The discipline requires more than a tool kit of methods such as checklists, surveys, assessment protocols, and analytical schemes, and so the profession is not hamstrung by a limited array of analytical tools for problem solving. The profession is more reliant on creativity in applying the right solution to real-life challenges.

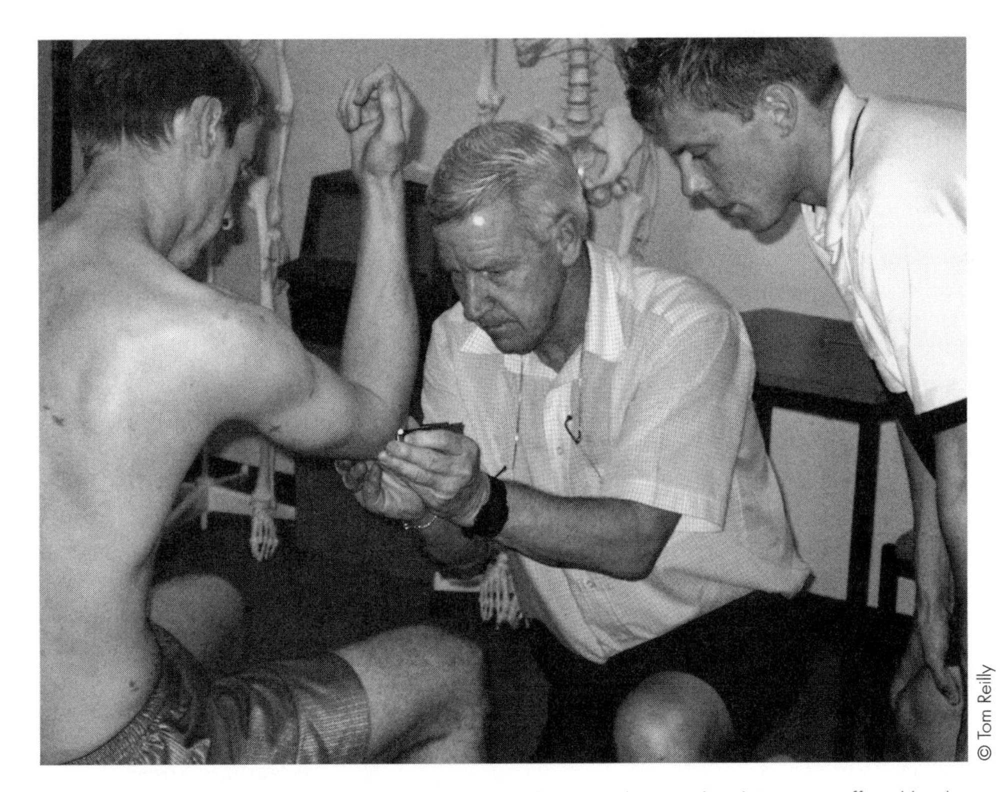

Figure I.5 Candidates practice anthropometric techniques during a level 1 course offered by the International Society for Advancement of Kinanthropometry.

Overview and Summary

Ergonomics emerged out of a necessity to understand how to optimize human performance and how to avoid breakdown of physical, physiological, and psychological resources. Ergonomics has become an influential profession worldwide, reflected by common principles in designing products and artifacts for human use in a wide variety of domains. These areas include not only work and industry but also domestic, leisure, and sport contexts. Although safety and well-being of workers take priority in occupational circumstances, dictated by legal requirements and employment policies, performance is often prioritized in elite sport. Nevertheless there must be a balance between risk and safety, personal achievement and health; this balance may be realized by ergonomics interventions, aided or implemented by allied professionals.

PART I

Risk Factors

IN ergonomics the primary focus is on the individual, whether he or she is an elite athlete, a participant in recreational activity, an industrial employee, or a self-employed person with a small business enterprise. The context in which that individual operates, the task engaged in, and the equipment used must all be considered if a harmonious relationship is to be established. There are likely also to be interactions with others in the team or in the workforce that need to be taken into account in a global ergonomics perspective.

The individual does not execute activities in isolation but in environments whose features may change in a nonsystematic manner. A host of environmental factors influence individual responses, mood states, and physical capabilities. Their effects may be complementary or opposing, additive or nonlinear, predictable or unknown. Part I considers the environment in broad terms and focuses on specific environmental variables as well. The more general view accommodates work ethics, lifestyle, and attitudes toward change. For example, redesigning tasks, workstations, training, and working practices calls for an appropriate balance between traditional ways and innovation.

The four chapters in part I provide a comprehensive outline of the risk factors that threaten equilibrium. The relationships among the individual, the activity, the technologies, and equipment are explored from a variety of ergonomic standpoints. Ergonomics provides a framework for identifying problems in the working of this relationship and for guiding the decision-making process that leads to a solution. Such an approach is not merely an application of standard checklists but requires comprehensive knowledge about human characteristics and human behavior.

Chapter 1 focuses on the physical properties of human structures. These are described in terms of the human sciences, emphasizing the interdisciplinary nature of the subject area. The chapter discusses how human characteristics can be quantified and to what extent humans can adapt to the external loads placed on them. The concepts and principles underpinning ergonomics are explained and the limits to human performance considered. Whereas the model of fitting the task to the person is a cornerstone of occupational ergonomics, the competitive athlete more readily

PART I

accepts the discomfort and the element of risk that are associated with exercise training. Indeed, a basic view of training is that the process is geared to fitting the person to the task, often a relatively inflexible task.

The health and safety of individuals are important criteria for the ergonomist. The formal regulatory requirements for safety at work apply to sport as well as to the traditional workplace. Nevertheless, risk is inherent in the majority of sports, its importance being related to the likely consequence of unplanned events. The training process itself may cause harm, particularly if recovery between strenuous exercise sessions is inadequate for the human body to return to homeostatic levels. These considerations form the content of the second chapter.

Chapter 2 presents causes of accidents and injuries, identifies external and internal factors in different activities, and highlights the role of human error. Negative predispositions toward injuries are explained in the context of accident proneness, and the use of screening procedures is explained in context as well. The critical incident technique for accident investigation is considered alongside epidemiological approaches. Postural analysis techniques are described and their relevance to physical activities placed in the context of preventing musculoskeletal problems. Injury prediction is reviewed, and indices for use in sport such as dynamic control ratios and leg-length discrepancy are evaluated. Ergonomic aspects of musculoskeletal disorders are related to characteristics of tasks, working postures, and faulty biomechanics. Human ethics procedures and formal risk assessment are explained in the context of individual protection. Consideration is given to high-risk sports and adventure activities such as high-acceleration rides and extreme sports. The concept of overtraining is presented and explained.

Chapter 3 is concerned with environmental stressors including heat, cold, altitude, hyperbaric conditions, air pollution, and noise. The human reactions to each stressor are considered. The text discusses tolerance levels, acclimatization processes, exposure simulations, and possibilities of acclimation for each environmental condition. Protective measures, including clothing, are considered as are means of environmental monitoring. The influence of seasonal variations on human activity and performance is covered at the end of the chapter.

Lifestyles are partly dictated by the natural harmony of activity and sleep cycles that are tuned to the alternations between daylight and darkness. As shown in chapter 4, although circadian variations are largely determined by endogenous rhythms, environmental aspects such as temperature and social activities fine-tune these cycles into a 24 hr period. Circadian rhythms are disrupted when sleep is curtailed or displaced as happens to shift workers. Daytime and nighttime patterns are also altered each year in practicing Muslims during the holy month of Ramadan. The syndrome of jet lag is associated with the desynchronization of circadian rhythms and the new local environment after crossing multiple time zones. Whether we travel for business, vacation, sport, or recreation, transmeridian flights and the accompanying experience of jet lag are part of contemporary lifestyles.

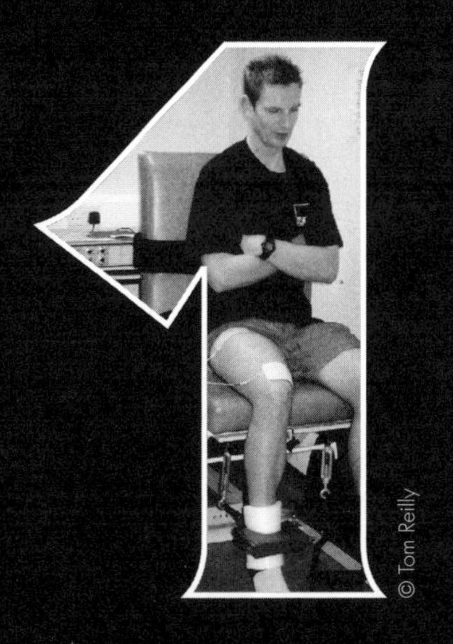

© Tom Reilly

Physical Properties of Human Structures

DEFINITIONS

attention—Selective choice of information from display.

blood lactate—Metabolite of anaerobic work produced within the active muscles and accumulating in the circulation.

electromyography—The recording of electrical activity resulting from contraction of skeletal muscle.

dynamometer—A device for measuring force produced by action around a joint.

flexibility—Range of motion about an anatomical joint.

force platform—A plate situated at ground level used to record the forces acting on contact with it.

hydrodensitometry—Measurement of body density using underwater weighing.

isokinetic dynamometer (device)—Device or machine used to measure dynamic force production or peak torque at preset angular velocities.

isometric force—Muscle tension at a specific angle of contraction, without any limb movement.

somatotyping—A formal system for quantifying physique or body shape

specificity—The principle of relating measures to the sport or action concerned.

one

A principle of ergonomics is that the target to be accomplished and the equipment to be used are designed with the operator in mind. In sport and exercise this means that the challenges inherent in the activity are within the capabilities of the individual and that any equipment used is suited to the athlete concerned. Competitive sport is unyielding in its requirements, so the athlete must possess or acquire the fitness required for the sport.

The term *fitness* is often used in a generic sense, because each sport (and activity) has its own specific requirements. A highly developed oxygen transport system is important in all endurance sports, but it must be complemented by specific muscular and neuromotor adaptations to training for the sport in question. Physical structure is at least as important as physiological function and psychological factors: For example, absolute aerobic power ($\dot{V}O_2max$) is critical for performance in rowing and absolute strength is important in scrummaging in rugby. The opposite applies in sports such as running where body mass must be lifted repeatedly against gravity and relative aerobic power is crucial.

Monitoring Activity Demands

The activity engaged in constitutes an external load for the athlete. The body reacts to the load in a specific way, and these responses are the internal load. Monitoring both the activity concerned and the responses to it provides insight into the consequences for the individual and the ergonomic implications of the work. A variety of methods are available to the ergonomist for such monitoring purposes.

Physiological Strain

A starting point in an ergonomics analysis is the quantification of load on the individual. The assumption is that the task imposes demands on the individual, whose responses can be used as indicators of physiological strain. Physiological criteria corresponding to exercise intensity include energy expenditure, oxygen uptake, body temperature, heart rate, and blood metabolite concentrations. Such variables have been shown to be associated with subjective perceptions of exertion, task difficulty, and thermal comfort as well as with biomechanical measurements that reflect the level of power output. When maximal physiological capabilities are also known, the physiological strain is expressed as relative loading. Where chronic overloading is possible during sustained periods of strenuous training or intensive competition schedules, endocrine responses and markers of immunosuppression have been used in attempts to explain underperformance (Gleeson et al., 1997; Gleeson, 2006).

Oxygen uptake and heart rate have traditionally been used to measure physiological strain in heavy occupational work. The availability of short-range telemetric devices has made the continuous recording of these responses possible in a range of field settings. Heart rate monitoring has been used conventionally to determine physiological strain in occupational contexts and to estimate the energy cost of specific activities. Although it can be maintained that these procedures are valid only in steady-rate exercise, the error in using heart rate to estimate energy expenditure during intermittent exercise of high intensity (such as soccer or field hockey) is within acceptable limits (Bangsbo, 1994). The recording of oxygen uptake during soccer training drills has yielded valuable information about metabolic loading related to competitive conditions (Kawakami et al., 1992). Continuous registration of heart rate during different training activities has generated information about their suitability for conditioning work or for recovery training. Sassi and colleagues (2005) used **blood lactate** and heart rate responses to a range of soccer drills to identify sessions that could be used as fitness stimuli and those that possessed purely tactical benefits. The physiological information can be used not merely as descriptive feedback on training inputs but also as a means of regulating the training intensity.

Traditionally, much ergonomics research was concerned with occupations that entailed heavy work. These jobs included forestry work, farm work, and work in the mining industry, where conditions were mostly hot and uncomfortable. The classification system designed by Christensen (1953) for assessing the severity of occupational work included body temperature as well as energy expenditure and heart rate. This system was developed for application to entire work shifts. With notable exceptions such as cycle road racing and ultramarathon foot races, the durations of sports contests are much shorter than a typical 8 hr work shift, but energy expenditure levels are much higher (table 1.1).

Once physiological systems are placed under strain, the body must be allowed to recover to the baseline state. This necessity raises questions about the optimal work-to-rest ratio. If rest periods are too long, the worker is underproductive; in contrast, the worker will underperform if fatigued from previous physical activity. These events have parallels in sport, especially sports that entail intermittent exercise, where the ability to recover quickly from all-out effort is a requirement.

Forces

The measurement of force provides information on the interaction of an individual with the environment. Several forces act simultaneously on a person to determine performance. Although some of the forces are known (e.g., gravity) and some can be computed (e.g., air resistance), the force that has the greatest influence on performance is the contact force between the individual and the environment. This contact force, usually referred to as the reaction force, often acts at the feet or hands but can in principle act at any point where the body makes contact with the surface. Specialized

Table 1.1 Energy Expenditure Levels

	Energy expenditure, kcal (kJ)/min	Heart rate, beats/min	Body temperature, °C
Too heavy	12.5 (52.3)	175	39.0
Very heavy	10.0 (41.9)	150	38.5
Heavy	7.5 (31.4)	125	38.0
Medium	5.0 (20.9)	100	37.5
Light	2.5 (10.5)	75	37.0
Very light			

Reprinted from T. Reilly, Introduction to musculoskeletal diseases: The Biomed IV Project. In *Musculoskeletal disorders in health-related occupations*, edited by T. Reilly (Amsterdam: IOS Press), pgs. 1-6, 2002, with permission from IOS Press.

measurement equipment, referred to as a **dynamometer,** has been constructed to monitor the reaction force in specific situations.

The simplest form of dynamometer entails measuring the tension force in a wire attached to the individual and is known as cable tensiometry. The technique was borrowed from the aircraft industry, where it was used to measure the strength of cables on the airplane. When this method was applied to human performance, it was originally used to measure isometric strength of a single joint. For example, to measure extension strength at the knee joint, the person was seated in a rigid chair with the ankle of the leg to be tested connected to a cuff, which in turn was attached by means of a cable to a load cell or strain gauge device. As the individual tried to extend the knee joint, the tension created in the cable was measured and recorded as a force. Later, the cable tensiometer was replaced by load cells and strain gauge assemblies that soon became the preferred means of recording forces.

Measurement of **isometric force** of a joint can provide useful data on strength capabilities of individuals and on how strength is influenced by muscular fatigue and other factors such as diet and heat stress. The force data can be further processed to obtain variables such as the rate of force development and rate of force decay. These, together with the peak isometric force, provide a range of variables that can be used to monitor a wide variety of individual and muscle performance characteristics and the relationship between muscle groups. The National Institute for Occupational Safety and Health (NIOSH) and other important groups have traditionally recommended an electronic load cell for isometric strength testing (Chaffin, 1975; NIOSH, 1977), although inferences may be limited to the angle at which measurements are made.

More sophisticated muscle function dynamometers have been developed commercially (e.g., Kin-Com, Lido, Cybex, Biodex) whereby the angular velocity of movement can be preset. These **isokinetic devices** are usually substantial pieces of equipment that were initially intended to provide a controlled environment for rehabilitation. Their measurement capability quickly led to their being adopted as a tool for measuring the force-generating capability of different muscle groups. The measurement principle is similar to the strain gauge device mentioned previously, but isokinetic devices are capable of measuring muscle strength (usually expressed as joint torque) during isometric, concentric, and eccentric modes and can be configured to measure many of the body's joints in both flexion and extension. These devices carry some measurement issues that users need to be aware of (Baltzopoulos and Gleeson, 2003), but contemporary software enables these devices to be used widely in the evaluation of sport performers. For example, there has been much interest in evaluating soccer players at different levels in terms of basic strength, bilateral strength, and strength asymmetries (Rahnama et al., 2003) as well as age-group soccer players (Iga et al., 2005) and the influences of soccer match play in inducing fatigue (Rahnama et al., 2006). Peak torque varies with the joint angle as well as the angular velocity. These factors must be taken into consideration when the effects of a training intervention on maximum voluntary strength are being assessed.

Multijoint strength cannot be measured by the dynamometers just described, and a force plate or platform is required. A **force platform** is a device that usually sits in the ground and can record the forces as contact is made on it, often with the feet but sometimes with the hands or other body part. The force platform is a sophisticated instrument that can directly measure up to six force variables (one vertical force, two horizontal friction force components, the friction torque, and two center of pressure locations). These force variables can be used directly or in combination to indicate different aspects of performance. The most informative force variable is the vertical force component, because this value usually is the largest. This force variable has been used to determine the forces attributable to walking (1.1 times body weight) and running (up to 2.5 times body weight) as well as the most demanding of sports such as triple jumping (up to 10 times body weight, Hay, 1992). The two horizontal friction forces can be used to record frictional resistance as an athlete changes direction during cutting movements or side stepping or to determine the influence of shoe sole or stud design on the performance of sport footwear (Lake, 2000). The friction torque is not widely used but has relevance, for example, in the cause of injury in cyclists (Wheeler et al., 1992) where high rotational torques have been associated with knee injury. The locations of center of pressure during locomotion have been used to identify characteristics of running technique (Cavanagh and Lafortune, 1980).

Force variables can be used to monitor behavior in similar contexts, as mentioned here, but their value is best appreciated when they are combined with a motion analysis system to provide information on internal joint forces. Automatic motion analysis systems are used to record three-dimensional data based on reflective markers attached to the subject. These data may be supplemented by concomitant force recordings. Such a process is typified by gait analysis, although the approach has been applied more widely to sport activities. Data on joint moments and forces are available for a variety of actions including running (Buzeck and Cavanagh, 1990), cutting, jumping (Lees et al., 2004), kicking a soccer ball, and many others. Furthermore, the mass distribution characteristics of an athlete can be calculated by incorporating anthropometric measurements, and a whole-body model can then be used to calculate time histories for joint torque. These techniques can be used to analyze sport skills, providing visual feedback on screen to athletes and coaches.

A pressure-sensing device provides information about the localized application of forces. This instrument is usually made of a series of small force-measuring cells (about 5 mm square) that give information about the force acting over a small area. When several of these cells are put together as a mat, the device is used to measure areas of high pressure; for example, a mat placed under the foot measures pressure as the heel makes initial contact with the ground until toe-off, where the pressure acts on the metatarsal heads. The regions of high pressure can lead to bruising and pressure sores, which can be prevented or alleviated with the use of custom-made orthoses, designed from the pressure data (Geil, 2002). A pressure mat can be placed in other interfaces between the body and the environment, for example, in the stump of an amputated limb to monitor the fit of the prosthesis or on the seat of a wheelchair.

Electromyography

The application of muscle force requires the contraction of muscle fibers. The small electric field produced as muscle fibers are activated can be monitored by means of **electromyography (EMG)**. The electrical field is detected either by surface electrodes placed on the skin above the underlying muscle or by indwelling electrodes inserted into the muscle through a needle. Both methods indicate muscle activity, but the latter method of measurement is the less popular because it is invasive and carries a risk that the fine wire making up the electrode becomes detached during use. This risk is enhanced during vigorous muscle contractions, where large changes in muscle fiber length occur. Nevertheless, in some cases indwelling electrodes provide the only way to monitor small or deep muscles (Morris et al., 1998). Surface EMG is the most popular approach, and many commercial systems are available that provide good preprocessed EMG signals for analysis.

Surface EMG signals can be used in a variety of ways, but care must be taken in their interpretation because the signal is susceptible to cross-talk from other active muscles besides those over which the electrodes are placed. It is also necessary to know whether a muscle contraction is concentric, isometric, or eccentric because the EMG signal has a different appearance under these different contraction conditions. One of the more basic uses of surface EMG is to identify the muscles that are active in the performance of a task and their timing pattern relative to one another (for a review, see Clarys and Cabri, 1993). The surface EMG signal can be further processed to gain insights into muscle function. One method is to rectify the electrical signal so that it has only positive components. Horita and colleagues (2002) used this method to detect the influence of muscle stretch on the stretch-shortening cycle in a jumping activity while the same research group studied alterations in the lower-limb muscles with increased running speed (Kyrolainen et al., 2005). The EMG data commonly are further smoothed to provide an "envelope" or integrated EMG signal that broadly reflects the action taking place. To relate the muscle activity to the motion being investigated, account must be taken of the electromechanical delay, that is, the time from activation of the muscle fibers to the time when they develop maximum force, which is around 30 to 120 ms and is dependent on the level of tension in the muscle prior to its activation. When relative activity between muscle groups is the concern, a method of normalization must be used. A maximum voluntary contraction (MVC) is often chosen, and although it too has limitations, it allows comparison between muscles, between different conditions, and between individuals. Once the best processing method has been determined, surface EMG can be used in a variety of applications.

Electromyography can be used to evaluate sports equipment. For example, Robinson and colleagues (2005) investigated the efficacy of various commercial abdominal trainers in comparison to traditional sit-ups or crunches. These authors monitored the lower rectus abdominis, upper rectus abdominis, and external oblique muscles over five different types of abdominal exercise including one using a commercially manufactured device. The authors reported significant differences between exercises: Three exercises (one involving a Swiss ball, another using raised legs, and another using a weight behind the head) resulted in greater muscle activity, whereas the commercially available abdominal roller showed less muscle activity, compared with the standard sit-up. Considering these findings, Robinson and colleagues recommended the commercially available device for inexperienced participants, because the device might enable them to perform a greater number of repetitions using the equipment than they otherwise would, which in turn would improve their motivation to exercise. For maximal loading, exercises other than the standard sit-up can be chosen. It was also apparent that the exercise involving the Swiss ball is an advanced exercise suitable only for those with a high level of training experience or those who require a high level of muscle training.

Onset of fatigue in active muscles is an indication that they are operating at or near their limits, and surface EMG is used to monitor the effects of fatigue. Robinson and colleagues (2005) investigated the influences of a typical 30 min circuit training program on muscle activity. When muscles were in the fatigued state, the normalized mean EMG for both exercises increased for lower and upper rectus abdominis but not for the external obliques. This differential result illustrated that the external obliques were not highly used during the fatigue condition of the experiment. The increase in the integrated EMG signal attributable to fatigue is thought to reflect the greater central effort made or the changes in muscle recruitment pattern needed when muscles become fatigued. Not all fatigue leads to a greater EMG signal. In a study of the effect of muscle fatigue in a simulation of the exercise intensity of soccer, Rahnama and colleagues (2006) reported an increase in EMG attributable to different running speeds (at 6, 9, 12, and 15 km/h) but a reduction from the start of activity to the end of the 90 min intermittent exercise protocol in some of the muscles monitored. This reduction was deemed to reflect the decline in muscle strength found in players as a result of game play (Rahnama et al., 2003).

Assessing Individual Characteristics

Databases accessible to ergonomists provide numerical information about human characteristics and activities and how they vary between individuals. These numbers are expressed as mean or average and as range or quartile. One difficulty is that having an average value for one characteristic does not guarantee being average on others. Hence it is necessary to consider individual variations. These variations are addressed for anthropometry, physiological capabilities, and performance measures.

Anthropometry

Anthropometry refers to physical measurements obtained from the human body using a systematic approach. Physical measures include linear variables that relate to the whole body, for example, stature, or a segment of it such as limb length. Height is an important dimension for sports such as basketball and in rugby union forward play, where the taller participants have an advantage. A second tier of anthropometric measures is related to body proportions, such as sitting height relative to weight. Provided overall stature is appropriate, a relatively large leg length bestows an advantage in high hurdling whereas a low center of gravity suits weightlifters. There is considerable evidence that Olympic-level athletes gravitate toward the competitive event for which they are anthropometrically best suited.

Combinations of anthropometric measures have health-related implications. The body mass index, or BMI (body mass in kilograms divided by height in meters squared), is used in epidemiological studies to identify individuals who are overweight or, worse still, obese. This index has been criticized as too crude, because it does not account for whether extra mass for height is attributable to muscle, bone, or fat. The body mass index is inappropriate for athletes who have a large muscle mass as a result of strength training, because they would be erroneously described as overweight.

The waist-to-hip ratio is an alternative index for use in population studies. The ratio differs between the sexes in that men tend to accumulate relatively more fat in the abdominal area, whereas women, particularly after menopause, gain weight around the hips. In the years after menopause, the relative predominance of hip adiposity is reduced as abdominal fat increases in women. A value of 0.9 is deemed acceptable for men, and the corresponding value for females is 0.8.

Body shape is independent of body size and is generally referred to as physique or **somatotype.** The system for recording body shape is known as somatotyping. It is based on three dimensions—endomorphy (or fatness), mesomorphy (muscularity), and ectomorphy (linearity). Each is rated on a scale from 1 to 7, but people with values >7 may be outside the standard somatochart, as shown in figure 1.1.The somatotype is determined from measurements of body size, limb circumferences, bone breadths, and skinfold thicknesses (Duquet and Carter, 2001). Again there is considerable evidence that elite performers in particular sports tend to congregate around the same area of the somatochart. Rather than constitute an analytic tool in its own right, the somatotype has a role as an accessory technique in ergonomics work. The plots display the variation in body shapes as well as central tendencies in the data.

Applications to Design

Designers may construct their products for the average user or according to parameters such as shoe, glove, or shirt sizes. Design principles can also accommodate sex differences and age in growing children. Sports equipment must be functional if players are to enjoy their sport and develop the skills necessary to compete at high levels. Physical size and strength differences between population groups have influenced equipment design, enabling the product to be fit for purpose. For example, although it has always been possible to purchase sport shoes designed specifically for children, it is now quite common to find sport shoes that are fabricated exclusively for females.

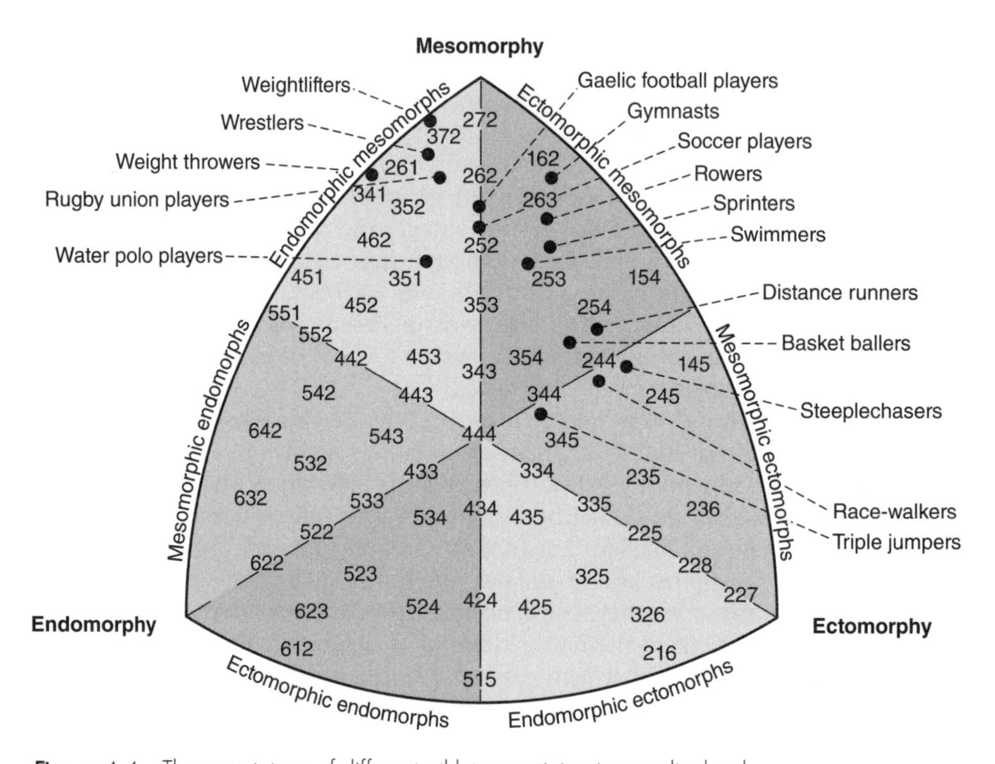

Figure 1.1 The somatotype of different athletes participating at elite level.
Reprinted, by permission, from T. Reilly, 1992, *Sports fitness and sports injuries* (London: Faber and Faber), 87.

Ethnic differences in foot shape have been reflected in the manufacture of footwear for different ethnic groups. Sports equipment sized to the individual is common in sports like golf, tennis, cycling, and soccer, and generic equipment such as headgear and footwear can be individually sized too. Manufacturers have relied on anthropometric databases that include measurements such as segment lengths, girths, joint mobility, reach, strength, handedness, and comfortable exertion levels to design their products. For high-level performance it is common to custom fit equipment to suit the requirements of individual players and, sometimes, the sponsor.

Formula One motor racing provides an example of how the workspace must be compatible with the individual's anthropometric characteristics. The layout of the car's controls must accommodate the reach envelope of the driver and allow room for the protective clothing and equipment worn. This example highlights how ergonomists consider the dynamic working situation before reaching design conclusions.

Body Composition

The body composition of athletes affects their suitability for their sport. Conceptual models of body composition embrace multiple factors but the simplest model incorporates two compartments: body fat and fat-free mass. In locomotory sports such as running and field games, extra fat deposits are disadvantages because they constitute dead weight during antigravity work. In contrast, fat deposits provide extra insulation to protect against hypothermia in long-distance swimmers, the layers of subcutaneous fat acting as a barrier to heat loss. Health is affected when body fat levels are increased, body fat constituting a risk factor for cardiovascular disease.

Techniques for body composition analysis have been described in detail elsewhere (Eston and Reilly, 2001; Heymsfield et al., 2005). Body composition refers to the separation of the body into its constituent compartments, the two-compartment model (fat mass, fat-free mass) being the most common. Methods are necessarily indirect because they rely on cadaver analysis for validation. Until recently, **hydrodensitometry** was used as the reference method for the development of doubly indirect techniques such as bioelectrical impedance, infrared interactance, and skinfold thicknesses. Body density is measured by underwater weighing of subjects using the principle of Archimedes and assuming a constant density for fat and fat-free tissue. These assumptions may break down in athletic subjects because of the accrual of bone with training or the demineralization of bones in amenorrheic endurance athletes. Air displacement is an alternative method but is not as reliable as the more conventional hydrodensitometry. Dual-energy X ray absorptiometry is considered the nearest technique to a gold standard of those available (table 1.2). It does not rely on the assumptions required for densitometry and can yield a three-compartment model (fat mass, bone mass, fat-free bone-free mass) and further detail.

Dual-energy X ray absorptiometry (DXA) was originally designed for screening purposes to identify osteopenia and osteoporosis in clinical contexts (see figure 1.2). Its main use in athletes was to examine bone mineral losses in female athletes experiencing secondary amenorrhea attributable to a combination of strenuous training and inadequate diet. Body mass can be divided into bone mass, fat mass, and lean body mass (consisting of fat-free and bone-free mass). The fat component is of particular interest to sport participants, and the technique has been used to demonstrate how

Table 1.2 Levels of Validation for Body Composition Analysis

Method	Level	Comment
Dissection	I	Direct
Potassium counting	II	Indirect
Total body water	II	Indirect
Medical imaging (DXA*, computed tomography, magnetic resonance imaging) (based on quantitative assumptions)	II	Indirect
Bioelectric impedance	III	Doubly indirect
Electrical conductivity	III	Doubly indirect
Infrared interactance	III	Doubly indirect
Anthropometry (calibrated against a level II method)	III	Doubly indirect

*DXA (dual-energy X ray absorptiometry) is a chemical technique.

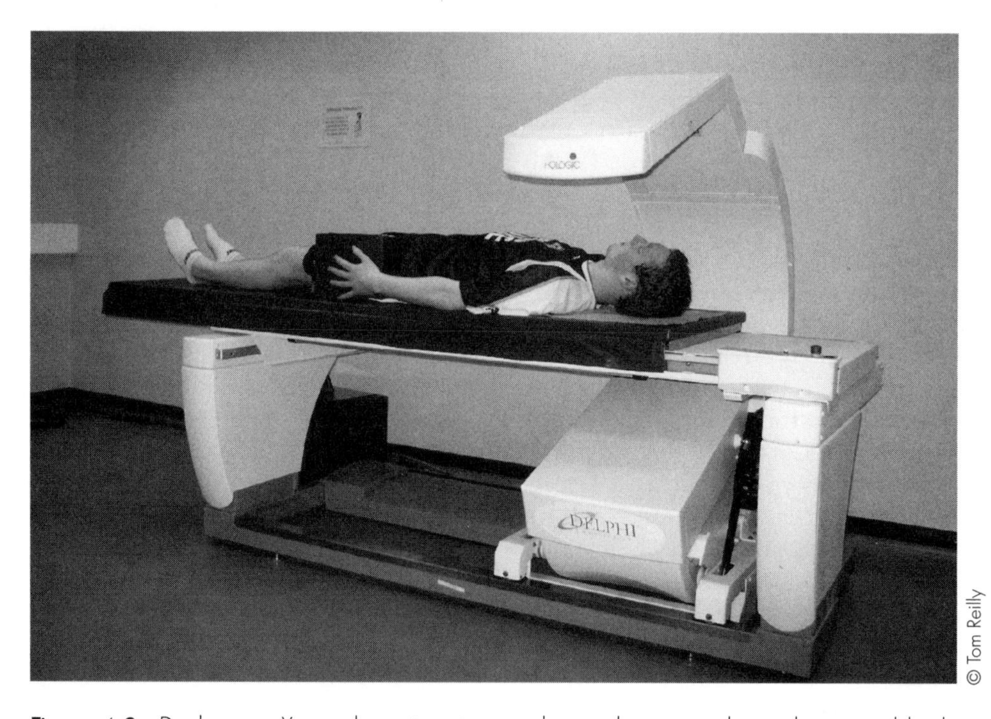

Figure 1.2 Dual-energy X ray absorptiometry may be used to assess bone density and body composition.

percent body fat varies in professional football players throughout the competitive season (Egan et al., 2006).

The most accessible method for estimating body fat is by means of skinfold thicknesses. This method is based on the assumption that the thickness of subcutaneous adipose tissue layers, determined at a number of sites on the body, is directly related to internal fat deposits largely in the viscera and around internal organs. This assumption has been questioned from results of cadaver studies (Clarys et al., 1987). Nevertheless, comparisons have shown good agreement with results derived using DXA (Egan et al., 2006). In a position statement, the Steering Groups of the British Olympic Association recommended that five skinfold sites be used and the total value derived to obtain an index of adiposity (Reilly et al., 1998). The anterior thigh was added to the four skinfolds used in the classical formula of Durnin and Womersley (1974)—biceps, triceps, subscapula, and suprailiac. A criticism of the latter was that the skinfolds were all based on upper-body sites.

Anthropometry has been used to estimate muscle mass, a compartment of the body that is especially relevant for sport participants. The classical anthropometric approach of Matiegka (1921) was revised as a result of the Brussels Cadaver Analysis project. Martin and colleagues (1990) showed that sport specialists at the international level could be distinguished on the basis of their muscle mass, normalized for body size. Similarly, games players have more skeletal muscle relative to their size than do reference groups (Reilly, 2003). Because the strength of skeletal muscle is directly related to its cross-sectional area, the increased muscle mass attributable to training-induced hypertrophy is associated with gains in strength. Cross-sectional area of muscle can be measured using imaging techniques ranging from ultrasound to magnetic resonance imaging, but such measurements are done mainly for experimental purposes.

Assessing Physiological Capabilities

A hallmark of sports ergonomics is the correspondence between the individual and the task or sport. This match becomes more refined as participants become skilled in specific sports and acquire physiological adaptations to their tailored training programs. Any method of testing individual capabilities should resemble key respects of the sport in question. Where possible, any test apparatus should be linked to the sport if inferences about fitness levels and training prescriptions are to be drawn from the observations.

Specificity

Assessment of physical and physiological capabilities is now a routine part of sport science work. For these assessments to be of use to practitioners, fitness testing should be conducted on a regular basis, frequently enough to provide individual feedback but not so often that it disrupts training and becomes a chore. Generic protocols that use standard ergometry are available for measurement of aerobic power, anaerobic capabilities, muscle performance, and other functional measures. Because these measures may lack **specificity** to the sport in question, a range of dedicated ergometers and tests have been designed to suit particular requirements. Exercise on these ergometers engages the most relevant muscle groups by mimicking the actions in the sport concerned. Consequently, it is possible to apply standard protocols for ski simulators, rowing and canoe ergometers, and other sport-specific devices. Sophisticated measurements of swimmers, kayakers, and rowers can be obtained from exercise in water flumes, where biomechanical analysis can complement the recording of physiological responses (see figure 1.3).

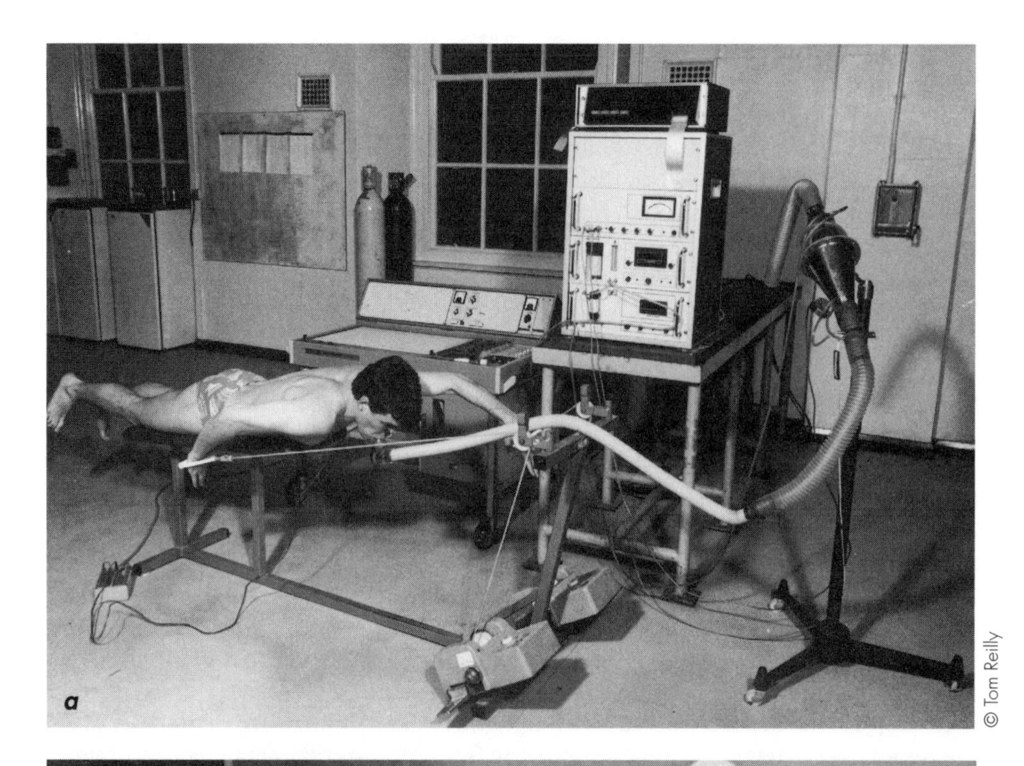

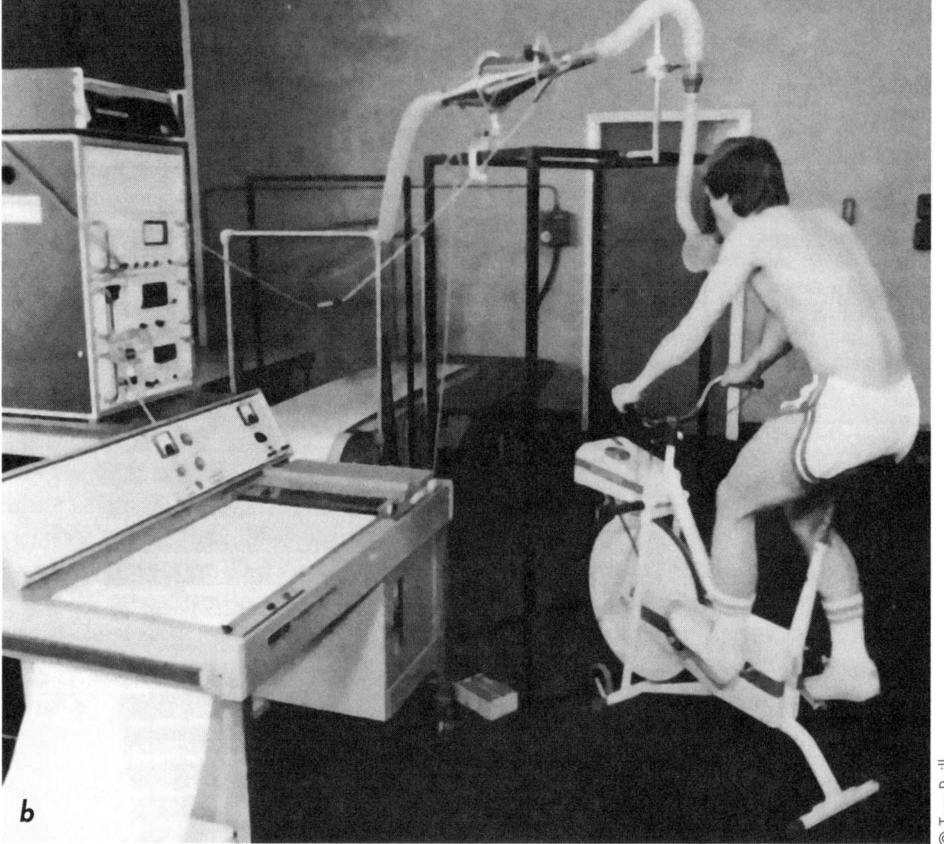

Figure 1.3 Examples of the early generation of ergometers for assessment of swimmers *(a)* and cyclists *(b)*.

Field Tests

Field-based tests have been used to enhance the ecological validity of fitness assessment and relate observations to competitive performance. Protocols have been designed that incorporate not only the locomotion patterns of the sport but also the essential skills (see Reilly, 2001). In these instances, the increased specificity comes at a price of missing important physiological information. There are concessions for reliability in such tests because environmental and surface conditions can vary when measurements are repeated in subsequent assessments.

The 20 m shuttle run test of Leger and Lambert (1982) is commonly used to assess aerobic fitness. It has been used to assess games players and for population studies, as in the Eurofit battery of tests (Reilly, 2001). An athlete's maximal oxygen uptake can be predicted from the number of shuttles completed, and a number of athletes can be assessed at one time. The limitation of this test is that performance is dependent on volitional effort, there being no specific criteria to indicate that maximal physiological capacities were actually attained.

The capability to reproduce high-intensity sprints is also important in sports in which exercise intensity varies intermittently. When the recovery periods are relatively short (<20 s), recovery depends on the athlete's oxygen transport system and ability to tolerate high levels of metabolic acidosis. The ability to reproduce all-out efforts is best assessed for team sports by means of repeated sprints (approximately seven) over a short distance (30 m) with a 25 s intermission between sprints. This type of test has high reliability and validity and can be executed with a series of timing gates (at 10 and 30 m) and on a nonmotorized treadmill using a standard protocol (Hughes et al., 2006).

Assessments can embrace the range of scientific disciplines and include psychological as well as physical and physiological methods. Indeed, fitness requirements for most sports tend to be multidisciplinary. The predictive power of any multi-item test battery is low when the characteristics required for success are complex. Reilly and colleagues (2000) showed that young soccer players being groomed for international level cannot be distinguished on the likelihood of future success, but they can be discriminated from subelite performers on a range of measures. Discriminant functions were found for aerobic power, speed, and agility and for anticipation and decision-making skills that characterize "game intelligence."

Physiological Measures

Physiological assessments rely on isolating specific functions using an established exercise protocol. Tests are available to measure aerobic and anaerobic capacities, submaximal and maximal responses; tests can be short term or progressive and incremental. The tests are conducted under controlled conditions, preceded by a warm-up, and conducted according to recognized procedures (Winter et al., 2006).

Aerobic Power and Capacity

Maximal aerobic power is indicated by the highest level of oxygen consumption that an individual can attain. It is referred to as maximal oxygen uptake ($\dot{V}O_2max$) and is determined by responses to a graded exercise test to exhaustion during which expired gases (O_2 and CO_2) are measured and minute ventilation is monitored. The attainment of maximal values is assessed at voluntary exhaustion using criteria that include a leveling off in $\dot{V}O_2$ despite an increase in work rate, a respiratory exchange ratio ($\dot{V}CO_2/\dot{V}O_2$) greater than 1.15, and peak lactate postexercise greater than 8 mmol/L. Because an improvement of about 25% in $\dot{V}O_2$max is considered to be a good training effect and the $\dot{V}O_2$max of an elite endurance athlete can be double the value of a sedentary individual, the genetic influence on this function is considered greater than the influence of training or environment.

The ability to maintain exercise intensity at a high fractional utilization of $\dot{V}O_2$max is important for endurance performance. This concept reflects aerobic capacity and is highly related to the so-called lactate (or anaerobic) threshold. The concentration of lactate in blood increases as exercise intensity increases. The point of deflection in the lactate response curve is taken to indicate the lactate threshold. The concentration of lactate in blood represents the balance between the production of lactate in active skeletal muscle and its clearance from the circulation. An alternative approach is to establish the work rate corresponding to a reference blood lactate concentration, such as the running velocity that induces a concentration of 4 mmol/L or V – 4 mM. The lactate curve shifts to the right with training, as shown in figure 1.4. The lactate threshold has been found to be a better predictor of marathon running performance than is $\dot{V}O_2$max (Jacobs, 1986).

Training effects are also marked by an improvement in mechanical efficiency, calculated by expressing efficiency as mechanical work performed as a percentage of energy cost. The efficiency can be measured easily when the work done (against gravity or a known resistance) is recorded, as in exercising on a cycle or rowing ergometer. For cycling, gross mechanical efficiency has been reported to be about 22% to 23% depending on the exercise intensity, whereas for weight training the figure is around 12% (Reilly, 1983). Calculation of net efficiency considers the oxygen consumption associated with resting metabolism. It is difficult to measure the mechanical work accomplished during running, and so the notion of running

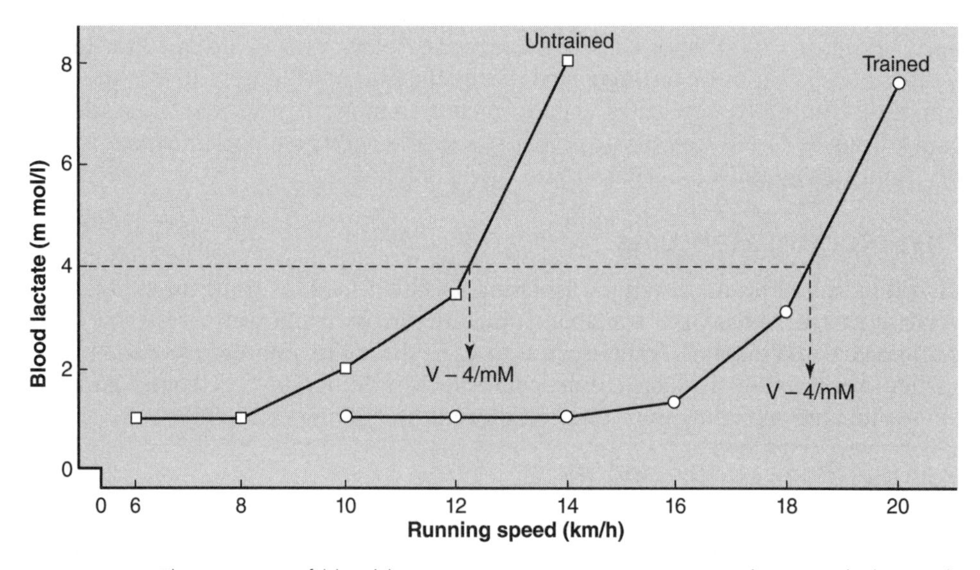

Figure 1.4 The response of blood lactate to progressive exercise is nonlinear and alters with training.

Reprinted, by permissions, from T. Reilly et al., 1990, *Physiology of sports* (London: E. & R.N. Spon), 138.

economy is applied. This concept refers to the oxygen consumption for a given running speed and is lower in well-trained athletes with good running technique than in average people.

Anaerobic Power and Capacity

Exercise at an intensity in excess of maximal oxygen uptake can be accomplished by using anaerobic sources of energy. Phosphagens and muscle glycogen (whose breakdown anaerobically leads to the production of lactate) provide the substrate for anaerobic metabolism and have high power but low capacity. The Wingate anaerobic test is the main laboratory-based method of measuring anaerobic power and anaerobic capacity. The test is performed on a cycle ergometer, instrumented to record flywheel revolutions and allow power output to be calculated. The average power output over the 30 s duration of the test is taken to represent anaerobic capacity, whereas power during the first 5 s is taken as peak anaerobic power. A fatigue index is determined from 5 s values of power over the first and last 5 s periods (Bar-Or, 1987). The test can be adapted for use on other ergometers, such as those used by rowers and kayakers.

Power output is determined on a treadmill when the belt is propelled by running on it, provided appropriate instrumentation is attached. The runner is attached to a load cell by means of a harness that allows the resistance to motion to be calculated. By knowing the length of the treadmill belt and counting the revolutions, the trainer or laboratory technician calculates power output from the forces registered. This system has proved to be a reliable method of assessing anaerobic power and capacity in single efforts and in repeated sprints (Hughes et al., 2006).

Anaerobic power is also determined using the stair-run test of Margaria and colleagues (1966). The individual sprints up two flights of stairs and the time is recorded by means of timing lights or photo cells. The resulting time, plus the subject's body mass and the vertical distance between steps, is used to calculate power. An alternative is to estimate power output in a vertical jump test, using a standard nomogram (McArdle et al., 1991). As an extension of this test, a sequence of jumps is attempted (over the course of 20 s) and the power decrement during this period used as an index of fatigue.

Muscle Strength

The traditional method of measuring muscle strength was by means of cable tensiometry (Clarke, 1967) or use of a load cell, strain gauge, or force transceiver. The technique was limited in that measurements were restricted to isometric actions: A series of recordings at different positions were needed if joint-angle curves were desired (see figure 1.5). There was also the question of the relevance of static actions to dynamic activities such as are found in sport and occupational settings. The same caution was applied to portable dynamometers such as used to assess back strength and leg strength (Coldwells et al., 1994), despite their value in field experiments (Reilly et al., 1998).

The previously described difficulties are overcome when isokinetic dynamometry is used. **Isokinetic dynamometers** permit the angular velocity of movement to be preset, and the force exerted is expressed as torque throughout the range of motion. The peak torque and the angle at which it occurs are determined. Eccentric as well as concentric modes of action can be used. Asymmetries between limbs and between

© Tom Reilly

Figure 1.5 Leg strength of the quadriceps is measured on a dynamometer at a specific joint angle.

flexors and extensors are determined to reveal imbalances in strength. Contemporary dynamometers have facilities for configuring different joints on the assembly for multiple assessments.

Isokinetic dynamometers are used to screen for muscle weaknesses and for imbalances between limbs or between muscle groups. Once deficiencies in muscle strength are identified, trainers can design programs for remedial action. One limitation of isokinetic machinery is that the maximum angular velocity is usually 400 °/s (6.97 rad/s), well below that recorded in competitive sport in skills such as kicking a football, drifting off in golf, or serving a tennis ball.

The main value of isokinetic dynamometry lies in monitoring the regain in strength during a rehabilitation program following injury. Part of the rehabilitation training can be performed on the dynamometer itself because the mode of muscle contraction and the range and speed of motion can be controlled. Using regular strength assessments, the trainer can determine when the injured athlete can safely return to full training and competition.

Mobility and Agility

Mobility or **flexibility** refers to range of motion about a joint. Range of motion in an intact joint is limited by soft-tissue opposition including ligamentous restraints or adjacent bone structures. Range of movement at a joint increases with flexibility exercises. Good mobility is desirable in sports such as gymnastics and the majority of track-and-field events. Lack of flexibility is associated with muscle tightness. Tightness in the hamstrings and adductors has been linked to injuries in soccer players, and a formal training program focusing on these muscles was found to reduce injuries in Swedish players (Ekstrand, 1982). Training to restore joint mobility to normal values is also a necessary part of rehabilitation.

The conventional means of measuring range of motion was by using flexometers or fluid-filled goniometers, which are manual methods. Electrogoniometry allows the range of movement to be recorded in real-life dynamic situations, including sport and industrial contexts (Boocock et al., 1994). Limb acceleration can be recorded and is especially relevant in sport skills where limb speed characterizes elite performers.

Agility refers to the ability to change direction quickly, a factor in swerving past opponents and avoiding tackles in games play. It is an important component of talent in young soccer players (Reilly et al., 2000). Because agility relies on balance and neuromuscular coordination, it has no specific physiological test. Rather, agility is assessed by means of performance tests that entail fast turns and zig-zag runs (Reilly, 2007; Reilly and Doran, 2003).

Reaction time is an important component of the ability to respond to stimuli in sports. Simple reaction time incorporates the sensory reception of a stimulus and the initiation of an appropriate response. Simple reaction time is measured by determining the time taken to respond to a light (or sound) stimulus by pressing a button with the index finger or a pedal with the foot. Complex reaction time is measured by selecting the appropriate response when an array of potential stimuli is presented. Reaction time varies between individuals but is faster for the arms than for the legs and is shorter for an auditory compared with a visual stimulus. In sprinting, electronic timing equipment built into the starting blocks measures a sprinter's reaction to the starter's pistol; a reaction faster than 100 m/s has been interpreted as a false start depending on the timing functions designed into the starting blocks by different manufacturers. The timing feature is intended to assist the starter in decision making but there are no definitive data to indicate certainty of a false start. The importance of reaction time is also evidenced by the circumstances of a penalty kick in soccer. From a distance of 10 m, the goalkeeper has little chance of reaching to a ball traveling at a velocity of 40 m/s, because movement time to reach the ball is also involved. Therefore, the goalkeeper must anticipate the direction of the ball to have a realistic chance of stopping the shot if it is well placed.

Whole-body reaction time refers to the ability to move the entire body quickly, rather than only one segment. Although simple reaction time has shown little difference between regular players and substitute players in top soccer teams in Japan (Togari and Takahashi, 1977), the first-team players had faster times for choice whole-body reactions. Linked with the ability to move the body quickly in the right direction is the ability to anticipate the right move. The capability to anticipate moves correctly was found to be a characteristic of elite, talented soccer players compared with those operating at a subelite level (Williams and Reilly, 2000). The talented players were able to assimilate information from a dynamic display and process the information more efficiently than their lower-level counterparts. The model of the stages involved in complex perceptual–motor processes (see figure 1.6) is equally relevant to the execution of sport skills as to occupational ergonomics.

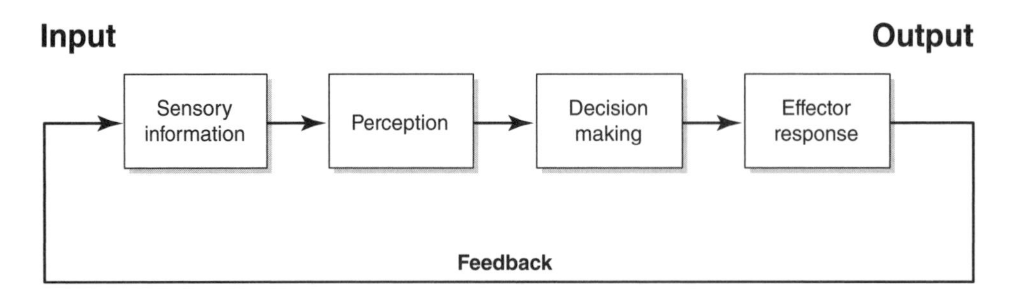

Figure 1.6 A systems model of the decision-making process in ergonomics and sport.

Assessing Mental Load

Many tasks within ergonomics have low loading on the traditional physical criteria yet are a source of stress. These tasks are analyzed according to the mental functions that are challenged and the individual's capacities. Mental workload refers to the degree to which task demands affect the individual's capacities for processing information. The term reflects the relative difference between the available and the required capacity of the information-processing system to perform a task at any one time. The implication is that the person becomes overloaded when the capacity required is greater than that available. Generally, a more demanding task calls for a greater mental effort to perform it adequately and consequently mental effort is used as an indirect measure of individual capacity. As mental effort is increased, there is an automatic decrease in the capacity remaining.

Mental Capacities

The mental load associated with cognitive function and decision making is much more difficult to quantify than is the physical component of work. Nevertheless, various methods have been used in an effort to quantify the relative loading on mental performance capability and to model the decision-making process. Such models separate sensory function and perception from decision making and effector responses, as shown in figure 1.6. The measurement of mental load is equally challenging in sport as in occupational settings.

Although there has been no parsimonious method of quantifying mental load, various approaches have been adopted. A first category includes physiological measures such as heart rate variability, electroencephalography, and electro-oculography (or other eye activity–related measures). The second category of methods includes subjective workload assessments. The third category contains discrete techniques for cognitive task loads. The NASA task load index, for example, has gained acceptance because of its potential applications to both individual and team workload assessments.

Some tasks require sustained attention and the ability to detect stimuli that occur at irregular intervals. When these signals are infrequent the individual may fail to detect them because of time on task, time of day, or inattention caused by distractions. The theory of signal detection was applied to classical vigilance tasks, in which performance was separated into two elements—one a sensory threshold and the other representing fatigue.

In other instances a person must react to novel circumstances or unforeseen changes in the immediate environment. The ability to react correctly will depend on being aware of the situation, assimilating a variety of stimuli quickly, and responding appropriately. The response may call on a combination of competencies, including social skills and avoidance of distractions. Various tools have been derived for quantifying situation awareness, based on ratings or subjective assessments.

Attention refers to selectivity of processing information and is an active rather than a passive process. Attention is not a unitary system but rather consists of

- focused attention, in which the individual tries to respond to one particular input out of the noise of inputs; and
- divided attention, which occurs when the person attempts to respond to two or more inputs concurrently. The task is to attend to both stimuli and react accordingly (Eysenck and Keane, 2001).

Three kinds of memory storage have been suggested:

- Sensory store—holds information briefly and is limited to one sensory modality
- Short-term sensory store—has very limited capacity
- Long-term store—essentially unlimited capacity

Short-term memory is the capacity to store (and recall) information for periods measured in seconds. It is usually measured in terms of its span, which is the longest sequence of items that can be reproduced correctly following a single presentation. For items in random order the span is about 7, hence the phrase "the magical number 7 plus or minus 2." Short-term memory is considered an area of cognitive processing and has been replaced by the notion of working memory. This concept has three components:

- A modality-free central executive that resembles attention
- An articulating loop that is speed based
- A visuospatial scratch pad, specialized for spatial and visual coding

In assessments of working memory, participants are required to manipulate information. For example, they may be asked to recall a string of digits in reverse order.

Dual Tasks

In many instances human capacities are challenged in combination so that errors occur attributable to overloading the weakest link. In view of this observation, human limitations are assessed when secondary tasks are introduced as dependent variables. Examples include the additional requirements of using mobile phones or monitoring global positioning systems when driving. A parallel is found in sport, especially in games, when psychomotor functions and cognitive demands are concomitant with high physiological loading.

Reilly and Smith (1986) showed that performance in a psychomotor task was adversely affected when accompanied by exercise in excess of about 40% $\dot{V}O_2$max. The warm-up effect of exercise was beneficial to psychomotor performance up to that intensity. A similar warm-up effect was observed in cognitive function, although performance did not deteriorate until exercise intensity exceeded 70% $\dot{V}O_2$max (figure 1.7). These findings have implications for sport, heavy industry, and military contexts

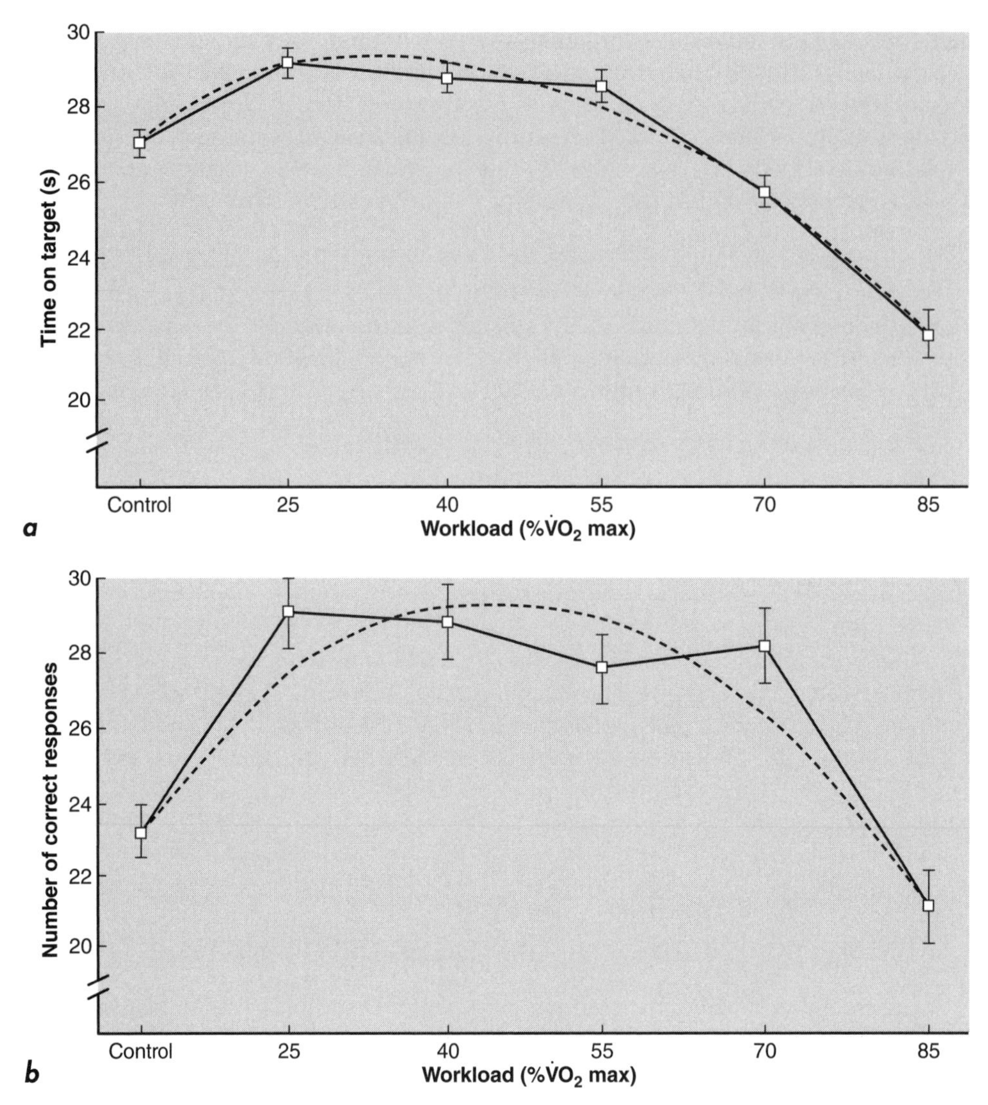

Figure 1.7 The relation between relative metabolic loading and performance in *(a)* a pursuit rotor task and *(b)* mental arithmetic. The curve fitted to the data is shown in each case.

Reprinted, by permission, from T. Reilly and D. Smith, 1986, "Effect of work intensity on performance in a psychomotor task during exercise," *Ergonomics* 29, 601-606.

where psychomotor tasks and mental decision making are often conducted under high metabolic loading.

Occupational Context

In occupational ergonomics the task is modified to suit the characteristics of the human operator. Considerable effort is also made to match the machinery and tools used to be compatible with human function. Task analysis is required to establish critical components of the operation where errors or inefficiencies might arise. A compromise to the comfort and well-being of the individual is avoided, the desired outcome being improvement in overall system performance.

Systems analysis is used in complex activities to optimize performance by separating functions between the human employees and machine operations. Increased automation has gone hand in hand with industrialization, leading to a decline in physical activity associated with industrial and office work. This trend has been linked with a sedentary contemporary lifestyle and has led to the promotion of active recreational programs to restore physical fitness and maintain health.

Similarly, in sport the participant must reach a level of physical fitness to meet the demands of the activity. Any gap between the level of fitness and demands of the activity must be breached by improving fitness through appropriate training. The alternative is to drop to a lower level of competition or to the reserve team. In team sports the fitness requirements may vary with the positional role, and the coach or manager has the option of changing the tactical role of a member with identified deficiencies. The alternatives are to improve fitness through training or drop to a lower level of play or engagement (see figure 1.8).

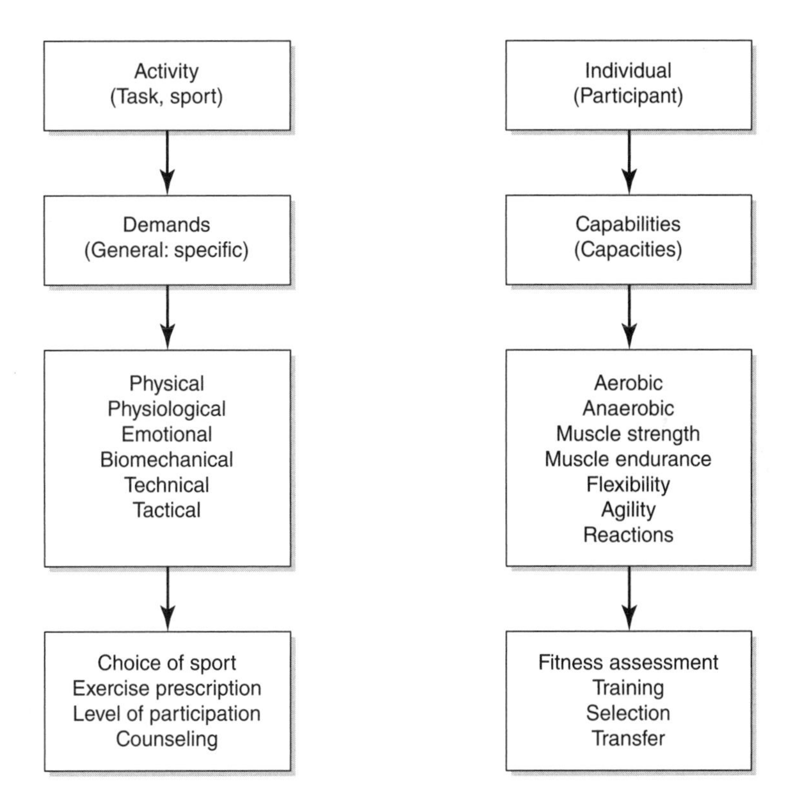

Figure 1.8 How demands of the sport are related to the capabilities of its participants.
Modified from Reilly 1991.

Overview and Summary

Physiological loading can be measured using reaction forces, energy expenditure, heart rate, and body temperature. The effectiveness of any training intervention can be evaluated by means of fitness assessments. Protocols for fitness tests have been standardized for sport science and for specific populations. Separate assessments are made for different physiological capacities. Mental loading is also assessed using various approaches. Using various methods of assessment ensures quality control in collection of fitness data and completes the feedback loop to the individual.

© Tom Reilly

Health and Safety

DEFINITIONS

dynamic control ratio—The eccentric strength of the hamstrings relative to the concentric strength of the quadriceps muscles.

ethics—A way to behave in accord with professional codes.

human error—The production of outcomes or actions not intended at the outset of any action.

immunosuppression—A physiological state whereby the body's defense mechanisms are weakened; usually occurs in an exercise context and is of short duration.

injury risk—The odds or likelihood of incurring trauma attributable to engagement in particular activities.

overload—A principle in training whereby the stimulus presented to the trainee is greater than what he or she is accustomed to.

risk assessment—A formal method of evaluating the likelihood and degree of trauma that is associated with specific activities.

safety—Principles and issues concerned with carrying out activities without incurring damage or personal harm.

specificity—The degree to which the activity or action in the context of training or testing corresponds to the sport concerned. It is the principle of relating measures to the sport or action in question.

MOST human activity involves some risk of an accident that harms the person undertaking the activity or others who are nearby. The resulting damage can be trivial or serious enough to threaten life. Sport, like industrial work and domestic activity, entails some risk of injury, although certain sports involve high-risk activity whereas others carry low risk.

Injuries

Ergonomists refer to a chain of actions that precede accidents. These actions are unplanned events that are attributable to error or an unintended outcome of behavior. This logic supports the use of an error-centered approach to examining injury causes. Not all errors lead to accidents, nor do accidents inevitably lead to injury, and so critical incident analysis is used to identify precursors of accidents by concentrating on instances where mistakes occurred but a serious incident was avoided (see figure 2.1). Safety officers use this technique to develop preventive strategies for aircraft, train, and boat travel as well as for the factory floor and the sports center. Some accidents have catastrophic consequences with tragic loss of life, mostly attributable to system failure and often triggered by **human error.**

The risk of incurring fatality has been calculated for many day-to-day activities such as crossing the road, traveling in a car or motorcycle, and traveling by airplane. These calculations are based on accident statistics that include recordings of deaths normalized for exposure rates. Risk in sport is estimated from information on the frequency of injury occurrence and the exposure rate of participants. The prevalence of injury refers to the known number of participants affected at one time, and the incidence refers to the number of new cases occurring in a given time frame. Exposure to **injury risk** in soccer, for example, is expressed as injuries per 1,000 hr of play or injuries per player exposure. In soccer, each game can be deemed as exposing 22 players to injury over the course of 90 min. In sports medicine, however, there is no definition for what constitutes an injury. In view of the lack of standardization in the presentation of injury statistics for this game, Fuller and colleagues (2006) provided a consensus statement for recording, analyzing, and reporting soccer injuries.

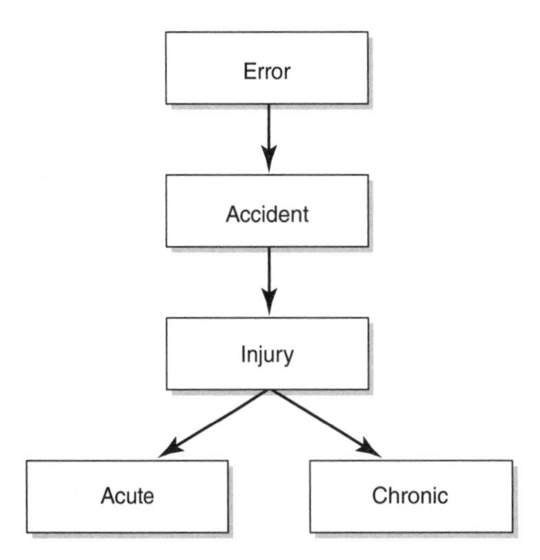

Figure 2.1 The causal chain from error to injury.

Injuries in industrial settings may be attributable to faults in workstation design or equipment or the condition of surfaces or floors. Factors leading to slips, trips, and falls have been a focus of attention in accident causation, as have work-cycle factors. Human error is considered a major causative factor in 45% of critical incidents in nuclear power plants, 60% of accidents in air flight, 80% of marine accidents, and 90% of road traffic accidents (Pheasant, 1991). There is a distinction between genuine errors and violation of safe working practices or the norms of safe behavior. According to risk homeostasis theory, individuals behave according to an internalized target level of risk that they attempt to maintain either by taking chances or avoiding risk as circumstances demand. Risk homeostasis explains why the use of protective equipment might promote more reckless behavior, either in driving on the public roads or playing contact sports. When individuals abandon operational procedures for **safety**, they can be considered negligent.

There are many parallels in occupational contexts and in sport for studying accidents and injuries. In both domains there is a reliance on the integrity with which human–machine systems, equipment, and work spaces are designed to suit the users. The combative component of sport is a distinguishing feature that places its participants at risk. The focus of this chapter is on health and safety considerations and injury occurrence in sport contexts with inferences for ergonomics at large where appropriate.

Each sport has its own unique distribution of injuries and mechanisms of occurrence. Roughly 70% of injuries to runners occur as a consequence of training, whereas this figure is 30% or less in soccer. The majority of soccer injuries are to the lower limbs, whereas in rugby upper-limb injuries are much more common. The distribution of injury location in runners is even more pronounced, because most injuries occur in the lower limbs, although low back problems are common (see figure 2.2). The wide variation in levels of risk associated with different sports was deemed by Fuller (2007) to influence participants' perceptions and acceptance of risk and consequently affects their choice of sport.

A major focus on ergonomics in the United States in the 1970s and 1980s was injury prevention associated with materials handling tasks. One injury prevention approach is documented by the NIOSH landmark document on preemployment testing for physically demanding jobs (NIOSH, 1977). Since that time employers have moved from preemployment testing to job redesign (NIOSH, 1981). The classic publication on the NIOSH lift is an example of job redesign: The weight of various tasks should be limited to fit the U.S. industrial worker

Figure 2.2 Patterns of injuries incurred by distance runners.

population. The revised equation (NIOSH, 1994) provided guidelines for a more widespread range of lifting tasks than the earlier versions, using methods for evaluating asymmtric lifting tasks and lifting objects with suboptimal couplings between the object and the worker's hands. The lifting index provides a relative estimate of the level of physical stress associated with a particular manual lifting task. The estimate is defined by the relationship between the weight of the load and the recommended weight limit; the recommended weight limit represents the weight of load that nearly all healthy workers could lift over an 8 h shift without an increase in the risk of developing lower-back pain.

This clear shift from selecting physically capable employees to redesigning jobs followed early research in both occupational ergonomics and competitive sports. The key factor that emerged from that body of literature is that exercise intensity or work severity is a major risk factor of injury; that is, the risk of injury increases when workers perform work tasks that approach the workers' maximum capacity. Similarly, recreational runners experienced more injuries when the frequency of training was increased above three times per week (Garbutt et al., 1988; Reilly, 1981).

Causes of injury in sport are subdivided into extrinsic and intrinsic mechanisms. The former include inappropriate footwear, dangerous terrain, climatic conditions, playing surfaces, and collision with others or with physical objects. Experiencing direct blows to the body, suffering a hard tackle, and being hit by the ball are sources of injury in contact games. Inadequate clothing or faulty equipment can lead to injury to the climber, and malfunctioning navigation equipment might lead to problems for the sailor. In contrast, intrinsic factors are attributable to poor joint flexibility, fatigue, incorrect biomechanics, muscle weakness, inability of musculoskeletal structures to tolerate high internal forces generated by activity, and an inability to make correct mental decisions during physical stress. A lack of fitness can cause injury, so improving one's fitness level is an effective preventive measure. Warming up properly also protects against injury. Injuries are reduced by increasing safety awareness and fulfilling risk assessment obligations.

The role of exercise intensity in injury has been linked to aerobic exercise. The American College of Sports Medicine (ACSM) has published position papers on the role of exercise in fitness, which for many years served as guidelines for exercise prescription (ACSM, 1978, 1990, 1998). The research foundation for the position statements came largely from the epidemiological research conducted by Pollock and his group (Pollock and Wilmore, 1990; Pollock et al., 1991). A key problem with high-intensity aerobic exercise programs is that the risk of injury increases. There is also an increased risk when the frequency per week does not allow adequate recovery between exercise days. The duration of exercise becomes a causal factor as fatigue sets in during prolonged sessions.

Predisposition to Injury

Individuals differ in their propensity to incur injury. An increased vulnerability can be attributable to behavior in training or competition or to inherent chronic defects or transient deficiencies. Participants carrying such innate risk factors into intensive sport, especially encounters that involve physical contact, are more likely to sustain injury than are their peers. Such deficiencies can be related to physical features or fitness levels. Psychological defects are relevant in the concept of the injury-prone athlete. Alternatively, the cause of injury may be attributable to overload, when the training stimuli are too high for the athlete. The overload can occur in a single session or be a cumulative effect of chronic overtraining.

Physical Predispositions

Injury can result from deficiencies in fitness or innate defects in the participant. Such deficits include muscle weakness, joint instability, limb asymmetry, or incomplete recovery from previous injury. Athletes who carry muscle weaknesses into competition are likely to experience situations where the muscle fails. Such weaknesses can be identified if athletes undergo regular profiling of their muscle strength capabilities. This profiling is available for individuals and teams with a systematized sport science support program. Muscle strength profiling should also disclose asymmetries between left and right limbs, the weaker of which is the side most likely to be affected in locomotor sports. Asymmetry is also reflected in improper hamstrings-to-quadriceps ratios. Some sprinters have disproportionately strong quadriceps relative to their hamstrings, causing injuries to the latter when they are overstretched when running at full speed. Soccer players have strong quadriceps but must balance this adaptation by undergoing corresponding training for the hamstrings. Athletes should pay attention to eccentric as well as concentric muscle contractions in training, in view of the eccentric role of the hamstrings in actions such as kicking a ball. The peak torque of the knee flexors in eccentric actions is compared to that of the knee extensors in concentric mode in computing a **dynamic control ratio.** The eccentric strength of the hamstrings should be equal to the concentric strength of the quadriceps to prevent injury (figure 2.3).

Strength angle profiles are determined using isokinetic force data at a selection of angles throughout the range of movement at a particular joint (Perrin, 1993). This kind of profile is especially relevant in avoiding reinjury, because reduction of strength may be evident only in a restricted range of motion. This deficiency can be corrected by undergoing isometric exercises for the range of motion where muscle strength was reduced.

In a study of Swedish soccer players, personal factors such as joint stability, muscle tightness, inadequate rehabilitation, and lack of training were deemed responsible for 42% of all injuries observed (Ekstrand and Gillqvist, 1982). In an extension of this research, Ekstrand (1982) reported that 67% of soccer players had tight muscles and such players were vulnerable to injury. Tightness was especially evident in the hamstrings and hip adductors. A program of flexibility training among Swedish professional soccer players over a complete season reduced the incidence of injury, leaving little doubt that flexibility protected these players against injury.

Stretching muscle prior to training and match play affects flexibility in the short term. Flexibility routines can be incorporated into the warm-up. Because flexibility is particular to each joint, the stretching routing should be developed for the sport or activity concerned (Reilly and Stirling, 1993). The incidence of injuries over a season was less in games players who paid attention to jogging, technique work (to rehearse game skills), and lower-body flexibility exercises than those who warmed up for the same duration but used a more general warm-up regimen. Warm-up is especially important in cold conditions to raise muscle and body temperature for the strenuous training drills to follow.

Between 17% and 30% of injuries are attributed to an incomplete recovery from previous injury at the same site. Secondary injuries of this type tend to be severe

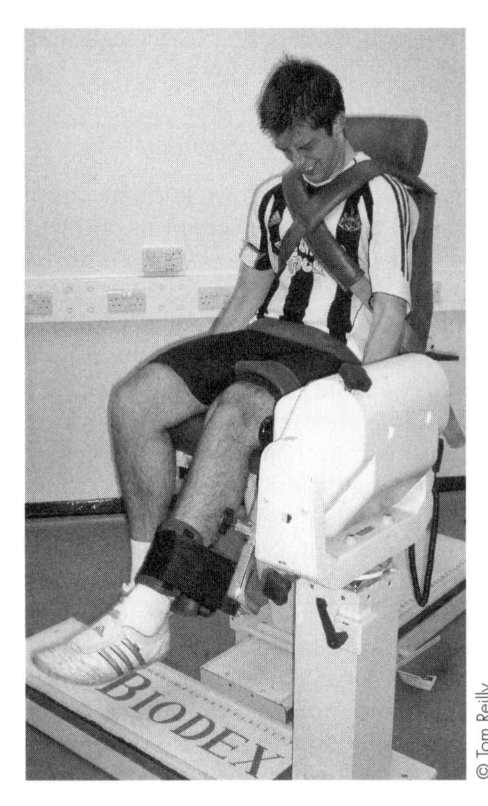

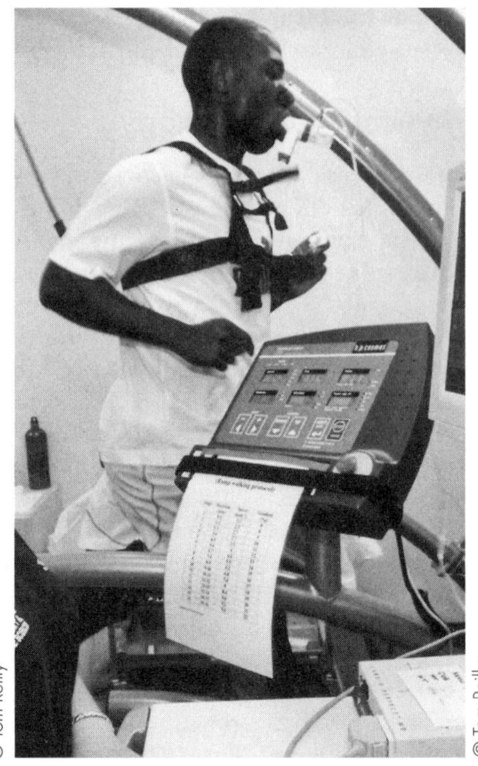

Figure 2.3 Laboratory setup for assessment of leg strength (left) and maximal oxygen uptake (right).

(Hawkins et al., 2001), emphasizing the importance of appropriate and complete rehabilitation. Such rehabilitative intervention should also be timely. Preexisting joint or muscle pain has been identified as an important correlate of injury (Dvorak et al., 2000). Wrapping injured areas was found to prevent reinjury.

Biomechanical imperfections may become apparent only when a participant engages in high training loads. *Morton's foot* refers to a disproportionately long second toe and has been implicated in injuries to road runners. Individuals who overpronate on landing are also subject to foot injuries. These athletes can reduce the risk of injury by using footwear with antipronation design features, whereby the manufacturer builds up the medial side of the shoe to compensate for pronation. Anatomical imperfections also are found in the knee joint, genu varum, and genu valgus, causing injuries when training loads are high.

Psychological Predispositions

It is thought that some individuals are injury prone, the susceptibility being linked to their psychological makeup. Psychological characteristics combined with behavior traits may cause these participants to incur injury more frequently than their peers. Personality may be related to a susceptibility to injury and illness, in that neurotic individuals tend to be cautious, indecisive, and easily stressed. There is some evidence that apprehensiveness is linked to injury, especially when players engaged in contact sports are not fully committed to tackles or offensive moves. In his review of the concept of injury proneness, Sanderson (2003) posited that stress is likely the intervening variable between personality and injury, in that a person's psychological makeup may cause him or her to experience stress in a wide variety of situations, including sport.

A selection of symptoms associated with psychologically vulnerable team players is shown in table 2.1. In developing profiles of accident-prone and overuse-prone young soccer players, Lysens and colleagues (1989) emphasized that psychological factors need to be considered alongside physical traits in determining these profiles. Some people become addicted to exercise, this condition being related to the release of brain endorphins that produce states such as the so-called runner's high. This addictive condition can be difficult to treat, often requiring a life-event catastrophe such as a marital breakdown for the person to realize there is a problem (Wichman and Martin, 1992). Accompanying the "addiction" is a reluctance to recognize when the training volume is sufficient.

Overtraining and Overreaching

Overload is an important principle in training and reflects the fact that biological systems must be presented with training loads beyond what they normally experience if the person is to improve performance. After adapting to a period of overload, the individual is capable of performing at a higher level, thereby achieving a goal of training. The load must then be increased to provide further overload so that the path

Table 2.1 Symptoms Associated With the Psychologically Vulnerable Player

Symptoms	Comments
Discrepancy between ability and aggressiveness	A player with modest ability who is overly aggressive is vulnerable.
Success phobia	Fear of failure is a common and well-understood phenomenon, but the incidence of fear of success should not be underestimated.
Uninhibited aggressiveness	The player presents a danger to himself and others.
Feelings of invulnerability	Such feelings are associated with reckless behavior.
Excessive fear of injury	The apprehension causes overly cautious play, paradoxically making injury more likely in, say, 50-50 tackles.
Extensive history of injuries	Repeated injury may indicate physical or psychological vulnerability.
Concealment or exaggeration of injuries	This indicates probable underlying psychological problems.
Marked anxiety proneness	The overly nervous player's performance is detrimentally affected, and injury is more likely.

to top fitness is a recurring cycle of overload, recovery, and adaptation. The difficulty arises when there is insufficient recovery between training bouts and performance capability begins to reverse rather than improve. This phenomenon is known as overtraining or overreaching.

Overload

Overload, a key principle in training, implies that physiological systems must be challenged beyond normal activity for adaptation to take place. Improvement in fitness accompanies this adaptation process. There is a continual need to reevaluate the training stimulus as the athlete becomes accustomed to challenging exercise, and the training level has to be progressed if further improvements are to be realized. Thus an upward spiral of overload, adaptation, and habituation occurs; training plans must include adequate time for recovery between sessions to permit physiological adaptation to occur and biological processes to regenerate so that the next training bout can be tolerated.

In this recurring cycle of training and overload, the athlete can reach a point where recovery is inadequate and fitness regresses rather than improves. The individual may feel stale, lose motivation, and lack vigor. Performance is impaired, and further increases in training become counterproductive. This syndrome of underperformance is often accompanied by reduced testosterone levels, increased cortisol concentrations, suppressed immune function, and general tiredness. The term *overtraining* is best reserved for physical injuries associated with repetitive loading of body tissues, whereas *overreaching* best describes the general breakdown of ongoing physiological adaptations to exercise.

The corollary of **overload** is "underload," the implication being that training effects are gradually lost if the training stimulus is not maintained at a high enough level or is abandoned. This situation arises during the off-season in professional athletes or following injury. To avoid loss of training effects, many athletes engage in maintenance programs of low-intensity exercise prior to formal preseason training, now a common practice in a number of sports. This strategy avoids the increased risk of injury when training is introduced abruptly. Many sports medicine specialists recommend that injured athletes engage in physical exercise in some guise, either using uninjured limbs or practicing non-weight-bearing exercise such as cycling or exercise in water. This focus on maintenance is an important aspect of rehabilitation.

A further principle in training theory is that of **specificity.** The inference is that training effects are restricted to the muscle groups and physiological systems that are used in a specific sport or activity. An obvious consequence is that overtraining also is specific to the soft tissues and anatomical regions that are most stressed in training or competition. Hence, some injuries are occupational hazards for specific sports and find their way into the terminology of injury. Examples are tennis elbow, jumper's knee, runner's knee, swimmer's shoulder, and footballer's ankle.

Overload Injuries

Overload injuries are associated more with repetitive stresses over time than with acute loading on soft tissues or joints. The occurrence of overload injuries in sport has parallels in occupational health. Examples are repetitive strain injury resulting in wrist and hand pain and the experience of low back pain attributable to manual materials handling.

Achilles tendinitis affects runners and games players. It is characterized by inflammation of the tendinous sheath and may force the athlete to reduce or abandon training temporarily. The injury results from logging excess training mileage, training on hard surfaces, wearing shoes with inadequate cushioning properties, or using faulty biomechanics. Training on a compliant surface such as grass can reduce symptoms, but recovery can be slow attributable to the relatively poor blood supply to the tendon.

Jumper's knee refers to a similar injury to the patellar tendon. The injury is manifest as anterior knee pain and tenderness around the patella and is aggravated when the knee extensors are contracted. Jumper's knee is observed in basketball players, volleyball players, and high jumpers, who commonly experience high forces through the knee joint. The condition includes tendinitis, degeneration, and sometimes partial rupture of the patellar tendon. The surface used for training and competition can be implicated in this injury; spring-loaded surfaces are optimal. Athletes and coaches should note the duration of training when sessions include repetitive jumping and plyometric regimens that entail multiple hopping routines.

Shin splints is a blanket term used to describe pain in the anterior lower leg. It mainly refers to an inflammation of the musculotendinous compartment in the medial margin of the tibia. It is an overuse injury, often affecting runners who train on hard surfaces and soccer players who use inappropriate studs for the type of pitch they play and train on. Athletes with severe shin splints may require surgery to reduce the pressure within the anterior tibial compartment.

Stress fractures are also considered to be caused by overuse. These include tibial stress fractures attributable to high mileage in runners, particularly when a large volume of training is conducted on the road. Female runners are especially vulnerable when secondary amenorrhea is associated with demineralization of the skeleton. Loss of bone mineral attributable to prolonged low estrogen levels causes bones to weaken and incur miniature fractures, which become debilitating.

Chronic exertional compartment syndrome is a reversible condition that results from repetitive activities or those that require a high level of exertion. The cause is increased pressure within a limited anatomical space that compromises the circulation and function of the soft tissues within that space (Dunbar et al., 1998). The most vulnerable compartments in athletes, especially runners, are in the lower legs, notably the anterior, peroneal (lateral), deep posterior, and superficial posterior compartments. Anterior compartment syndrome is more common in runners than in games players. Compartment syndromes are also found in military recruits forced to endure long marches and were originally called *march gangrene*. Persistent complaints and debilitating pain on exercise accompany this morbidity.

Overuse injuries are caused by continued or repetitive actions or exposure of a structure to high loads. These injuries occur as a result of training errors, biomechanical abnormalities, inadequate or appropriate footwear, or unfavorable terrain. Training errors include inadequate warm-up; excessive training regimens; abrupt increases in the duration, frequency, or intensity of training; and inadequate rehabilitation from injury. Biomechanical abnormalities include leg-length discrepancies, soft-tissue inflexibility, incorrect alignment, and joint stiffness. Problems with footwear include poor shock absorption qualities, poor grip, and poor fit. Terrain such as hills or uneven road surfaces can be responsible for injuries as well. Stress fractures, which are microfractures in cortical bone resulting from excessive tensile loading, are classified as overuse injuries (Corrigan and Maitland, 1994). Such injuries are common in the metatarsals, fibula, and tibia.

Exposure to whole-body vibration is a recognized risk factor for low back pain and other disorders. Workers who use vibrating tools such as pneumatic drills and chainsaws for extended periods are particularly vulnerable. Seated humans are more sensitive to vertical vibration at 4 to 5 Hz and more sensitive to horizontal vibrations at lower frequencies. The biomechanical response of the seated operator has conventionally been assessed using kinetic (impedance, apparent mass) and kinematic approaches (Mansfield and Maeda, 2007). Whole-body vibration is indicated from measurement of acceleration of the seat, frequency weighting, and scaling of vibration according to the axis involved. Mansfield and Maeda showed that measuring multi-axis vibration, as mostly occurs in work environments, is more representative than measuring vibration in a single direction attributable to interactions between the different axes.

Vibration is experienced in a range of sports, including all the human–machine events like car racing, motorcycling, and road-race cycling. Such responses are attenuated by designing damping elements in the machine. A paradox is that local stimulation of musculoskeletal structures causes adaptations that are beneficial to performance, provided the frequency and amplitude of vibration are appropriate. Consequently, vibration platforms have gained acceptability as training aids in both professional and recreational sport and are considered elsewhere in this book in a training context.

Immunosuppression

A long-held belief among sport practitioners is that athletes are vulnerable to illness and infection **(immunosuppression)** when they are engaged in strenuous training programs. As a consequence, these athletes tend to pick up common colds and especially upper respiratory tract infections more frequently than sedentary individuals. Only recently has this belief been confirmed with evidence that exercise can have an acute effect on immune function, but the relationship is not linear. The J-shaped curve in this relationship indicates that moderate exercise boosts whereas strenuous exercise impairs the immune system. Whether the exercise is continuous or intermittent at the same overall work rate does not seem to matter (Sari-Sarraf et al., 2006).

The immune system is designed to protect the body against invading pathogens and harmful microorganisms. It has both innate and acquired mechanisms. The innate system includes white blood cells and the cells of the tissue macrophage system, whereas acquired immunity includes antibodies consisting of globulin molecules (B-cell immunity) and activated lymphocytes (T-cell immunity). The innate immune system develops through the thymus gland to form T-lymphocytes and through bone marrow as B-lymphocytes. The T-cells account for about 75% of mononuclear leukocytes in peripheral blood and, in addition to forming subsets important in immunological surveillance, secrete a number of regulatory proteins including the cytokines. The B-lymphocytes can be triggered to differentiate into plasma cells that produce five immunoglobulin subclasses. One of these, salivary immunoglobulin A (s-IgA), is commonly used as a mucosal marker of immunity. The cytokines have a role in orchestrating immune activity and comprise four main groups: 10 interleukins (notably IL-6), tumor necrosis factor, three interferons, and colony-stimulating factors.

The suppression of immune function following strenuous endurance exercise is short lived, lasting 3 to 5 hr (Nieman and Bishop, 2006). This period of vulnerability to bacterial and viral infection is referred to as the *open window* (see figure 2.4). Over these hours, there are changes in both the innate immune system and acquired immune mechanisms. The acute depressive effect has consequences for several immune cell functions including reduced cytokine production from lymphocytes, decreased cytolytic activity of natural killer cells, lowering of s-IgA, and numerous other changes (Gleeson, 2006). The potential immune responses are weakened with a consequent increased risk of incurring an infection. The effect is accentuated by an increase in circulating hormones during endurance exercise—these include cortisol, adrenaline, noradrenaline, and prolactin, the increase in cortisol being modulated in part by cytokines such as IL-6.

There is evidence of a cumulative effect of exercise on the immune responses, attributable to an inadequate recovery of immunological status following exercise. Immunosuppression has been found in young soccer players completing two matches within 48 hr and youth players engaged in training camp activities for 10 days (Reilly and Ekblom, 2005). Immune malfunction also occurs when athletes undertake training sessions twice in one day. From their studies of competitive swimmers, Dimitrou and colleagues (2002) concluded that the optimal time of day for training, with a reduced immunosuppressive effect, is the evening. This recommendation was based on low resting and postexercise cortisol levels and peak saliva flow rate.

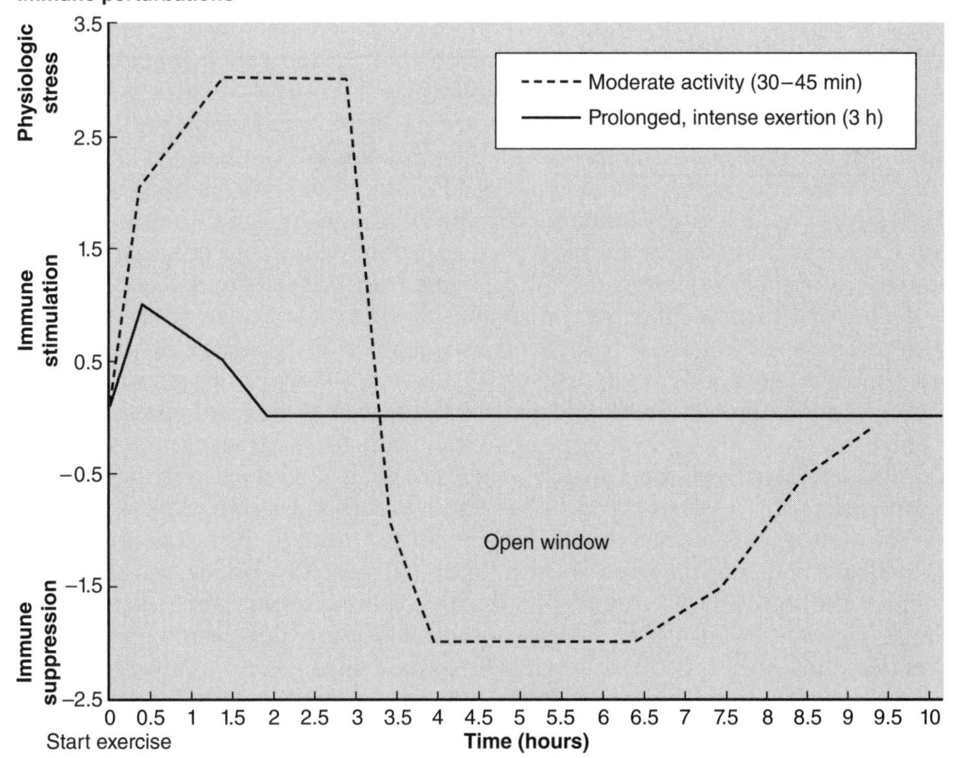

Figure 2.4 The open window theory of immune perturbations following strenuous exercise.

A link has been suggested between overtraining and chronically suppressed immune system function. Gleeson (2006) concluded that although athletes in periods of intensified training are not clinically immune deficient, it is possible that resistance to common minor illnesses is compromised by small changes in several immune variables. Underperformance linked with excessively intense training is attributable to lowered energy stores, and normal performance can be recovered by lowering the training load, allowing longer time for rest between training bouts, and increasing carbohydrate intake. Even in the absence of an underperformance syndrome, carbohydrate supplementation reduces the immunosuppressive effect of exercise.

Risk Assessment

Safety is a fundamental goal in the field of ergonomics. Individuals who are harmed, either by their occupational tasks or by practicing their sport, are of little immediate use to their organizations or their teams. Activities can be scrutinized to consider sources of injury or harm **(risk assessment).** Such scrutiny takes a variety of forms ranging from formal assessment to analysis of critical incidents. Some leisure activities have inherent risk but are also fun and exciting, which must be balanced against the stress these activities induce. These factors are considered in the sections that follow.

Formal Risk Assessment

Safety is a fundamental criterion in ergonomics and is underlined in the regulatory framework for assessing risk. Risk assessment procedures are mandatory in the workplace in many countries. Employers may have other legal obligations, such as prescribed by the COSHH (Control of Substances Hazardous to Health) regulations in the United Kingdom when substances hazardous to health are being handled. The regulations are known to exercise scientists as part of documenting their laboratory activities and conducting them in a safe manner. Such formal procedures are rarely monitored in a real-life sports environment although participants in high-risk sports usually adopt an intuitive approach to assess the risk to themselves and their colleagues.

Formal risk assessment entails an analysis of the task or activity concerned and quantification of the risk involved. The consequence of an accident is described and rated, and the likelihood of its occurrence is estimated. The activity is then scored as to the severity of the possible injury and the probability of its occurrence to produce a risk score. These scores are totaled to give an overall figure for the operation in question. The aggregate score represents an overall index of the risk involved, and the detailed information helps to locate the most hazardous parts of the activity (see table 2.2).

The desired outcome of a risk assessment is a reduction in the risk of injury. The individual completing the risk assessment protocol must leave on file a record of the content. The documentation must show that the risk assessment was completed, that the risks were evaluated, that preventive or control measures were identified, and that these were implemented. This record is reviewed and updated on a regular basis. This record keeping is especially important for sport science support work in laboratory or field conditions as well as for experimental investigations. This forward planning is usually done in conjunction with preparing documents for formal ethical approval for the project concerned.

Table 2.2 Details Required for Rating and Generic Assessment of Risk

Risk rating details		
What is the frequency that an accident or incident can happen? [A]	☐ Common × 6	
	☐ Regular × 5	
	☐ Frequent × 4	
	☐ Occasional × 3	
	☐ Possible × 2	
	☐ Improbable × 1	
What is the severity of the accident or incident? [B]	☐ Multiple deaths × 6	As defined by the Reporting of Injuries, Diseases and Dangerous Occurrences Regulations of 1995 (United Kingdom)
	☐ Death of one person × 5	
	☐ Major injury to several people × 4	
	☐ Major injury to one person × 3	
	☐ Minor injury × 2	Requires treatment
	☐ Trival injury × 1	No treatment required
Risk rating [A] ___ × [B] ___ = ___		

Risk assessment rating		
1-6 low	Trivial to tolerable risk	Trivial: No action is required other than to record incident if near-miss determined.
		Tolerable: No additional controls are required. Consideration may be given to a more cost-effective solution. Monitoring is required to maintain controls.
7-12 medium	Tolerable to moderate risk	Moderate: Efforts should be made to reduce the risk, but the costs of prevention should be carefully measured and limited. Where moderate risks are associated with extremely harmful consequences, further assessment may be required to establish precise likelihood.
13-36 high	Substantial to intolerable risk	Substantial: Work should not be started until the risk has been reduced. Considerable resources may be required. Where the risk involves work in progress, urgent action should be taken.
		Intolerable: Work should not be started (or continued if started) until the risk has been reduced. If it is not possible to reduce risk even with unlimited resources, the work must remain prohibited.

The processes associated with risk assessment were illustrated by Leighton and Beynon (2002), who described their approach to risk assessment for musculoskeletal disorders in health care professionals. The investigators conducted an epidemiological survey of occupational injuries to highlight the injuries incurred in a sample of health care specialists, the anatomical regions mostly affected, and the causes of the injuries. This survey helped to identify the health specializations most at risk and to focus on specific activities. Then the investigators shadowed a sample of workers over a typical work cycle and conducted postural analysis of selected activities. The outcome was that manual handling, such as moving a patient, was highlighted as a high-risk activity, and preventive practices were recommended.

Surveys and Checklists

Survey methods can be used in the early stages of an ergonomics project to identify any hazards associated with task design, equipment use, and workstation layout. An experienced ergonomics team may have a validated checklist that could be applied to an occupational or a sport and recreational setting. Although the checklist is a useful tool in identifying potential hazards and eliminating them, the research team must be alert to novel features of any situation.

The Nordic Musculoskeletal Questionnaire (Thomas, 1992) is a tool for quantifying the period prevalence (12 months), point prevalence (7 days), and intensity of musculoskeletal problems (such as aches, pain, discomfort, numbness, or tingling) in different anatomical areas. These areas include neck; shoulders; elbows; wrists and hands; upper back; lower back; hips, thighs, and buttocks; knees; and ankles and feet. Responses to detailed questions about neck, shoulder, and low back indicate the respondent's risk for problems in these areas, absence attributable to such problems, and effects on work or leisure activities.

Critical Incident Analysis

An error-centered approach is a classical feature of ergonomics research. Elimination of errors should enhance performance and increase effectiveness. A reduction of errors should also reduce accidents and injury. Concentration on critical incidents where errors occurred is a productive means of reducing injuries and increasing awareness of safety.

The critical incident technique has been used in occupational and industrial ergonomics with a view to enhancing preventive programs. Mismatches between controls and visual displays have been highlighted in instances where near-accidents have been reported. In competitive sport, this technique has been used to analyze errors to correct defects in performance, whether these errors have been forced by pressure from the opponent or have occurred without such pressure.

Rahnama and colleagues (2002) used a computerized analysis to investigate critical incidents as precursors of injury in soccer. About 18,000 actions were studied with respect to their potential to cause minor or major injuries (figure 2.5). The most likely actions to cause injury were tackling or charging an opponent or suffering a tackle. The risk was highest during the first and the last 15 min of a game. The risk was elevated also in specific attacking and defending zones of the pitch where possession of the ball is most vigorously contested. Although these trends are of obvious interest to medical and support staff, coaches and trainers should also be aware of them.

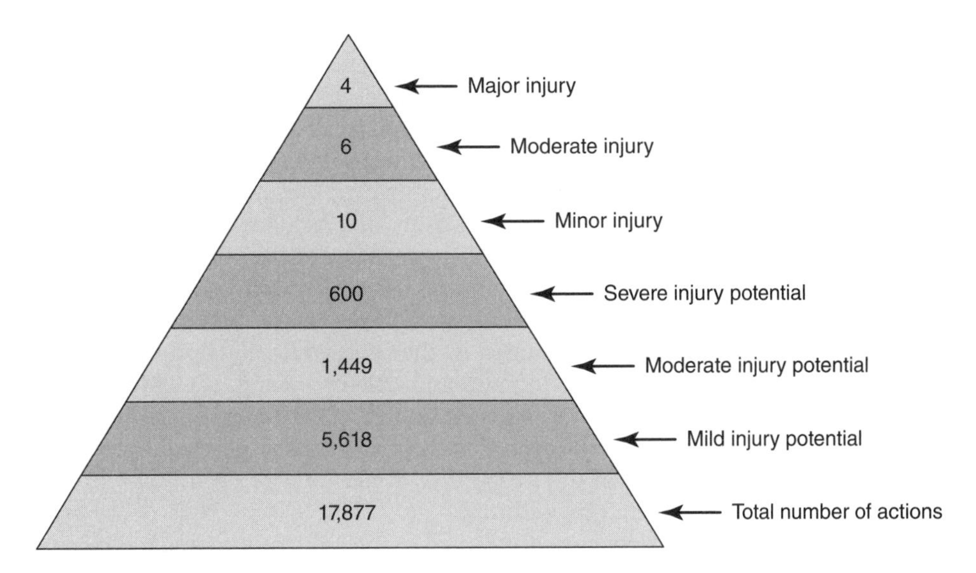

Figure 2.5 Total actions studied, with the number of events in the injury potential and actual injury categories.

Reprinted, by permission, from N. Rahnama, T. Reilly, and A. Lees, 2002, "Injury risk associated with playing actions during competitive soccer," *British Journal of Sports Medicine* 36, 354-359.

Risk Taking and Fun

Ergonomics can be applied to leisure and domestic contexts as well as sport and industrial settings. Leisure activities range from active outdoor adventures to the more organized open amusement parks. In the former case, the participants rely to a large extent on experiencing reasonable weather and negotiable terrain. In the latter case they rely for their safety on the proper functioning of the machinery and the attention of its engineers. In both cases, participants experience awareness of the stressful nature of the environment and are conscious also of the risks involved.

In an attempt to separate the biological correlates of fear and fun, Reilly and colleagues (1985) studied psychological and physiological responses of a group of females to a high-acceleration ride in an amusement park. The mixture of thrill and anxiety was reflected in the widespread screaming of those undertaking the ride. The emotional responses often reach hysteria in such circumstances. The heart rates reached an average of 170 beats/min, peaking when the traveling cars accelerated under the effect of gravity. Gravitational forces exceeded 3.5 *g* during the ride, and blood concentrations of cortisol, adrenaline, and noradrenaline increased approximately twofold. Noradrenaline and adrenaline were most highly related to the feeling of fear, whereas the change in fatty acid levels was correlated with fun. The mean heart rate was reduced by 15 beats/min when the ride was repeated after a 20 min interval. Although accidents during such high-thrill rides are rare, they do occur, especially in temporary fairground facilities.

Participants in free-fall activities such as parachuting and bungee jumping testify to their exhilarating effects. Heart rates of about 160 beats/min have been observed in parachutists awaiting the drop from their aircraft. The emotional tachycardia is less pronounced in experienced than in novice participants. The fear associated with the uncertainty of the experience is reduced in repeated encounters.

A relatively high risk of serious injury does not necessarily deter individuals from taking part in certain sports. An encounter with danger presents a challenge to some participants, and people with certain personality types are characterized as risk takers. In some activities the consequences of an accident can be fatal, for example, high-speed racing on land or at sea and expeditions to the high mountains or the Arctic or Antarctic regions. Nevertheless, climbers will attempt to reach the high mountainous peaks and sailors broach the broad seas simply "because they are there."

Planning in detail, conducting trial runs, and using checklists reduce the risk of accidents and give confidence to the participant. Adoption of standard procedures and adherence to codes of practice outlined by the sport governing body are important safeguards that can make sports like sub-aqua diving, caving, parachuting, and water-skiing enjoyable and relatively safe recreational pursuits. Attention to the proper functioning of equipment is essential and machinery must be maintained thoroughly. Participants must be schooled in the proper use of personal protective equipment and in the conduct of emergency procedures.

Engineering design for safety has contributed to the reduction of tragic accidents within motor racing. The progressive improvements have been reflected in the design of racing cars and motorcycles, fire-protective clothing, track design, and race procedures. Protective criteria must also be extended to consider the safety of spectators whether in Formula One circuits or the slopes of downhill ski runs.

Many recreational activities find their attraction in attempts to overcome nature, including the force of gravity. There are elements of risk in parachute jumping, bungee jumping, sky diving, caving, and breath-hold diving, and where accidents occur they tend to be fatal. The safety of the participant is entrusted to the care of the trainer and the reliability of the equipment being used. It is only to be expected that participants feel anxiety and anticipation before participating in those high-risk activities.

Spectator Safety

Special attention is directed toward crowd control and spectator safety when large attendances congregate for major sporting events. Modern sport arenas are designed with the comfort and safety of spectators in mind as well as the quality of the playing environment. Seating, space for sitting comfort, ease of access to catering and restrooms, and entry and exit from the stadium are all important considerations.

Issues arise also from crowd control and patterns of pedestrian traffic to and from sporting events. In team sports the situation can be complicated by antagonism between supporters of rival teams, which has been a source of trouble in soccer in particular. The phenomenon of "soccer hooliganism" gave rise to frequent skirmishes between fans outside and inside the grounds, the social origins of which were described by Dunning and colleagues (1988). Alcohol was thought to have been a factor in crowd violence and hooliganism on the soccer terraces (Reilly, 2005). This belief led to the banning of alcohol at soccer and cricket grounds in Britain in the mid-1980s. The restrictions were extended to other sporting events in the years that followed, before being relaxed as crowd surveillance methods became effective.

Major disasters have occurred at sport stadiums around the world, attributed to causes such as overcrowding, unsafe facilities, and accidents such as fires. Soccer hooliganism was implicated in the deaths of Juventus fans at the Heysel Stadium

(Brussels) in 1986 when some of them died in terrace crushes. Overcrowding was implicated in the deaths of 96 Liverpool supporters at Hillsborough (Sheffield) in 1989. Improvements in stadium designs have reduced the risk of such disastrous crushes and have improved emergency management. Surveillance of spectators with closed-circuit television, control of pedestrian traffic outside the grounds by horse-mounted police, and good stewarding inside the stadium have increased safety. The traffic flow of visiting supporters to car parks, coaches, and train stations also must be considered. Event organizers are responsible for determining the number of tickets to issue and the method of allocation.

Inevitably when large numbers of people congregate, health-related problems occur among individual spectators, especially when conditions are exceptionally hot or cold. First-aid facilities must be in place with clear guidelines for dealing with emergencies. In major venues, these include medical emergencies such as cardiac arrests, which require that defibrillation equipment be available. Such provision is also necessary at outdoor music festivals, horse-racing venues, and similar large gatherings.

Edwards (1991) focused on disaster plans for air shows, because these events can attract large numbers of the viewing public. The necessity for disaster drills and coordination between all rescue agencies was emphasized. Rescue personnel and command posts need to be highly visible to be effective. Good communication includes having dedicated radio frequencies at site and contact between security and paramedical and medical personnel. Edwards recommended identification of separate sites for treating injuries according to their severity. The principles of disaster planning apply across many scenarios where large crowds of spectators are assembled.

During events in enclosed stadiums and for some events in open terrain, the safety of participants as well as spectators must be considered. Boxers and wrestlers are elevated from spectators and separated from them by the ropes of the ring. Safety nets on downhill ski courses prevent the participant from going out of control and also provide a barrier for crowd protection. Injuries to participants may occur because of the surroundings immediately around the playing area. In games players, these features might include the quality of the surface adjacent to the pitch, the placement of advertising, and the proximity of the boundary between spectator areas. A risk assessment of the surroundings reduces the chance of injury during events.

Human Ethics and Risk

It is universally accepted that individuals have an inalienable human right to protection against harm by others. This principle is enshrined in international legislation, and even in human warfare there are agreed-upon rights allowed to captives and prisoners of war. Most professions have codes of conduct that bind their members in their formal behavior toward the communities they serve. The British Association for Sport and Exercise Sciences, for example, has a code of practice for each of its accredited practitioners—biomechanics, physiology, psychology, performance analysis. The longest established code of practice is probably the oath of physicians that prohibits the release of confidential information about the patient.

It is helpful to distinguish between morals and ethics. Morality refers to a set of beliefs or principles that guide each individual: These principles can be determined by religious practices or can reflect a purely individual philosophy that may be intuitive. A moral stance is adopted when a sports club turns down sponsorship from a tobacco company, for example, on the ground that members cannot support a company whose products harm human health. **Ethics** refers to a way of behavior and in instances such as experimental work on human subjects must be under a recognized form of control. In their quest to solve real-life problems, ergonomists may conduct laboratory investigations using human subjects. These investigators need to be aware of the limits to which they can go in the use of human volunteer participants in their studies and the obligations they must accept.

Philosophers have not agreed on the universality of ethical principles, illustrated in the extremes of the categorical imperative on one hand and situation ethics on the other. The former represents the views that there are absolute directives that indicate whether a particular action is right or wrong, and always so. An opposite viewpoint is that whether actions are right or wrong depends on the circumstances. It can be argued that the code adopted by mountaineers on the high peaks whereby a climber dying of hypothermia is abandoned by a colleague in order to preserve the latter's slim chance of survival is an example of situation ethics as the lesser of two evils. The alternative is that both climbers would die on the mountain.

The rights of human subjects in experimental work were first endorsed in the Helsinki Convention for the Use of Human Subjects in 1949. This reflected the necessity for a formal body at institutional level to oversee the recruitment of subjects and the conduct of research on them, especially in view of the litigious consequences of anything going wrong. Research ethics committees are designed to protect the integrity and dignity of the volunteer subject or participant and to ensure that he or she comes to no physical or psychological harm. The risk to the research worker is also a concern in certain studies.

Ergonomists and sport scientists must get ethics approval before they can embark on any experimental work. This requirement applies in North America and in the majority of European nations. Ergonomics interventions based on observations of human participants must also have formal ethical support. The ergonomist working independently of a university is placed differently with respect to this rigid requirement but is ultimately bound by the profession's code of practice. Most academic journals will not accept work for publication without an explicit statement about the ethics involved, and personal indemnity may be compromised without ethical approval.

To receive approval from a research ethics committee, a research project must be thoroughly planned and must conform to an accepted research design. It would not be deemed ethical to conduct work on participants when the research design is flawed. The application for committee approval requires the applicants to address a number of features, including the project, the investigators, the procedures, the participants, and the handling of the data collected. The originality of the work and the benefits to the community must be addressed. Information is also provided about the competence and experience of the investigators and the discomfort to participants caused by the experimental procedures. These might range from noninvasive procedures such as monitoring surface electromyography to recording rectal temperature or performing muscle biopsies. The handling of body fluids—such as sweat, blood, urine, and saliva—is subject to special provision within health and safety regulations.

Applicants requesting approval from a research ethics committee must explain the recruitment of participants and how information about the project is presented to them. The details are provided to participants in a participant information sheet, and participants must provide written signed consent for their participation. A screening questionnaire such as the PAR-Q (Kordich, 2004) is useful in identifying individuals who require an additional medical screen prior to participation in the project. Special provision is made for vulnerable groups, including pregnant women, children, elderly people, people who are ill or disabled, prisoners, and minority groups. Care is also taken that the researcher is not vulnerable at the site where observations are being made, for example, a research student and novice climber proposing to accompany an experienced group on a dangerous climbing maneuver.

Exercise scientists must take due care when conducting physiological tests on athletes or members of the general population for purposes of fitness assessment. Some form of pretest screening must be used, such as a questionnaire. Even when maximal functional assessments are being made, the participant must be assured that he or she can withdraw from the study or test without any recriminations. Active athletes are usually accustomed to exercising to exhaustion, and such tests are not an inordinate difficulty for them. Blanket ethical approval can be given for the conduct of procedures such as determination of $\dot{V}O_2max$, peak isokinetic torque, or lactate threshold, but such approval does not negate the need for informed consent. Without having ethical approval for work conducted on humans, the research group is unlikely to find a publication outlet for the findings and, what is more, may be deemed personally liable should harm occur to the participant.

Overview and Summary

The competitive nature of sport implies a risk of injury to participants. The likelihood of injury varies with the activity and sometimes with the level of competition. Training programs for sport should incorporate preventive elements that reduce the risk of soft-tissue injuries. Careful planning of training regimens is needed so that the overload involved does not itself induce musculoskeletal damage or lead to overtraining. Risk is an attraction inherent in adventure activities, the self-imposed challenge being to overcome environmental adversities. Experience provides insights into the possibilities of unforeseen accidents. Guidelines for health and safety are important in raising awareness of hazards to be faced and taking measures to reduce the risk of accidents.

© Tom Reilly

Environmental Stress

DEFINITIONS

acclimatization—A physiological process that allows the individual to adapt to specific environmental stresses.

air ions—Ions produced by energy sources that displace one electron from a molecule of one of the more common atmospheric gasses.

hyperthermia—Overheating of the body so that core temperature approaches the outer limit of its normal narrow range of elevation.

hypothalamus—Part of the brain that regulates body temperature.

hypoxia—Relative lack of oxygen, caused at altitude by the decrease in ambient pressure.

muscle glycogen—Source of energy used in exercise of high intensity.

precooling—Strategy to reduce body temperature before exposure to hot conditions.

presbycusis—Damage to hearing associated with the aging process.

three

SPORT is played throughout the world and in many different environments. The environment is sometimes unsuitable for sport because of seasonal variations in climate that interrupt the competitive program. In cold and wet weather, it becomes impossible to maintain playing pitches to an adequate standard and the conditions are too harsh for the comfort of players. At the other climatic extreme is the difficulty of coping with high heat and humidity. The hottest part of the day can be avoided by starting competitions in the morning or late evening. This timing is not always practical for international tournaments, and athletes from temperate climates may have to compete in unfamiliar conditions. They will be unable to cope well unless they have prepared for the climate.

Athletes are sometimes required to compete or train at moderate to high elevations above sea level. Training camps for top teams are regularly held at altitude resorts, and this environment provides a particularly novel challenge to sea-level inhabitants. Endurance athletes can be hampered by altitudes of 2 km and above, whereas climbers and mountaineers encounter more extreme altitudes.

Acclimatization enables the human body to adapt partially to environmental challenges. Many physiological functions can adjust to new environmental stressors. The processes of adaptation differ in time scale and degree according to the stresses involved, which can interact with each other. Environmental circumstances can be so extreme that the body cannot acclimatize and the time period for safe exposure is limited.

The major environmental variables that affect sport performance and training are considered in this chapter: heat, cold, hypoxia, ambient pressure, air quality, weather conditions, and noise. The biological background is provided prior to a description of the consequences of environmental conditions for athletes and tourists.

Thermoregulation

The temperature of the human body is relatively constant, being regulated about a value of 37 °C. Core temperature refers to temperature within the body's central organs, usually measured as rectal, tympanic, esophageal, or intestinal temperature. Oral and axillary temperatures are less reliable and are rarely used in studies of athletes.

The body can be thought of as a shell surrounding a warm body core. Mean skin temperature is usually about 33 °C, representing a gradient of about 4 °C from core to shell. The temperature of the shell responds to changes in environmental temperatures, and the normal temperature gradient from skin to the air facilitates a loss of heat to the environment.

The human body exchanges heat with the environment through different routes to achieve equilibrium. The heat balance equation is expressed as

$$M - S = E \pm C \pm R \pm K$$

where M = metabolic rate, S = heat storage, E = evaporation, C = convection, R = radiation, and K = conduction.

Heat loss and heat gain must be in balance for thermal equilibrium. Metabolic processes produce heat, the basal metabolic rate being about 1 $kcal \cdot kg^{-1} \cdot hr^{-1}$. One kilocalorie (4.186 kJ) is the energy required to raise 1 kg of water 1 °C. Energy expenditure during strenuous exercise can increase this value by a factor of 15 to 16, but only 20% to 25% of the energy expended is reflected in external power output. The remaining 75% to 80% is dissipated as heat within the active tissues, causing heat storage within the body. The body has physiological mechanisms for losing heat, thereby avoiding overheating. The thermal state of the body must be maintained in circumstances where heat is lost very rapidly to the environment, such as occurs in very cold weather.

Neurobiology and Thermoregulation

Specialized nerve cells within the **hypothalamus** regulate body temperature. The neurons in the anterior portion constitute the heat loss center, initiating responses that lead to a loss of heat (figure 3.1). The posterior hypothalamus contains the heat gain center that operates to preserve body heat when the environment is cold. Skin

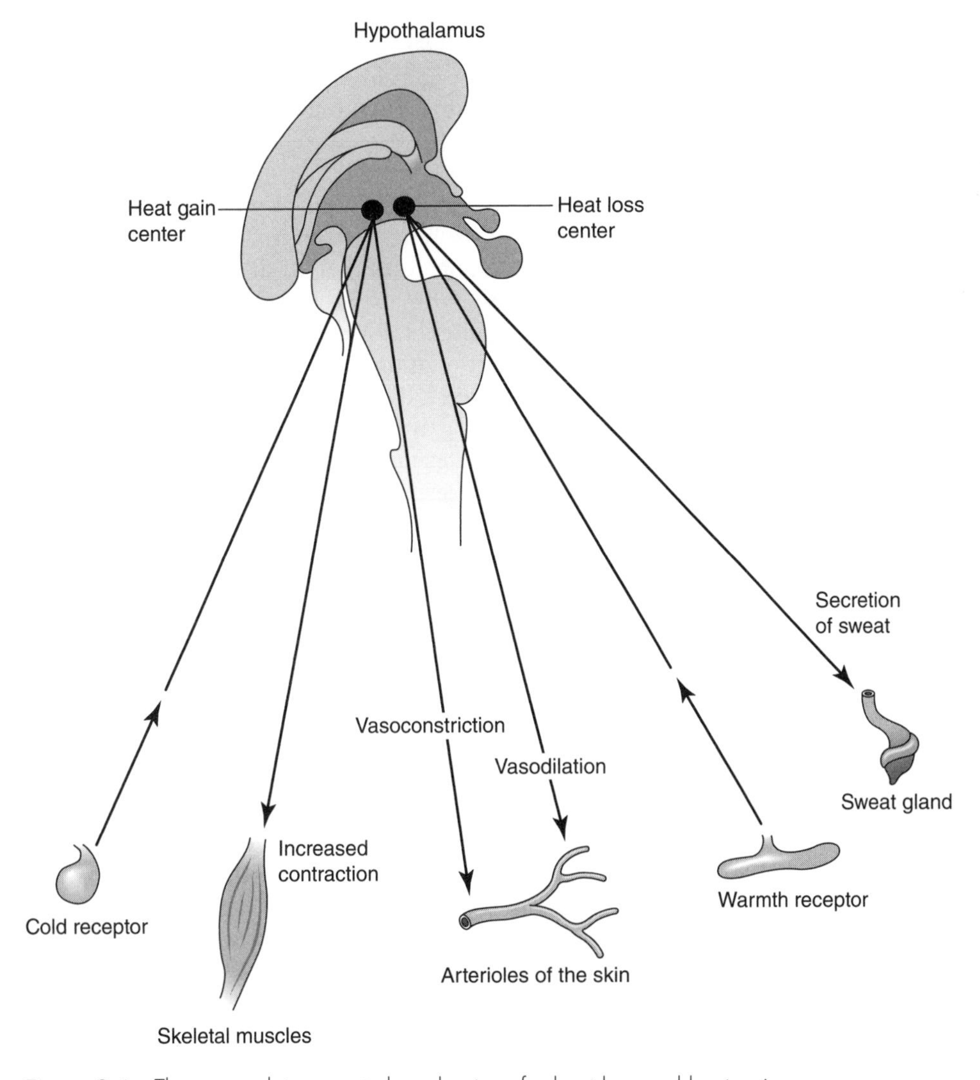

Figure 3.1 Thermoregulatory control mechanisms for heat loss and heat gain.
Reprinted from *Sport, Exercise, and Environmental Physiology*, T. Reilly and J. Waterhouse, pg. 16, Copyright Elsevier 2005.

receptors for heat and cold provide afferent pathways from the periphery to these control centers.

When conditions are hot, peripheral vasodilation causes a redistribution of blood to the skin where it can be cooled. Subsequently, eccrine sweat glands secrete a solution onto the skin surface for evaporative cooling. Conversely, in cold conditions vasoconstriction reduces blood flow to the skin, allowing shell temperature to decrease but protecting core temperature.

The hypothalamic cells that control thermoregulatory responses sense the temperature of blood that flows through them. These cells receive signals from warmth and cold receptors located in the skin. The heat loss and gain centers thereby receive information about both the body's internal thermal state and environmental conditions.

Heat Stress

The degree of **hyperthermia** and the rate of increase in core temperature limit performance when conditions are hot (Drust et al., 2005). Hyperthermia induces fatigue indirectly by reducing blood flow in critical areas (e.g., brain, splanchnic tissues) or acts directly on the central nervous system (CNS) to reduce arousal and the ability to activate muscle.

A direct effect of high core temperature on the CNS can be more important than the effects on cardiovascular and blood pressure responses. Attainment of a critical internal body temperature around 40 °C has been associated with reduced voluntary activation during isometric (Morrison et al., 2004; Nybo and Secher, 2004; Todd et al., 2005) and shortening and lengthening muscle contractions (Martin et al., 2004) in various muscle groups, with this activation failure being located at or above the cortical level (Todd et al., 2005). Other changes indicative of alterations in CNS function have been observed via electroencephalogram (Nybo and Secher, 2004).

It has been suggested that the brain regulates the number of motor units that are recruited or derecruited during prolonged exercise to maintain body temperature below a critical level (Marino, 2004; Morrison et al., 2004), preserving cell function and avoiding catastrophe. An electromyogram can reveal central activation in the reduced work rate that characterizes fatigue. A superimposed twitch technique can separate central and peripheral factors in such cases.

Neurobiological mechanisms underlie hyperthermic fatigue, irrespective of whether a critical core temperature is reached or there is a gradual anticipation of its occurrence. Alterations in central serotonergic and dopaminergic activity (Cheung and Sleivert, 2004; Cheuvront et al., 2004) can be involved; prolactin, a surrogate for serotonin, is elevated after both passive and active heat exposures (Low et al., 2005a; 2005b). Catecholaminergic neurotransmission also seems to be important in propagating the inhibitory signals that may arise from the CNS during hyperthermia (Watson et al., 2005a). Other neurotransmitters such as glutamate and γ-aminobutyric acid are affected because of the increased ammonia concentrations associated with heat stress (Mohr et al., 2006). These changes are compounded by altered permeability of the blood–brain barrier, which also is associated with hyperthermia (Watson et al., 2005b).

Heat and Exercise Responses

The temperature within the active muscles and the core temperature both increase during exercise. Skin temperature is elevated further when the exercise is performed in hot conditions. Blood flow to the skin is increased to facilitate the loss of heat: The body surface can lose heat to the environment (by convection and radiation) attributable to the warm blood being diverted through its subcutaneous layers. As the cardiac output approaches or reaches maximal values during intense exercise, the increased cutaneous blood flow can compromise blood supply to the active skeletal muscles. In such instances, the athlete has to reduce the exercise intensity or take longer recovery periods if the exercise is intermittent.

The optimal ambient temperature for marathon running is about 14 °C. The work rates of games players during matches and in training are directly affected by high ambient temperatures. The distance covered in high-intensity running during soccer match play in 30 °C was 500 m compared with 900 m when the temperature was 10 °C lower (Ekblom, 1986). This lowered work rate reflects changes in the overall pace of play. The amount of increase in core temperature is affected by the exercise intensity and the competitive level. In ambient temperatures of 20 to 25 °C, rectal temperatures of Swedish First Division players averaged 39.5 °C compared with 39.1 °C for players of lower divisions (Ekblom, 1986).

Dilation of peripheral blood vessels increases blood flow to the skin. The increased vasodilation reduces peripheral resistance and causes blood pressure to decrease; the protection of blood pressure from too large a decrease limits the amount of vasodilation that occurs. The kidney hormone renin stimulates angiotensin, a powerful vasoconstrictor, and this response offsets a decrease in blood pressure. The decline in blood pressure is a risk when prolonged training or competition is conducted in the heat and is related to collapse following exercise in hot conditions.

As core temperature increases the sweat glands are stimulated, evaporative sweating being the major route of heat loss to the environment during intense exercise. These glands are noradrenergic and secrete a dilute solution containing electrolytes and trace elements. Heat is lost to the environment only when the fluid is vaporized on the body's surface, no heat being exchanged if sweat drips off or is wiped away. Evaporative heat loss is decreased when the relative humidity is high, because the air is already highly saturated with water vapor. Hot, humid conditions are especially detrimental to performance and increase the risk of heat injury.

Two liters of fluid or more can be lost in 1 hr of exercise in the heat; this figure varies with climatic conditions and between participants. Individuals who sweat profusely may be dehydrated after 60 to 90 min, whereas those who sweat little will be at risk of hyperthermia. A fluid loss of 2% body mass is thought sufficient to impair performance. A fluid loss of 3.1% body mass was reported during a soccer match at 33 °C and 40% relative humidity and also when ambient temperature was 26.3 °C but humidity was 78% (Mustafa and Mahmoud, 1979). Soccer players training in the evening when the temperature was 32 °C experienced a net loss of body mass of 1.6% ± 0.6%, despite having free access to a sports drink for the 90 min of training (Shirreffs et al., 2005).

Reduced plasma volume attributable to sweat loss compromises the supply of blood available to the active muscles and to the skin for cooling. The endocrine glands and kidneys attempt to conserve body water and electrolytes, but the needs of thermoregulation override these mechanisms and the athlete may become dangerously dehydrated through continued sweating. The main hormones that protect against dehydration are vasopressin (antidiuretic hormone), which is produced by the pituitary gland, and aldosterone, which is secreted by the adrenal cortex and stimulates the kidneys to conserve sodium.

Athletes should be adequately hydrated before playing and training in the heat. Water can be lost through sweat at a faster rate than it can be replaced by means of drinking and subsequent absorption through the small intestine. Thirst is not a precise indicator of the dehydration level. Athletes should consciously try to drink regularly, about 200 ml every 15 to 20 min if possible, when training in the heat. The primary need is for water because sweat is hypotonic. Electrolyte and carbohydrate solutions can be more effective than water in enhancing intestinal absorption.

Many components of sport performance are impaired once core temperature increases above an optimal value of about 38.3 to 38.5 °C. Progressive levels of dehydration also cause performance to deteriorate, and cognitive as well as physical and psychomotor aspects of skill can be affected. Reilly and Lewis (1985) reported that cognitive function was best maintained when an energy drink was provided to subjects compared with water only, which itself was superior to no fluid provision (see figure 3.2). The investigators required subjects to add columns of two-digit numbers presented as one slide at a time as quickly as possible while placing equal emphasis on accuracy. Similar trends were observed for both the amount correct and the number of sums attempted.

The relative exercise intensity (%$\dot{V}O_2$max) rather than the absolute workload determines thermal strain. The higher are the maximal aerobic power ($\dot{V}O_2$max) and maximal cardiac output, the lower is the thermal strain on the individual. A well-trained cardiovascular system helps the athlete cope with the dual roles of supplying oxygen to the active muscles during exercise and serving thermoregulatory needs. Such a person will acclimatize more quickly than one who is unfit. Training improves exercise tolerance in the heat but does not eliminate the necessity for heat acclimatization.

A metabolic consequence of exercising in hot conditions is that **muscle glycogen** is used more rapidly than normal. Intermittent exercise at 41 °C for 60 min increased muscle glycogen utilization compared with exercise at 9 °C (Febbraio, 2001). During such activity, there is a corresponding shift in the respiratory exchange ratio and a decrease in the use of intramuscular triglycerides. Glycogen content in brain astrocytes is also likely to decrease. The outcome of these changes is that fatigue occurs earlier than normal.

Thermoregulation During Exercise

The problems experienced by elite performers in maintaining thermoregulation and avoiding extreme dehydration differ in important ways from those of recreational participants. Elite athletes benefit more from training adaptations with an expanded plasma volume that allows more blood to be shunted for peripheral convective cooling. Augmented sweating in elite athletes leads to a progressive reduction in body water during exercise. Conversely, the lower-level competitor experiences less heat storage and may overcompensate for fluid losses by drinking copiously at all available opportunities. In such cases, hyponatremia can result and has been reported among the slower finishers in the Boston marathon, for example (Almond et al., 2005). Hyponatremia has been a concern in underground miners working long hours in very hot conditions and drinking frequently. It has also been a concern in ultramarathon events and Ironman triathlons (Laursen et al., 2006) that last for up to 12 hours in warm conditions that do not cause thermoregulatory failure. To prevent hyponatremia in these circumstances, Montain and colleagues (2005) produced a quantitative

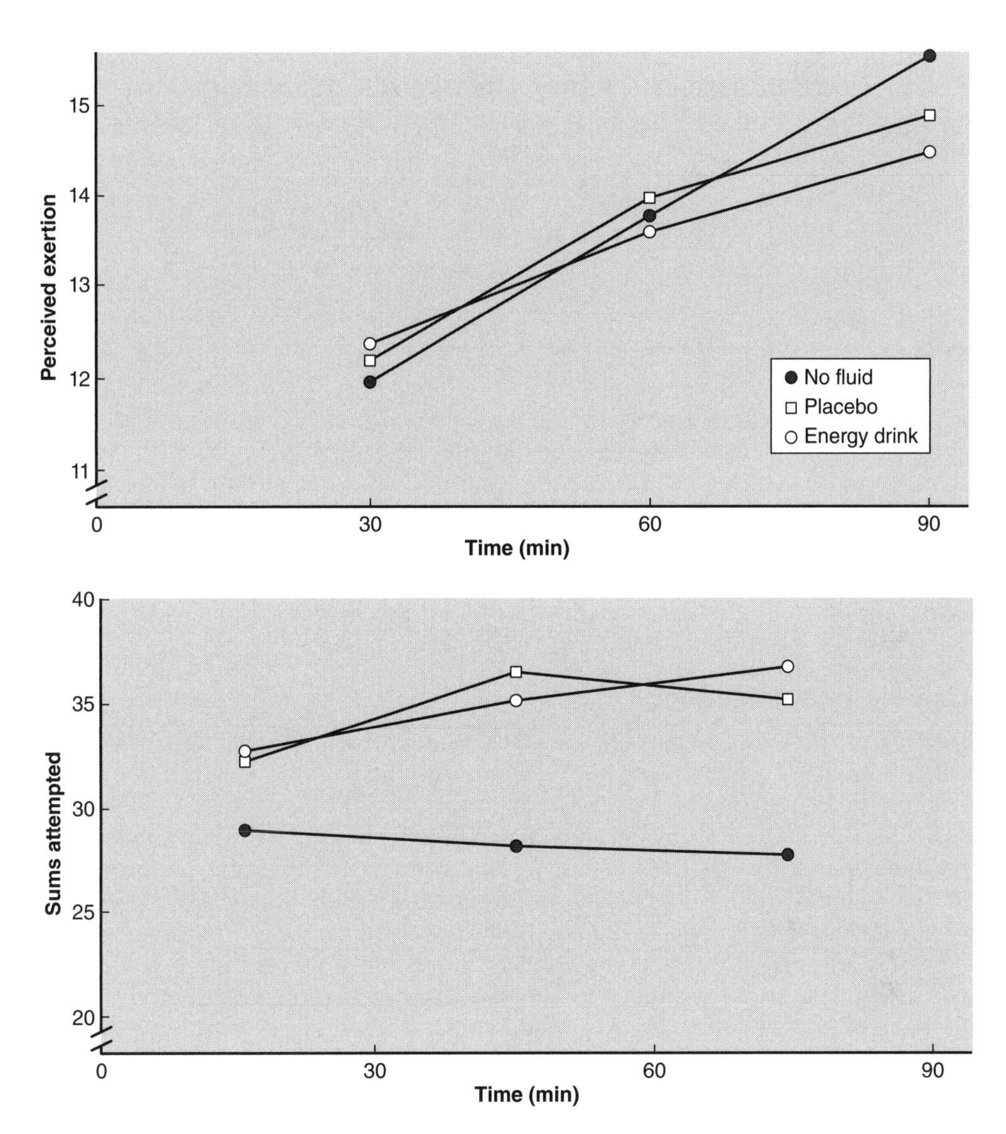

Figure 3.2 Rating of perceived exertion and mental performance under conditions of no fluid, a placebo, and an energy drink.

Reprinted, by permission, from T. Reilly and M. Williams, 2003, *Science and soccer*, 2nd ed. (London: Routledge), fig. 12.3, p. 169.

model to prescribe intakes of electrolyte-containing beverages that would minimize drinking relative to sweating rate.

High central body temperature can impair endurance performance before muscle glycogen stores are depleted (Nybo and Secher, 2004), suggesting that a critical core temperature limits performance in the heat. An opposing view is that blood flow to the active muscle is compromised because of the requirement to maintain blood pressure. During self-paced activity, the exercise intensity may be down-regulated to avoid reaching a critical brain temperature (Marino, 2004). There is likely an interplay between various biological triggers of hyperthermic fatigue (Cheung and Sleivert, 2004), and both central and peripheral mechanisms are implicated. There may not

Mechanisms Associated with Lowering of Body Temperature as an Ergogenic Maneuver

Precooling

- Increased time to critical t score
- Increased work rate
- Decreased sweat production
- Improved thermal comfort
- Altered thermal sensibility
- Delayed fatigue triggers

be a fixed set point for thermoregulation but rather one that is variable with training, acclimatization, menstrual cycle phase, and time of day (Waterhouse et al., 2005). Assumption of a ceiling in body temperature has led to promotion of preventive measures such as heat acclimatization, **precooling** before exercise, and administration of sports drinks and pharmacological agents. The mechanisms underlying these maneuvers have been partly explained (see highlight box above) and recent findings have contributed to the understanding of cellular and neuroendocrine adaptations accompanying improved heat tolerance (Reilly et al., 2006).

Clothing and Precooling

Participants must choose appropriate clothing when exercising in hot conditions. Light, loose clothing helps create convective air currents to cool the skin. Clothing of natural fiber such as cotton (or at least a cotton–polyester mix) is desirable under warm and radiant environmental conditions. When training takes place in hot conditions, a cooling vest can be effective in reducing heat stress and maintaining performance. One study showed that a lightweight cooling vest worn during the rest and warm-up prior to exercise and during the cool-down after exercise provided a significant thermoregulatory advantage (Webster et al., 2005). The advantage was evident in a decrease in core temperature, skin temperature, and sweat rate and in enhanced recovery of the thermoregulatory system postexercise. Endurance performance was impaired when the cooling vest was discarded after the warm-up. A commercial cooling suit designed for applications in sport may not be available to many teams in hot weather, and other strategies must be used to keep the body from overheating during exercise.

Precooling is used to reduce body temperature and enhance the body's capacity to store heat during the subsequent bout of exercise (Drust et al., 2000). This strategy is particularly useful in sports that extend over long periods under high environmental temperatures and where rules permit limited opportunity to consume fluids.

Numerous cooling modalities promote both local (e.g., ice packs) and systemic (e.g., ice packs, cold air, water immersion, and ice vests) reductions in body temperature (Castle et al., 2006; Grahn et al., 2005). Because of its high rate of cooling, cold-water immersion is frequently used to evaluate the physiological effects of precooling on exercise performance (Castle et al., 2006; Hasegawa et al., 2006a). Availability of facilities and time constraints inherent in using cold-water immersion restrict its application in many field settings. Precooling in competitive situations relies on minimal disruption to the athlete's preparation for competition.

Cooling vests worn during the precompetition warm-up enhanced time to exhaustion during constant-rate exercise (Hasegawa et al., 2006b; Webster et al., 2005) and improved performance during self-paced trials (Arngrimsson et al., 2004). Similarly, application of ice packs to the lower limbs resulting in both local and general changes in body temperature improved sprint performance during 40 min of intermittent exercise in the heat (34 °C) (Castle et al., 2006). Hand cooling also may have value in reducing whole-body temperature (Grahn et al., 2005; Hsu et al., 2005). Immersion in water for 30 min and subsequent use of an ice jacket were found by Quod and colleagues (2008) to be more effective than use of an ice jacket alone in reducing rectal temperature and improving cycling performance in a time trial over approximately 40 min.

Following precooling maneuvers, the body's capacity to store heat is enhanced for the subsequent bout of exercise because of the lowered starting value. The resulting thermal advantage is characterized by reductions in core body temperature, skin temperature, subjective estimates of thermal strain, and whole-body sweat loss (Arngrimsson et al., 2004; Castle et al., 2006; Hasegawa et al., 2006a, 2006b; Webster et al., 2005). A decrease in cardiovascular (Arngrimsson et al., 2004; Hasegawa et al., 2006a, 2006b) and metabolic strain (Hasegawa et al., 2006a) and a reduced perception of effort (Hasegawa et al., 2006b) are also consistent with performance improvements.

The benefits of precooling include more than enhancing performance by increasing heat storage capacity and the time to reach a critical thermal limit (figure 3.3), because studies using self-paced exercise models indicate core temperature values at fatigue that are below previously suggested critical limits (Arngrimsson et al., 2004; Castle et al., 2006). Precooling strategies lead athletes to select higher metabolic work rates during exercise in the heat (Arngrimsson et al., 2004; Castle et al., 2006) and therefore attenuate the decline in CNS drive likely under such conditions. The naturally lowered temperature in the morning compared with the evening cannot explain the ergogenic effects of extrinsic cooling (Waterhouse et al., 2005). Local and systemic cooling strategies may have important interactions that are not yet clear.

Heat Acclimatization

Acclimatization refers to adaptations of physiological systems to the natural climate. The term *acclimation* refers to physiological changes that occur in response to experimentally induced changes in one particular factor (Nielsen, 1994).

Heat acclimatization leads to an earlier onset of sweating (sweat produced at a lower increase in body temperature) and a more dilute secretion. The heat-acclimatized individual sweats more at a given exercise intensity than one who is not acclimatized. Distribution of blood to the skin achieves more effective cooling after a period of

Physiological Changes Attributable to Acclimatization

- Increased blood volume
- Increased sweat rate*
- Decreased sodium and chloride content of sweat and urine
- Decreased glycogen utilization

*This includes the onset of sweat at a lower temperature and the greater distribution over body surface.

acclimatization, although the acclimatized athlete depends more on evaporative sweat loss than on blood distribution (see highlight box on p. 53).

Appreciable heat acclimatization takes place within 10 to 14 days of initial exposure. Further adaptations enhance the athlete's capability to perform under heat stress (Nielsen, 1994). Ideally, athletes should be exposed to the climate of the host country for at least 2 weeks before the competitive event. Alternatively, an acclimatization period of about 2 weeks is recommended before the event with subsequent shorter exposures as training is tapered before competition. If these methods are not feasible, heat acclimation can be attempted before leaving for the host country. Acclimation refers to the adaptations that occur in artificial environments such as a heat chamber whereas acclimatization occurs from exposure to natural conditions. This goal may be achieved prior to exposure to the competitive environment in various ways (see Reilly, 2003):

1. Seek exposure to hot and humid environments; train at the hottest or the most humid time of day at home.
2. Seek access to an environmental chamber for periodic bouts of heat exposure. Exercise rather than rest under such conditions for the exposure to be effective. About 3 hr per week exercising in an environmental chamber provides a good degree of acclimatization (Reilly et al., 1997).
3. Wear heavy sweat suits or windbreakers in training to keep the microclimate next to the skin hot. This practice adds to the heat load imposed in cool environments and induces a degree of adaptation to thermal strain.
4. Learn from experiencing exposure to heat. Gauge how exercise performance is affected and how to pace the effort so that the conditions can be tolerated.
5. Make repeated use of a sauna or Turkish bath, although this passive procedure is only partially effective.

Training can be undertaken in the cooler parts of the day at first so that an adequate workload can be achieved and sufficient fluid taken regularly. The athlete should sleep in an air-conditioned environment if sleeping is disturbed by the heat; to achieve acclimatization, he or she should spend part of the day exposed to the ambient temperature other than in air-conditioned rooms. Attention should be directed toward restoring body fluids, because thirst is itself not a precise indicator of rehydration needs.

Glycerol has been found to increase fluid retention and reduce cardiovascular strain during exercise when ingested with water 2 hr before exercise (Anderson et al., 2001). The authors concluded that this practice of using glycerol (1.0-1.5 g/kg) to increase water retention preexercise improves performance in the heat by mechanisms other than alterations in muscle metabolism. The ergogenic effect of taking glycerol with a large volume of water has not been replicated. Armstrong (2006) concluded that use of glycerol cannot be recommended, especially if hydration can be maintained during exercise.

Body weight should be recorded each morning and athletes should try to compensate for weight loss with adequate fluid intake when attempting to acclimatize to heat. Alcohol is inappropriate during training because it is a diuretic, which increases urine output. For individuals in training, urine volume should be as large as usual and the urine should be pale rather than dark color. Sport scientists working with athletic groups can collect urine samples in the morning and use a standard color chart to assess hydration status. Although there is no ideal measure of hydration status, the most suitable methods are osmolarity, specific gravity, and conductivity (Pollock et al., 1997).

The positive features of prolonged acclimatization strategies can be negated by the potential impairments in training quality associated with initial exercise in high ambient temperatures. Some training sessions should be scheduled for the cooler times of the day to allow the required level of work to be completed. This can complicate the development of acclimatization protocols because some adaptations show time-of-day specificity (Maruyama et al., 2006). Furthermore, the loss of heat acclimatization is accompanied by physiological readjustments, the nature of which are still poorly defined.

Sunburn

Sunbathing after arrival in a hot climate does not help acclimatization. Although a suntan will eventually protect the skin from damage caused by solar radiation, acquiring a suntan is a long-term process and is not immediately beneficial. A sunburn can cause severe discomfort and a decline in performance. Individuals thus should use an adequate sunscreen when they are exposed to solar radiation.

The electromagnetic spectrum just beyond visible light, with a wavelength of 400 to 320 nm, is known as UVA. The next part of the spectrum, from 329 to 280 nm, is UVB; it has more energetic photons and is mainly responsible for sunburn. Sunscreens are designed to protect against sunburn, but skin cancers, particularly melanoma, are related to exposure to UVA. Sunscreens thus should offer some protection against both radiation forms: Oxidative damage attributable to free radical production has been implicated in some melanomas. Individuals with pale skin and red hair are thought to be more susceptible to skin cancers. Repeated exposure is less risky than occasional exposure, possibly because of the acquisition of a protective suntan. Covering exposed skin with shirts and sun hats is the preferred preventive measure for visitors.

Heat Injury

Hyperthermia (overheating) and loss of body water (hypohydration) lead to abnormalities that are referred to as heat injury. Progressively they may be manifest as muscle cramps, heat exhaustion, and heat stroke. Cramps occur more frequently in individual events like distance running than in game play but can be observed in soccer matches or training sessions in the heat.

Heat cramps are associated with loss of body fluid, particularly during exercise in intense heat (Reilly, 2000). The electrolytes lost in sweat cannot adequately explain the occurrence of cramps, which seem to coincide with low energy stores as well as reduced body water levels. The muscles used in the exercise are usually the ones affected, mostly the leg (upper or lower) and abdominal muscles. Stretching the affected muscle helps to relieve the cramp, and sometimes massage produces a good outcome.

A core temperature of about 40 °C is deemed characteristic of heat exhaustion. There is an accompanying tachycardia, a feeling of extreme tiredness, dizziness, and breathlessness. Symptoms may coincide with a reduced sweat loss but they usually arise because the skin blood vessels are so dilated that blood flow to vital organs is reduced.

Core temperatures of 41 °C or higher are observed in individuals suffering heat stroke. Hypohydration, caused by loss of body water in sweat and associated with a high core temperature, can threaten life. Cessation of sweating, confusion, and loss of consciousness are characteristic of heat stroke, which is a true medical emergency. Treatment is urgently needed to reduce body temperature. There may also be circulatory instability and loss of vasomotor tone as the regulation of blood pressure begins to fail.

An index of heat stress requires that a number of environmental variables be measured. The dry bulb temperature alone is insufficient because the relative humidity indicated by the wet bulb temperature is relevant for evaporative heat loss. The radiant heat load is monitored by means of a globe thermometer. A weighted combination of these measures is incorporated in the most widely used heat index, the wet bulb and globe thermometer (WBGT) Index. This measure has been adopted in industrial, military, and sport contexts.

Thermal comfort is achieved as a subjective state of satisfaction within a narrow range of thermoequibrium. The requirements for thermal comfort specified as an international standard (ISO 7730) indicate that the operative temperature should be 20 °C to 24 °C, the mean air velocity be below 0.2 m/s, and relative humidity be in the range 30% to 70%. For such conditions to be met, the percentage of people expressing thermal discomfort should be less than 10%. An assumption is that the person is sedentary but occasionally engaged in light activity.

Cold and Exercise Responses

Winter sports are often held in near-freezing conditions. Core temperature and muscle temperature may decrease, and exercise performance is increasingly affected. Muscle power output can be reduced by 5% for every 1 °C decrease in muscle temperature below normal levels (Bergh and Ekblom, 1979). A decrease in core temperature to hypothermic levels is life-threatening: Fortunately, the body's heat gain mechanisms arrest the decline, and true hypothermia is rare. In most cases, athletes experience thermal discomfort and arrest the decrease in body temperature by adopting appropriate behavior.

The posterior hypothalamus initiates a generalized vasoconstriction of the cutaneous circulation in response to cold, a response mediated by the sympathetic nervous system. Blood is displaced centrally away from the peripheral circulation, causing the temperature gradient between core and shell to increase. The decrease in skin temperature reduces the gradient between the skin and the environment, which protects against a large loss of heat from the body. Blood returning from the limbs is diverted from the superficial veins to the venae comitantes lying adjacent to the main arteries. Arterial blood is cooled by the venous return almost immediately as it enters the limb by means of countercurrent heat exchange.

A decrease in limb temperature impairs motor skills. Muscular strength and power output are reduced as the temperature in the muscle decreases and the conduction velocity of nerve impulses to the muscles is delayed. The sensitivity of muscle spindles also declines, impairing manual dexterity. Thus athletes must preserve limb temperature during competitive sports, in particular when manual dexterity must be maintained.

The body's autonomic nervous system stimulates shivering in response to the decrease in core temperature (see highlight box on p. 57). Skeletal muscles contract involuntarily to generate metabolic heat. Shivering tends to be intermittent and may persist during exercise if the intensity is insufficient to maintain core temperature. Shiv-

Selected Effects of Cold on Human Physiological Responses and Performance

- Increased peripheral vasoconstriction
- Increased metabolism
- Increased periodic shivering
- Decreased muscle spindle sensitivity
- Decreased nerve conduction velocity
- Decreased manual dexterity
- Decreased muscle strength and power
- Increased accident risk

ering may occur during pauses in training when conditions are cold or compounded by sleet or rain. Coaches should be alert to such behavioral signs of hypothermia in young players.

Shivering, fatigue, loss of strength and coordination, and inability to sustain work rate are all early symptoms of hypothermia. Once fatigue sets in, shivering decreases and the condition can worsen to include collapse, stupor, and loss of consciousness. This risk applies more to recreational than professional athletes, because the former might not be able to sustain a work rate to keep themselves warm in extreme cold.

Athletes can protect themselves against cold, so cold is less problematic than heat. The temperature in the microclimate adjacent to the skin can be maintained by appropriate clothing and use of more than one layer. Team players might respond positively to cold conditions by maintaining a high work rate. Alternatively, they may be spared exposure to cold by using indoor training facilities where these are available.

Natural fiber (cotton or wool) is preferable to synthetic material for cold conditions. Sweat produced during exercise in these conditions should be able to flow through the garment. The best material will allow sweat to flow out through the cells of the garment while preventing water droplets from penetrating the clothing from the outside. Fabric that becomes saturated with water or sweat loses its insulation and cause body temperature to decrease quickly in cold and wet conditions, so the optimum is a combination of natural fibers with synthetic material.

The trunk should be well protected when training is conducted in cold weather. Warm undergarments can be worn beneath a full tracksuit. Dressing in layers increases the insulation provided, and the outer layers can be discarded as body temperature or ambient temperature increases. A T-shirt underneath a team jersey helps to preserve warmth, but some participants may need a full tracksuit.

In cold and wet weather, the outer layer of clothing should be capable of resisting both wind and rain and the inner layer should provide insulation. The inside layer should also wick moisture from the skin to promote heat loss by evaporation. Polypropylene and cotton fishnet thermal underwear has good insulation and wicking properties and so is suitable to wear next to the skin. Individuals clad for Arctic conditions are shown in figure 3.3.

Immediately before competing in the cold, sport participants should stay as warm as possible. A thorough warm-up (performed indoors if possible) helps in this regard. Cold conditions are thought to increase the risk of muscle injury in sports that involve intense anaerobic efforts, and a systematic warm-up affords some protection in this

Figure 3.3 Close-up of cold-protective equipment *(a)* without loading and *(b)* in use with accompanying load carriage and *(c)* in group excursion with Arctic adventurers.

respect. Individuals may need to wear more clothing than they normally do during field games.

Aerobic fitness does not directly protect against cold. Nevertheless, it will enable individuals to keep more active and not succumb so easily to fatigue. Individuals with a high aerobic fitness are able to maintain activity at a satisfactory level to achieve heat balance. In contrast, people with poor endurance are at risk of hypothermia if the pace of activity falls too low. Individuals with low levels of subcutaneous adipose tissue are poorly insulated and may feel the cold more than others. They may be obliged to stay more active than their better-insulated counterparts and generally possess the aerobic fitness to do so. Nevertheless, shivering during activity signals the onset of danger and is a warning of impending hypothermia.

The severity of cold conditions is indicated by the wind-chill index. Whereas air movement helps to alleviate heat strain in hot conditions, it can accentuate heat loss when conditions are cold. The wind-chill index is used by climbers to estimate the risk of hypothermia in the mountains. The cooling effects of combinations of certain ambient temperatures and wind speeds are expressed as temperature equivalents and are estimated by means of a nomogram. The wind-chill index is used by sailors and explorers, and by people engaged in outdoor recreational activities in the winter. The index incorporates air velocity and ambient temperature; the clothing requirements for protection against hypothermia can be estimated from its values. The values calculated correspond to a caloric scale for rate of heat loss per unit of body surface area and are then converted to a sensation scale up to a point where exposed flesh freezes within one minute (Reilly and Waterhouse, 2005).

Altitude

The major stress associated with altitude is related to the lowered density of the air caused by decreasing ambient pressure. The result is a reduction in the oxygen that is available to the body's tissue, a condition known as hypoxia. There are compensating mechanisms that help in tolerating altitude, but their time course varies according to the physiological variable concerned. The deleterious consequences of altitude depend on the heights reached. Responses characteristic of altitude sickness are associated with different heights. Altitude training is an essential part of preparing to compete at altitude but can also induce physiological adaptations that benefit later performance at sea level. Cold and inclement weather present additional stresses at very high altitudes to which sojourners and mountaineers never fully acclimatize.

Physiological Adjustments

As altitude above sea level increases, the barometric pressure decreases. At sea level the normal pressure is 760 mmHg, at 1,000 m it is 680 mmHg, and at 3,000 m it is about 540 mmHg. Altitude conditions are referred to as hypobaric (low pressure), and the main physiological problem associated with this environment is **hypoxia,** or relative lack of oxygen.

Oxygen constitutes 20.93% of normal air, the partial pressure of the oxygen at sea level being 159 mmHg (20.93% of 760). The partial pressure of oxygen decreases with increasing altitude: This decrease corresponds to decreasing ambient pressure while the proportion of oxygen in the air remains constant. There are fewer oxygen molecules in the air at altitude for a given volume of air and so a smaller amount of oxygen is inspired for a given inspired volume, which reduces the amount of oxygen delivered to the active tissues.

The alveolar oxygen tension (P_AO_2) is of critical importance in the uptake of oxygen into the body through the lungs. Water vapor pressure in the alveoli is relatively constant at 47 mmHg as is the partial pressure of carbon dioxide (PCO_2) of 35 to 40 mmHg. The decreased alveolar tension at altitude causes a less favorable gradient across the pulmonary capillaries for transferring oxygen into the blood. Exercise that depends on oxygen transport mechanisms is impaired at about 1,200 m once desaturation occurs. The oxygen dissociation curve of hemoglobin (Hb) is sigmoid-shaped and is affected by ambient pressure (figure 3.4). Red blood cells are normally 97% saturated with O_2 but this figure decreases when P_AO_2 levels decrease to a point corresponding to this altitude (1,200 m). The O_2-Hb curve is not affected much for the first 1,000 to 1,500 m of altitude, but as the pressure drops further, the curve becomes steeper and the oxygen supply to the body's tissues is increasingly

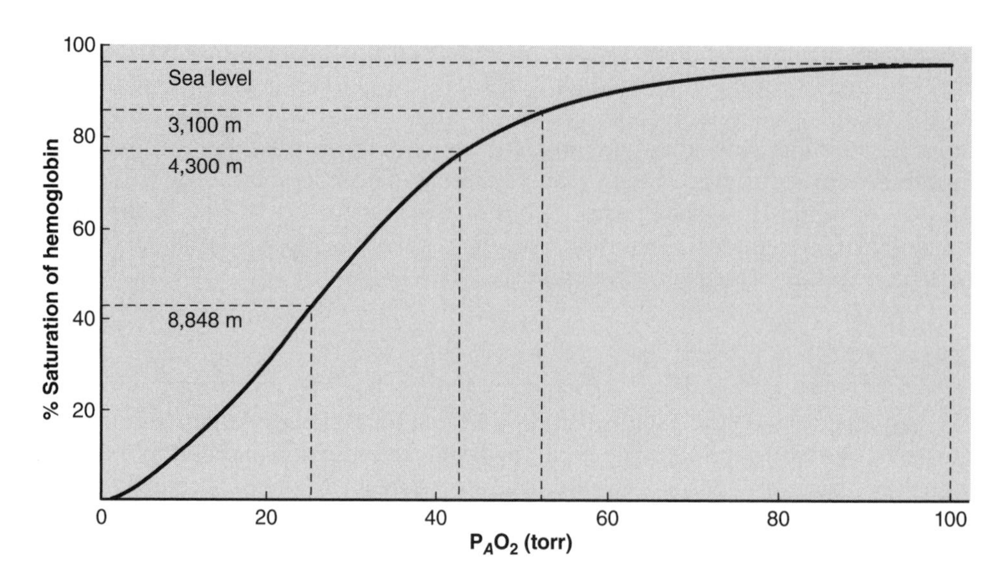

Figure 3.4 The oxygen dissociation curve of hemoglobin for a blood pH of 7.4 and body temperature of 37 °C. The horizontal lines indicate percentage hemoglobin saturation of arterial blood for the different altitudes and the vertical lines indicate partial pressure of oxygen (PO_2).

Reprinted from *Sport, Exercise and Environmental Physiology*, T. Reilly and J. Waterhouse, pg. 54. Copyright 2005, with permission from Elsevier.

impaired. At an altitude of 3,000 m, arterial saturation is reduced to about 90%, the corresponding value for the altitude of Mount Everest being less than 30%.

Increased ventilation is the body's acute physiological compensation for hypoxia. The depth and frequency of breathing both increase, the hyperventilation increasing the amount of CO_2 that is released from blood passing through the lungs. Because CO_2 is a weak acid when dissolved in body fluid, elimination of CO_2 leaves the blood more alkaline than normal attributable to an excess of bicarbonate ions. The kidneys compensate by excreting excess bicarbonate, thereby restoring the normal pH level of the blood over several days. The decrease in the body's alkaline reserve reduces the buffering capacity of the blood for tolerating additional acids (such as lactic acid diffusing from muscle to blood during exercise).

Production of 2,3-bisphosphoglycerate (2,3-BPG) by the red blood cells increases at altitude. This increase helps the unloading of oxygen from the red blood cells at tissue level. The increased number of red blood cells enhances the oxygen-carrying capacity of the blood. This process begins within a few days at altitude, stimulated by the kidney hormone erythropoietin, which causes the bone marrow to increase its production of red blood cells. This adaptation requires that the body's iron stores be adequate, and supplementation of iron intake may be needed before and during the stay at altitude (synthetic versions of erythropoietin have been used for blood doping, a procedure banned in sport). There is an apparent increase in hemoglobin in the early days at altitude, due to hemoconcentration and a transient decrease in plasma volume. A true increase in hemoglobin may take 10 to 12 weeks to be optimized.

After a year or more at altitude, the increases in total body hemoglobin and red cell count still do not match values observed in natives of high altitude. Individuals born at sea level would never theoretically be able to compete in aerobic events at altitude on equal terms with those native to altitude. Strategies must be devised to allow sea-level natives to demonstrate their skills as well as prepare physiologically by acclimatizing to altitude.

Exercise at Altitude

Despite the acute physiological adjustments to hypoxia that occur, athletes experience difficulty in exercising at altitude compared with sea level. This difficulty, and potential health-related problems, caused FIFA in 2007 to consider a ban on any international soccer matches being played at altitude greater than 2,800 m; the proposed ban was not implemented because a few countries would have been disadvantaged by it. Changes in maximum cardiac output and in oxygen transport lead to a decrease in maximal oxygen uptake ($\dot{V}O_2max$). At an altitude of 2,300 m, corresponding roughly to Mexico City, the initial decline in $\dot{V}O_2max$ averages about 15%. After 4 weeks at this altitude, $\dot{V}O_2max$ improves but still remains about 9% below its sea-level value. For sea-level dwellers, the initial decline in $\dot{V}O_2max$ is about 1% to 2% for every 100 m above 1,500 m (see Reilly, 2003).

Participants in individual endurance events operate at submaximal intensity for a continuous period whereas games players have short episodes of maximal anaerobic efforts in between submaximal exercise periods. Maintaining a fixed submaximal exercise intensity is more difficult at altitude than at sea level. The highest level of endurance exercise that can be sustained is determined by the intensity at which lactate begins to accumulate in the blood. This lactate threshold is lowered at altitude although the percentage of $\dot{V}O_2max$ at which it occurs is unchanged. As active muscles rely more on anaerobic processes to help cope with the relative hypoxia, games players need longer low-intensity recovery periods during match play, following their all-out, high-intensity efforts.

Heart rate, ventilation, and perceived exertion are all increased above the normal sea-level values at any given submaximal exercise intensity (Bangsbo et al., 1988). Therefore, the exercise intensity that can be sustained is reduced. Participants should be prepared to pace their efforts more selectively during exercise at altitude and accept a lower work rate during training. These modifications are especially important in the first few days at altitude and differ between individuals according to their level of aerobic fitness, prior acclimatization, state of health, and previous experience of altitude. Physiological factors such as pulmonary diffusing capacity, total body hemoglobin, iron stores, and nutritional state are also influential.

With progressive adaptation to altitude, the heart rate during submaximal exercise decreases compared with that on initial exposure and may approach sea-level values after 3 to 4 weeks. The skeletal muscles also adapt, but improvements in maximum blood flow and oxidative metabolism require many months at altitude. There are also some changes in enzymes associated with anaerobic metabolism. The buffering capacity of muscle is enhanced with a prolonged stay at altitude. This adaptation, along with changes in activities of enzymes associated with anaerobic glycolysis, complements the adaptations that occur in oxygen transport mechanisms. Altitude conditions may be favorable when training is geared toward improving running speed, because faster than normal velocities can be reached given the reduced air resistance against which the body moves.

The reduced air density at altitude has other implications for sport participants. When speed training is conducted at altitude, the recovery period between sprints should be lengthened. Furthermore, objects can be thrown through the air more easily than normal. This phenomenon has implications for playing strategies, including the practice of set pieces in soccer, long drives in golf, and throwing events in track and field.

The reduced aerobic capacity at altitude affects the quality of training in endurance athletes. The "live high, train low" model advises periodically returning to low altitude for intensive sessions but sleeping at altitudes of 2.2 to 2.8 km (Levine, 1997). This method is thought to enable physiological adaptations and maintain the quality of training.

Preparing for Altitude

Athletes scheduled to compete at altitude or to stay at an altitude training camp must consider the physiological consequences of such a visit. Individuals suffering infection or having low iron stores are unlikely to benefit from altitude training because their ability to increase red blood cell production is limited. Strenuous training should not be attempted for at least 2 days after arrival at altitude until the risk of developing acute mountain sickness has passed: This syndrome is characterized by headaches, nausea, vomiting, loss of appetite, sleep disturbances, and irritability. These problems can be encountered at altitudes above 2,500 m but are mostly associated with higher altitudes. Training sessions should be reduced in intensity to the same perceived exertion as experienced at sea level; full workouts are not advisable until 7 to 10 days after arrival. Recovery periods between repeated sprints should be lengthened when exercise is intermittent. This applies to conditioning work, interval training, and game practices.

The air at altitude tends to be drier than at sea level, and the body loses more fluid by means of evaporation from the moist mucous membrane of the respiratory tract. This loss is accentuated by the hyperventilation response to hypoxia. The nose and throat become dry and irritable, which can cause discomfort. It is important to drink more than normal to counteract the fluid loss. Drinking fluids helps to offset the decrease in plasma volume that is a characteristic response to altitude (Ingjer and Myhre, 1992). Sports drinks that contain carbohydrate can be helpful, given the increased reliance on glycogen as a fuel for exercise. A greater than normal proportion of carbohydrate should be consumed, especially in the first few days at altitude.

About 14 days should be allowed before competition for acclimatization to altitudes of 1,500 to 2,000 m and 21 days before competition at 2,000 to 2,500 m. These periods may be shortened if the athletes have had previous exposure to altitude in their buildup to the competition. Individuals without previous exposure to altitude need about 1 month to adapt to locations above 2,500 m and may lose fitness in the process.

Where a prolonged stay at altitude before competing is impractical, partial acclimatization can be achieved by frequent exposures to simulated altitude in an environmental chamber. Continuous exercise of 60 to 90 min or intermittent exercise of 45 to 60 min performed four or five times a week at simulated altitude of 2,300 m has proved effective in 3 to 4 weeks (see Reilly, 2003).

Portable simulators that induce hypoxia are available to wear as a backpack. These devices lower the inspired-oxygen tension and accentuate exercise stress but also increase the resistance to breathing. They do not promote the kind of adaptations that are experienced at altitude or that result from sustained exercise in a hypobaric chamber. However, the athlete may gain psychological benefits from experiencing the stress of hypoxia that these simulators provide. Normobaric hypoxic chambers and hypoxic tents have become available to athletes for training and rehabilitation. These devices are likely more beneficial during rehabilitation, because the relative circulatory strain is greater than a comparable work rate under normal sea-level conditions (see figure 3.5). These devices also have obvious value in providing individuals with experience of exercise under hypoxic conditions.

Extreme Altitude

High altitudes attract mountaineers who are driven to climb to their peaks. Tolerance of extreme altitudes is likely to be beyond the capacity of many individuals, even those who are partially acclimatized. A height of about 5.9 km is the limit of human habitation. The highest permanent human settlement in the Andes is at 5.5 km, and above this height acclimatization is replaced by a steady deterioration. Chilean mine workers experience loss of appetite and difficulty sleeping above this level; similar symptoms are displayed by mountain climbers.

Climbers may suffer from altitude sickness that can take various forms (see table 3.1). Many of these conditions are life-threatening (Houston, 1982). Altitude tourists are likely to suffer at lower altitudes, lacking a systematic approach to acclimatization and the experience to deal with altitude stress.

Figure 3.5 Normobaric hypoxic chambers provide a means of simulating altitude conditions.

Table 3.1 Altitude Sickness

Form	Symptoms
Acute hypoxia	Mental impairment and usually collapse after rapid exposure above 5,500 m (18,000 feet).
Acute mountain sickness	Headache, nausea, vomiting, sleep disturbance, dyspnea at greater than 2,500 m (8,000 feet): common and self-limited.
High-altitude pulmonary edema	Dyspnea, cough, weakness, headache, stupor, and rarely death. 3,050 m (10,000 feet) and above: requires rapid descent or early treatment.
High-altitude cerebral edema	Severe headache, hallucinations, ataxia, weakness, impaired thinking, stupor, or death. Above 3,550 m (12,000 feet): uncommon. Descent is mandatory.
Subacute and chronic mountain sickness	Failure to recover from acute mountain sickness may necessitate descent.
	Dyspnea, fatigue, plethora, and heart failure may develop after years of asymptomatic residence at altitude.
Chronic conditions worsened by altitude	Sickle trait, chronic cardiac or pulmonary disease.
High-altitude deterioration	Long periods spent above 5,500 m (18,000 feet): insomnia, fatigue, weight loss and general deterioration.
	Deterioration is more rapid at higher altitudes.

The dangers of climbing the high peaks are illustrated by the fact that 1 in every 16 climbers ascending Mount Everest dies in the attempt. Survival relies on detailed planning of the climb, aerobic fitness, an experienced leader, appropriate equipment and clothing, and favorable weather conditions. The gradual debilitating effects of struggling with hypoxia, loss of appetite, and muscle wasting mean that the time window available to conquer the high peaks is limited.

Air Quality

Air quality is important for comfort, health, and sometimes performance. Impurities can result from a variety of pollutants, indoors and outdoors. Air content can trigger adverse reactions in some individuals. Pollution has been a concern at recent Olympic Games, especially Athens 2004 and Beijing 2008.

Air Pollution

Inhaled air can contain impurities that impair the health as well as the performance of athletes. At rest, air breathed in through the nose is filtered and many pollutants are prevented from reaching the airways. During exercise there is a shift to oral breathing, and the scrubbing action of the nasal passageways is bypassed. When the air is polluted, lung function can be adversely affected, depending on the concentration of the pollutants and the sensitivity of the individual who is exposed.

Caution is needed when athletes compete or train in cities with high levels of air pollution or train close to roadside traffic. Recreational participants need to pay attention to the proximity of local pollutants such as smoke from factories. People with asthma should be especially alert to possible adverse reactions when pollution levels are high or conditions are foggy.

Pollutants are described as primary and secondary, depending on whether they retain the form in which they were emitted from source or are formed through chemical reactions between source and target (Reilly and Waterhouse, 2005). Sulfur dioxide (SO_2), carbon monoxide (CO), nitrogen dioxide (NO_2), benzene, and particulate matter like dust and smoke are the main primary pollutants. Particulates less than 10 microns in diameter, referred to as PM-10s, enter the airways relatively easily. Secondary pollutants include ozone (O_3) and peroxyacetyl nitrate, formed as a result of ultraviolet radiation affecting NO_2 and hydrocarbons. Elevated ozone concentrations cause symptoms such as coughing, chest pain, difficulty in breathing, headache, eye irritation, and impaired lung function, all of which are likely to affect exercise performance (Florida-James et al., 2004).

The World Health Organization sets standards for health-based air quality in the case of pollutants such as ozone, and so their concentrations are monitored regularly in the major cities worldwide. National standards apply in other cases such as PM-10s. Individuals show some degree of adaptation to ozone, reflected in the reduced sensitivity of those habitually exposed to it (Florida-James et al., 2004). For individuals with previous exposure to ozone, performance is only partially recovered when concentrations are high. Antioxidant supplements can ameliorate some adverse effects of ozone by counteracting the oxidative stress it causes. People with asthma may experience worsened symptoms when PM-10 levels are raised, attributable to oxidative stress–mediated inflammation (MacNee and Donaldson, 1999).

Air Ions

Air ions are produced naturally by events such as the shearing of water droplets in a waterfall and the rapid flow of great volumes of air over a land mass, by solar and cosmic radiation, and by a variety of radioactive sources. The ions are produced by these sources of energy that displace an electron from a molecule of one of the common atmospheric gases. The molecule is left with a positive charge while the displaced electron is normally captured by another molecule to which it imparts a negative charge.

Air ions are biologically active; negative air ions are thought to promote beneficial mood states whereas positive air ions have been shown to cause characteristic complaints from subjects (Hawkins and Barker, 1978). Negative air ion generators are commercially available for office and domestic environments to enhance mood and performance. Negative air ions have been reported to produce positive effects on psychomotor and mental functions (Hawkins and Barker, 1978).

Favorable effects of air ions also extend to physical exercise. Perceived exertion during an incremental exercise test was lowered by negative air ions compared with positive ions (Sovijari et al., 1979). Inbar and colleagues (1982) reported that the elevation in heart rate and core temperature during exercise was less pronounced on exposure to negative air ions compared with neutral air.

Brain serotonin, body temperature, and oxidative metabolism may mediate the effects of air ions. Reilly and Stevenson (1993) reported decreases in rectal temperature, oxygen uptake, and heart rate at rest on exposure to negative air ions; an effect on serotonin was suggested attributable to the alterations observed in circadian rhythm characteristics. Serotonin levels in the brain are known to be affected by air ions, and such alterations would explain the psychological benefits of negative ions and the adverse effects of positive air ions on mood states. Large doses of negative air ions in vitro have promoted efficiency of aerobic and metabolic processes (Krueger and Reed, 1976). Reilly and Stevenson (1993) concluded that factors other than thermoregulatory mechanisms mediate the influence of air ions on tissue metabolism. The effects they observed on metabolism and heart rate at rest tended to disappear under exercise conditions.

Allergens

Allergic reactions of the airways begin in the nose. Allergens such as pollens or house dust alight first in the nose, where they are filtered out. The inflammatory response caused by mast cell discharge blocks the nose and leads to a shift to oral breathing. The allergens thereby achieve direct access to the lower respiratory tract, where an inflammatory response can trigger an asthma attack. Allergens from a different flora can also trigger allergic reactions, causing irritation of mucous membranes of the eye as well as those of the respiratory tract.

Allergy is a state of altered reactivity in the host that results from interactions between an antigen and an antibody (Harries, 1998). Antigens stimulate the production of antibodies by penetrating the mucous membrane of either respiratory or gastrointestinal tracts. An antigen must be soluble in water, and immunoglobulin E (IgE) is the antibody associated with the allergy. When the IgE molecules on the surface of mast cells recognize the antigen, calcium channels in the mast cell membrane are opened and histamine is released. The appearance of neutrophils signals the inflammatory response that develops some hours later. These separate events explain the timing of both the immediate acute and the late asthmatic responses. Continued exposure to antigens like ragweed or grass pollens causes symptoms to persist.

Ragweed and grass pollen act as antigens that stimulate the production of antibodies after penetrating the mucous membrane of the respiratory tract. Some people are allergic to antigens, the reactions beginning in the nose where allergies such as pollen are first filtered out. The inflammatory response causes a blocked nose and promotes oral breathing, which gives the allergies direct access to the lower respiratory tract. Here the same inflammatory response causes asthma instead (Harries, 1998). Athletes vulnerable to allergies should seek medical advice for therapy.

Hay fever triggered by high pollen counts creates discomfort in many people. Antihistamines alleviate the symptoms in many cases. Without such prophylactic measures, afflicted individuals can find their competitive performance impaired by allergic reactions.

Noise

Many sports are associated with noise, whether induced by participants, equipment, machinery, or the roar of spectators. Sources of noise in ordinary life include thunder claps, motor vehicles, subway trains or jet engines, factory machinery, or radios and sirens. Noise may be regarded as unwanted sound having a nuisance effect on the listener. In communication theory noise is deemed to interfere with the signal being transmitted, and so engineers concentrate on improving the signal to noise ratio. In a human factors context, whether in sport or industry, the sensory mechanisms for transmission and reception are relevant considerations.

Sound pressure is expressed in dynes per centimeter, but sound is measured in decibels (dB) or bels (10 dB = 1 bel). Frequency is indicated in hertz (Hz), representing the number of waves arriving each second. Pitch indicates the subjective sensation of frequency. Sound is generated by vibrations from some source and transmitted through the atmosphere to the ear. For frequencies of 20 to 15,000 Hz, sound is heard by the human ear. Intelligible speech lies within the bandwidth of 620 to 4,800 Hz, but the main frequencies used in speech fall in the range of 300 to 3,500 Hz. Ultrasound indicates mechanical vibration in excess of 20,000 beats.

The decibel is a physical measure of sound intensity on a logarithmic pressure ratio scale. The scale and its zero point are chosen to correspond to subjective phenomena, although the scale is physical in nature and based on amplitude. Arbitrary zero on the decibel scale is the faintest sound the human ear can hear at 1,000 Hz; this value represents 0.00002 N/m^2 (0.0002 $dyne/cm^2$).

The ear converts the pressure waves of sound into neural signals for transmission to the brain. The auditory system converts pressure changes in the air, via pressure changes in a liquid medium, into pulses that contain information on pitch and loudness. This feat is accomplished by a series of transformations into (a) mechanical movement, (b) hydrostatic pressure, and (c) an electrical or neural signal. These transformations occur in the outer, middle, and inner ear.

Sounds pass from outer ear to middle ear to inner ear in a millisecond, and so the middle ear has a natural reflex to protect against sudden load noises. The tensor tympani and stapedius muscles hold the connections of the ossicles rigid and the eardrum stiff, which reduces the propagation of loud sound waves into the cochlea. When these muscles, attached respectively to the malleus and stapes, are stimulated to contract, they dampen movements of the ossicular chain and so decrease sensitivity of the hearing apparatus. This action protects the ear from damage, except in cases of sudden and unexpected loud noise. The response time of these muscles is of the order of 100 ms, so some damage to sensitive aural structures can occur before the auditory attenuating reflex becomes active.

Noise as a Pollutant

Although environmental noise is mostly accepted in the community as inevitable, it may be considered as an irritating pollutant. Its disturbing characteristics are acknowledged when communities object to increasing aircraft traffic by extending terminal runways. Complaints of disturbance are also directed at noisy neighbors, revving motorcycles, or loud music. Background noise at nighttime induces frustration when it disrupts or prevents sleep, especially in competitive athletes. Noise as an alarm signal calls for immediate attention for safety reasons, but frequent false alarms are annoying.

Unwanted noise is annoying, can damage hearing, and may have other long-term effects on health. Long-term exposure to traffic noise may account for 3% of deaths from ischemic heart disease in Europe (Coghlan, 2007). It was also suggested that

2% of Europeans suffer severely disturbed sleep attributable to noise pollution and 15% suffer severe annoyance.

Noise may cause ill health by accentuating the normal endocrine responses to stress. Modern sources of public noise are not always accounted for in laws to combat noise pollution, nor are there widely available resources for monitoring, regulating, or researching noise. This deficiency and the increasing complaints about noise as a pollutant and risk to health have led to initiatives for concerned actions, at least among European countries (Prasher, 2000).

Types of Deafness

Sensorineural loss leading to impaired hearing is distinguished from conduction loss. Middle ear deafness is normally caused by a stiffening or damping of the ossicles that leads to a moderate degree of hearing loss. An infection of the middle ear, perforation of the eardrum, or infection of the eustachian tube can also be responsible. Low tones rather than high tones are impaired by the resulting increase in resistance. Provided the inner ear is intact and functions normally, antibiotics and surgical repair of the drum can return hearing to normal. Amplification by means of a hearing aid will also be successful.

Nerve deafness affects the perception of all sounds or only a range of the spectrum. Damage occurs either to special receptor cells or to the nerve trunks supplying input to the brain's hearing center. Infections such as measles and mumps can cause cell damage in the inner ear. Head injuries, especially skull fractures, can destroy nerves, and exposure to high noise levels is particularly destructive to the receptor cells in the inner ear.

Temporary threshold shifts result from exposure to high noise levels. An increase in the hearing threshold takes place 0.15 octave higher than the noise tone. In this type of nerve deafness, hearing loss occurs first in high tones over about 4,000 Hz, which passes unnoticed, being above speech frequencies. The repetition rate of impact noise must be longer than 0.1 s to allow the middle ear muscles time to contract and protect the tympanic membrane and attached ossicles, a response not possible when the noise is unexpected. In firearm discharge, the intensity of noise peaks within 0.2 to 2 ms, too short for the protective reflex mechanisms to act.

Dullness of hearing can be associated with a ringing or buzzing sensation known as tinnitus. Once the receptor cells have degenerated, recovery of hearing is not possible. The cells affected no longer generate electrical impulses and so no signals are transmitted to the hearing center. Amplification devices then provide no benefit.

Presbycusis refers to a loss of hearing attributable to age rather than noise-induced damage. There is increasing deterioration at the higher frequencies (400 Hz or above), and losses do not interfere with speech communication until about age 65 years. The normal human speech range spans 300 to 3,500 Hz, but speech for both males and females tends to predominate at about 500 Hz. Cutting off the top of the speech frequencies (1,000-3,000 Hz) results in whispers being inaudible and difficulty in discriminating between specific words. Loss of hearing attributable to age is additive to that produced by loud noise.

Recreational Activities and Sports Involving Noise Hazards

Noise is implicit in the vast range of sports and recreational activities undertaken by individuals. This section identifies some of these activities and discusses the nature of the noise stress. Methods of avoiding damage to hearing attributable to noise exposure are covered, including the use of protective equipment.

Shooting

Many sport governing bodies advise participants to use ear protection to reduce the risk of hearing damage. A 0.22 rifle (one of the less noisy weapons in sporting use) can produce peak sound pressure levels in the 130 to 160 dB range. Despite the short duration of these peak sound pressure levels, they represent high values of impulsive sound and significant noise levels. Shooting regularly with a 12-bore shotgun damages the ears of recreational participants. Wearing ear protectors is the only way of avoiding risk.

The head provides some degree of protection for the ear that is averted in rifle shooting. Turning the head, as when firing a shoulder-held weapon, provides a "head shadow" for one of the ears. This shadow can produce attenuations of 25 to 30 dB at frequencies above 1,000 Hz but is negligible below this frequency. Pistol shooters keep the head squarely facing the target, placing both ears equally at risk.

Motor Racing

Race track meetings, hill climbs, motocross, grass-track racing, and other racing events not held on public roads are invariably noisy. Without the screeching of engine or tires, these sports might not be as attractive to spectators. Noise levels reach values above 100 dB, but the duration of exposure may be relatively short for spectators.

Drivers, mechanics, and track officials are most at risk, although spectators positioned near the track may also receive high doses. Mechanics and officials should wear ear protectors, like those developed for noisy industrial situations. Earplugs made of "glass down" are suitable, and safety helmets should be worn by competitors.

Motorcycling

In the United Kingdom, the noise level permitted for an 8 hr working day under the Health and Safety at Work Act of 1989 is 85 dB. Motorcyclists experience values in excess of this level. Noise from the machinery is satisfactorily controlled by the Auto-Cycle Union in the United Kingdom, and the main hazard is associated with turbulent airflow round the rider's helmet. This low-frequency "wind noise" amounts to 90 dB when the machine is driven at 56 km/hr.

Hearing loss spreads to frequencies on either side of the 3,000 and 6,000 Hz region as the damage progresses. Grand Prix motorcyclists (20 of 44 examined) displayed hearing losses greater than expected for age-matched controls (McCombe and Binnington, 1994). The hearing deficit tended to increase with racing experience. Just fewer than 40% of the riders wore earplugs regularly, most competitors discarding them to pick up other environmental sounds, in particular the engine and exhaust of the motorcycles.

Indoor Pool Swimming

Public swimming pools with inadequate acoustical treatment can produce noise pollution. Sound levels during busy periods can reach peak levels of 100 dB, and 90 to 92 dB levels are common. Sensible precautions are to avoid lengthy stays during busy periods and to wear rubber ear protectors designed to keep out water. These protective devices must fit tightly to be effective.

Snowmobiling

Snowmobiles provide a convenient form of transport over terrain covered in snow. These vehicles are used by mountain rangers for transportation and by competitors in snowmobile racing. Snowmobiles can produce up to 136 dB at full throttle (26 horsepower engine). Temporary threshold shifts lasting 4 to 14 days are found in both racing drivers and riders. Sound pressure levels caused by snowmobiles exceeded the Damage Risk Conditions of the American Committee of Hearing and Bioacoustics by 10 dB for the speech frequency range (500-2,000 Hz). Data suggest that drivers should not exceed 11 min of continuous snowmobiling. Unrestrained snowmobile operation would constitute a significant noise dosage for those exposed.

Do-It-Yourself Activities

Power tools, as used in "do-it-yourself" tasks, are noisy and can be hazardous if used for lengthy periods. Noisy machines include some motorized lawnmowers, chainsaws, electric hedge trimmers, circular saws, sanding machines (especially when used on metal surfaces), hammers (in certain circumstances), and percussion tools. When these tools are used outdoors, the only person at risk is the operator. Ear protectors are recommended by the Noise Advisory Council as essential when using chainsaws. Risks are greater when power equipment is used indoors—in workshops and garages with little sound-absorbing material. For the average do-it-yourselfer, machines should be used only for short periods of time.

Loud Music

Age and taste determine whether music is perceived as too loud. Discotheques, clubs, and pop concerts have been criticized for highly amplified music that may involve a small but definite hearing risk. Peak noise levels of 120 dB are common at discos, and average levels often exceed 105 dB. According to data prepared for industrial noise exposure, a person who spends 15 min/day exposed to continuous noise of 105 dB is at equal risk of hearing damage as a person spending 8 hr at 90 dB. Young people going to clubs or discotheques once or twice a week have their hearing rested in the interim. Individuals who are subjected to high noise levels during the day and further noise in the evening are at increased risk. Noise levels affect workers in nightclubs (disc jockeys and audiovisual technicians), who may get little respite from high noise doses during their shift. Tinnitus, a lingering, ringing sensation in the ears, is a common complaint among crowds at outdoor music festivals: In the majority of sufferers, the symptom disappears within 2 days.

Safety recommendations for attending night clubs, discotheques, or pop concerts include standing away from loudspeakers and occasionally adjourning to a quiet room or away from outdoor loudspeakers. This advice applies particularly if tinnitus is experienced.

Protection and Noise Control

The appropriate means of engineering protection are controlling the source of noise or isolating it. The individual may be protected by use of ear devices.

Control at Source

Proper design of equipment, regular maintenance, attention to lubrication, and use of rubber mountings are helpful in controlling noise at source. Hard walls, floor, and ceilings reflect sound and increase noise and should be addressed at the design stage. These problems can also be treated by creating sound traps, especially by means of rough, porous surfaces. Thick carpets and padded furniture can also help control noise.

Isolation of Noise

Noise can be isolated by acoustical treatment of the environment, using sound absorbers and strategic layout of equipment. These remedies correct for errors of omission at the design stage. An example would be the faulty location of plant and equipment-generating noise within sport facilities. Effective design would isolate such sources from employees and facility users.

Ear Protection

Effectiveness of ear protection is limited to 35 dB at 250 Hz and 60 dB at higher frequencies. A wide range of safety devices are available.

- *Prefabricated ear plugs* are available in various sizes, made of soft, flexible material. The plugs must provide a snug, airtight, and comfortable fit. They should be nontoxic and have a smooth surface for cleaning. Soft plastic models are available in five sizes. Noise reduction depends on fit and can vary by 10 dB. The plug may be sited accurately by putting it into the external ear, effectively opening and straightening the ear canal.
- *Malleable and disposable ear plugs* made from glass wool, wax, cotton, or a mixture of these can be formed into a small cone by hand, and the apex is then inserted into the middle ear. This type of plug can provide similar reduction as the prefabricated type if made and sited correctly. Glass wool or glass down is the material for these plugs. Glass down is a form of glass wool in which fine fibers form a material of downlike softness suited to the delicate skin of the middle ear. Glass wool is not recommended in areas with high intermittent noise levels or where it is necessary to remove and reinsert protective devices regularly during work periods. Cotton wool by itself provides little attenuation; if waxed it is a little better but still unsatisfactory.
- *Individually molded ear plugs* are made of silicon rubber and are molded in a permanent form within the ear canal so the plug conforms to the shape of the canal. When correctly made and situated, this device is superior to the prefabricated design. The degree of protection depends on expertise of molding, and usually five or six separate fitting attempts are required. The silicone rubber is supplied with a curing agent, and the two are then mixed to a consistency resembling that of putty before being inserted into the ear. No jaw movements are allowed during a 10 min period of curing. On removal, the plug is in a permanent form and is comfortable to wear.
- *Ear muffs* are rigid cups that cover the external ear and are held against the head by a spring-loaded, adjustable band. The seal to the head is made with soft cushions that envelop the whole ear. A fluid seal is a plastic ring containing fluid such as glycerin that fits around the ear and minimizes sound leakage. To overcome the high-frequency resonances within the cup, the space is filled with absorbent material (plastic foam). The seal material must be nonirritant and nontoxic and unaffected by sweat. Ear muffs provide a greater protection than earplugs, but their noise reduction depends, first, on the cup–head seal (this seal is reduced when muffs are worn

over long hair or spectacle frames) and, second, on the force of the cups pressing against the side of the head. Their maximum noise reduction is 35 dB at 240 Hz, 60 dB at higher frequencies. Ear muffs fit most people, are suitable for people frequently moving into and out of high noise levels, and can be worn by individuals for whom ear plugs are unsuitable because of infection of the outer ear canal.

- *Helmets* prevent sounds from reaching high levels attributable to conduction via skull bones. Helmets are used in circumstances where the head must be protected from physical injury. The best protection is afforded by a combination of helmet and ear plugs. Kirk (1993) monitored helmet-mounted ear muffs in loggers over a 12-month period and concluded that the devices were a safe and effective form of protecting against chainsaw noise.

Pleasurable Effects of Sound

Some sport contexts generate noise that becomes an emotive element of the sport. Examples are the screeching of tires in motor racing, the revving of engines in motorcycling, the starter's gun in sprinting, the sound of a driver making contact with a golf ball, or a cricket ball being hit sweetly by the batsman. Whether sound is distracting or alerting and acceptable depends on its meaning to the individual, as illustrated by the use of music in sport.

Music can influence the mood of spectators at sporting events. Loud popular music is used to arouse the audience and evoke emotions prior to the start of competition. Spectators at World Cup soccer tournaments anticipate the rhythmic sound of samba music to accompany the performance of the Brazilian national team.

Background music is used in leisure environments to soften the atmosphere and create a feeling of relaxation. Music is used by individual athletes in the dressing room before a competition to help prepare mentally for the forthcoming contest. Athletes use their own headphones and select their music to suit the desired mood.

Individual responses to exercise can be affected by accompanying music. Besides increasing arousal in a competitive context, music may be a distracting influence on strenuous training, dissociating one's perceptions to external rather than internal stimuli associated with perceived exertion. Alternatively, under light to moderate exercise intensities as in walking or jogging, listening to music through headphones can have a relaxing effect and aid compliance to the exercise program. The impact of music on the performer seems to depend on its type and the performance context.

Slow, soft music tends to decrease physiological responses to submaximal exercise, particularly evident in a lowered heart rate. Furthermore, exercise to exhaustion has been increased when accompanied by soft, slow music compared with fast, loud music or to a control condition. Fast rock music elevates heart responses to submaximal exercise. Although soft, slow music has a favorable effect on perceived exertion during exercise of light intensity, the influence of accompanying music tends to disappear when maximal exercise is undertaken.

Overview and Summary

The environment in which athletes train and compete has implications not only for performance but also for health and safety. The quality of the playing surface, for example, forces a choice of appropriate footwear so that performance can be executed without increased risk. Playing conditions sometimes exceed safety thresholds, and climatic conditions may compromise the safety of participants.

Air pollution can endanger health as well as threaten performance. Ozone concentration may exceed acceptable limits in some of the world's major cities (Athens, Beijing, Mexico City, Los Angeles). Fog worsens pollution but training is usually curtailed in foggy conditions for reasons of visibility.

Accurate assessment of environmental variables is needed to calculate the risk of heat injury. The main factors to be considered are the dry bulb temperature, relative humidity, radiant temperature, air velocity, and cloud cover. The most widely used index in sports contexts is the wet bulb temperature, which takes both ambient temperature and humidity into account. The wind-chill index is used to determine risk in cold conditions. Apart from the chilling effect of the wind, blustery conditions make outcomes more difficult to anticipate and skills become more erratic as a consequence.

The novel environmental challenge—hypoxia, temperature, travel, pollution, weather—calls for preparation and planning. Weather conditions are not always predictable and even can vary widely during the course of competition. An awareness of the dynamic biological adjustments that the body makes allows the sport participant to minimize the adverse effects and discomfort associated with environmental variables.

4

Circadian Rhythms

DEFINITIONS

circadian rhythms—Cyclic variations that recur over each perioc about 24 hr.

electroencephalography (EEG)—Technique used to monitor brai waves, notably the relative degrees of arousal during sleep.

exogenous factors—Environmental factors such as natural daylig ambient temperature, meal times, and social activity that fine-tu the innate rhythms to a 24 hr period.

jet lag—Symptoms of disturbed or desynchronized circadian rhy attributable to traveling across multiple time zones by air flight.

Ramadan—The holy month practiced by devout Muslims in whic food and drink are avoided between sunrise and sunset.

sleep–wake cycle—The natural alternation of rest and wakefulne that accompanies the daily fluctuations of darkness and daylig

four

ATHLETIC contests are played at various times in the day, ranging from morning starts in marathon races to nighttime football matches under floodlights. Similarly, some employees start work early in the morning whereas others work late into the evening. These times can disrupt normal diurnal rhythms because they are out of synchrony with the typical time for work or training. Competitive performance in sport depends on a host of factors, including physiological and psychomotor variables. The work rate in games is correlated with maximal aerobic power, and sprint performance is influenced by anaerobic power (Carling et al., 2005). These performance measures may themselves be affected by diurnal variation, changes that occur within the normal daytime hours.

Circadian rhythms refer to cyclical changes within the body that recur around the 24 hr solar day. Core temperature shows a cycle every 24 hr and is regarded as a fundamental marker of the body's circadian rhythm. Observations on rectal temperature can be fitted with a cosine function and the time of peak occurrence identified. The time that the peak occurs is referred to as the acrophase and is usually found between 5 and 6 p.m (17:00 and 18:00 hr). Many measures of human performance follow this curve in body temperature (Drust et al., 2005). These measures include components of motor performance (such as muscular strength, reaction time, jumping performance, and self-chosen exercise intensity), factors that are important in many sports (figure 4.1). Tasks related to soccer such as dribbling, juggling, and chipping the ball also show a time-of-day effect (Reilly et al., 2007a).

The **sleep–wake cycle** is another relevant biological rhythm to consider. This cycle is linked with the pattern of habitual activity, that is, sleeping during the hours of darkness and working or otherwise staying awake during daylight. Arousal states vary between these times, tending to peak just after midday at the time that circulating

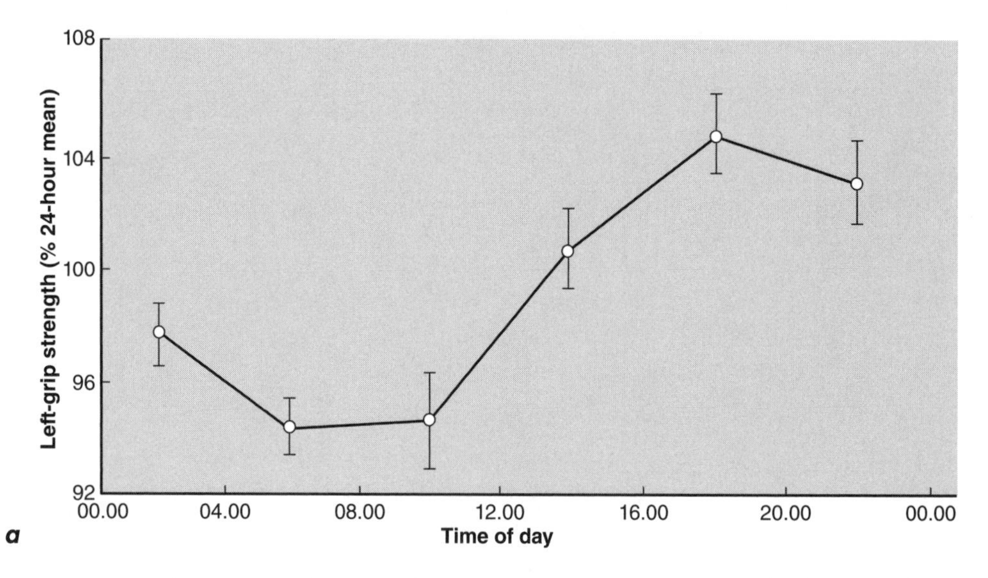

Figure 4.1 The circadian rhythm in self-determined work rate corresponds closely to that of core body temperature. Other measures such as left-grip strength, whole-body flexibility, and leg strength follow the same form as self-chosen work rate.

Reprinted from T. Reilly, G. Atkinson, and J. Waterhouse, 1997a, *Biological rhythms and exercise* (Oxford, UK: Oxford University Press), 42, 44, 52, 53. By permission of Oxford University Press.

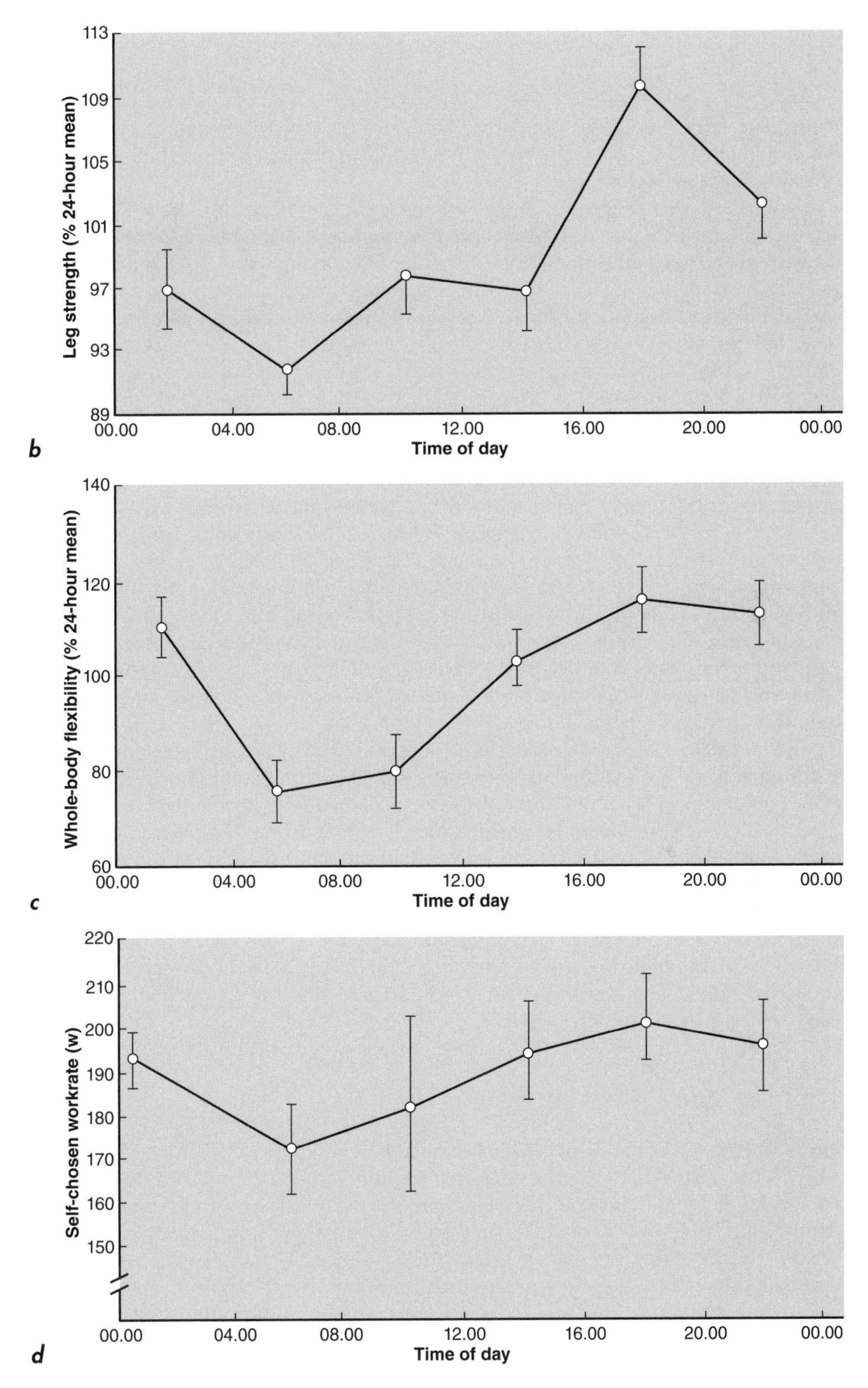

Figure 4.1 *(continued)*

levels of adrenaline are at their highest. A sports team forced to compete at a time of day its members would normally be inactive would not be well equipped, biologically or psychologically, to compete.

Endogenous and exogenous factors influence circadian rhythms, depending on the degree to which specific rhythms are governed by environmental signals. The **exogenous factors** include natural and artificial light, temperature, type and timing of meals, and social and physical activity. Endogenous rhythms imply internal body clocks, the suprachiasmatic nucleus cells of the hypothalamus being the site of control of circadian rhythms. These cells are linked by neural pathways to the pineal gland, and timekeeping functions have been attributed to the pineal, its hormone melatonin, and related substances such as serotonin. The most relevant circadian rhythms for human performance seem to be the body temperature curve and the sleep–wake cycle.

The harmonious coexistence of distinct circadian rhythms cannot be assumed when the normal sleep–wake cycle is disrupted. Such perturbations occur as a result of changes in domestic circumstances that interrupt normal sleep, when anxiety prohibits restful sleep, and when a person does nocturnal shift work. These perturbations also apply to travelers on long-haul flights over multiple meridians and to a lesser extent to Muslims fasting during the holy month of **Ramadan,** when eating and drinking are eschewed from sunrise to sunset (Reilly and Waterhouse, 2007). The consequences of a perturbed sleep–wake cycle are usually apparent in mood, alertness, and performance (Bonnet, 2006). The effects of these disruptions may be more pronounced in athletic activities, particularly adventure events where the amount of time allocated for sleep is minimized.

The importance of good-quality sleep for sport participants is acknowledged by practitioners (e.g., Reilly, 2006). Insights can be gained into the role of sleep by looking at the consequences of disruptions to the sleep–wake cycle and how individuals cope in such conditions. This chapter considers the effects of training at different times of day before reviewing the disruptions to circadian rhythms caused by transmeridian air flights. The effects of total sleep deprivation, chronic sleep loss, and partial sleep allowances are considered, and results are reviewed from both laboratory and field studies. The circumstances of nocturnal shift work are then reviewed along with remedies for counteracting any performance impairments. Sleep disruption in the context of individual differences is discussed, and guidelines are provided for coping with necessary breaks to normal sleep.

Training and Time of Day

Physical performance measures usually demonstrate a peak in performance that occurs close to the acrophase of the circadian rhythm in body temperature. On this basis the ideal time for exercise would be about 5 to 6 p.m., assuming the environmental temperature is within the comfort zone (Reilly, 2009). There is probably a window of some hours during the day when maximal performance can be achieved. The optimal point can be realized with appropriate warm-up and physical and mental preparation. Consequently, events commencing before 3 p.m. and after 7:30 p.m. do not necessarily entail suboptimal performance, particularly because muscle and core temperatures rise during the course of sustained exercise. Particular consideration to warm-up is needed late in the evening, say 8 p.m., in cold conditions.

The time of habitual training and the time at which competitive events are held often do not coincide. The majority of professional football teams train in the morning, starting at 10 or 11 a.m. Strenuous physical conditioning exercise is best conducted in the early evening, the time at which many amateur teams train. Joint stiffness is greatest in the morning and so special attention should be given to flexibility exercises during warm-up prior to morning training sessions. When athletes have to compete at a time of day to which they are unsuited, simply training at that time in the few days beforehand seems to be helpful.

Endurance performance is not necessarily hampered by a morning start, provided that exercise is not overly high in intensity at the start. Reilly and Garrett (1995) allowed subjects to pace themselves over an extended exercise test at two different times of day. During the morning, subjects began slowly but increased the intensity of exercise progressively throughout. In contrast, they started at a higher work rate in the evening but by the end of the 90 min period were operating at a lower intensity than they had done in the morning. Sweating takes place at a lower body temperature in the morning than in the evening. In hot conditions, the increase in core temperature may reach a critical upper value more quickly in the evening compared with the morning, when the environmental temperature is also likely to be cooler. Thus, the best time of day for exercise depends on the tasks to be performed, the goals of the session, and the environmental conditions.

Circadian variations in skills related to soccer were examined by Reilly and colleagues (2007a). Muscle strength and body temperature conformed to the typical circadian rhythm with a peak denoted for about 6 p.m. The soccer-skills tests also demonstrated a diurnal variation. The tasks that required the greater degree of motor control, juggling and chipping tests, tended to peak earlier than those like dribbling speed that involved more gross motor functions. This separation quite likely reflects the existence of more than a single circadian rhythm, whereby functions related to the nervous system tend to reach a peak earlier in the day than those linked with changes in body temperature (Reilly and Waterhouse, 2005). This separation may also entail a mental fatigue effect associated with the length of time awake. It seems that skills may be best acquired in midday sessions just as the curve in arousal approaches and reaches its high point. Consequently, there is a case for young professionals to do their skills work at light intensity in morning sessions. The more intense exercise can be retained for a later session following lunch and a recovery.

Sleep–Wake Cycle

The sleep–wake cycle is fundamental to humans, activity being associated with the hours of daylight and sleep with the hours of darkness. This daily recurrence is linked with the responses of the pineal gland to the environment. The pineal hormone melatonin is secreted at dusk, but its production is inhibited by light. Once melatonin is secreted, its vasodilatory properties cause body temperature to decrease. Another effect is that it causes drowsiness, preparing the body for sleep.

Sleep itself is divided in separate stages, based on **electroencephalography (EEG)**. Typically in an 8 hr sleep, an individual goes through five or six cycles, made up of stages 1 and 2, slow-wave sleep (stages 3 and 4), and REM, or rapid-eye-movement, sleep. It is mainly during REM sleep that dreaming occurs. The first half of a night's sleep tends to contain relatively more non-REM sleep, REM sleep becoming more prominent in the second half of the night (see figure 4.2).

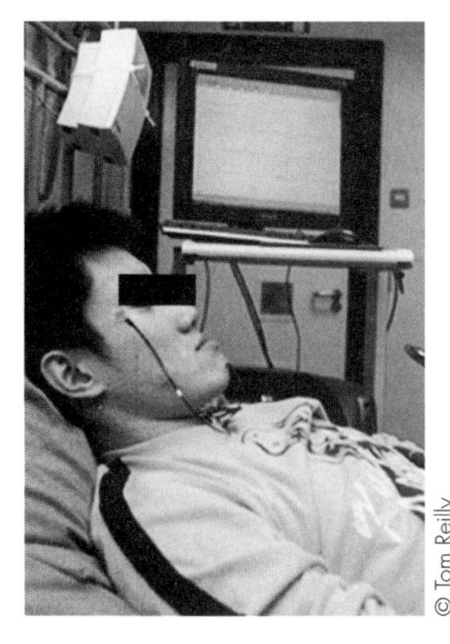
© Tom Reilly

Figure 4.2 Electrodes and display unit used to monitor electroencephalography.

It is not clear why the body needs sleep. Many functions are consolidated during sleep and many restorative processes occur. It is clear that neural mechanisms benefit from sleep, as do many aspects of immune function. Many individuals, workers and athletes, believe that good sleeping habits promote their well-being and that lack of sleep compromises it.

Travel Fatigue and Jet Lag

Elite athletes are regularly called upon to travel large distances to participate in international or interclub competitions. Teams may also participate in closed-season tournaments or friendly games overseas as part of preseason training. Such engagements are made possible by the speed of contemporary air flight. Although international travel is routine nowadays for recreational purposes, it is not without attendant problems for the traveling athlete, which should be recognized in advance.

Many athletes have their regular routines disrupted when they travel abroad. They may be particularly excited about the trip or worried about planning for the departure. Depending on the country to be visited, visas and vaccinations may be required. Professional teams usually have arrangements made for them by their administrative and medical staff. These arrangements extend to coping with formal procedures at departure and disembarkation and avoiding any mix-ups in dealing with ground staff and security controls.

Having arrived safely at the destination, the athlete may suffer travel fatigue, loss of sleep (depending on flight times), and symptoms that have come to be known as **jet lag.** This term refers to the feelings of disorientation, light-headedness, impatience, lack of energy, and general discomfort that follow traveling across time zones (see highlight box on p. 81). These feelings are not experienced with traveling directly northward or southward within the same time zone when the passenger simply becomes tired from the journey or stiff after a long stay in a cramped posture. Jet lag may persist for several days after arrival and can be accompanied by loss of appetite, difficulty in sleeping, constipation, and grogginess. Although individuals differ in severity of symptoms they experience, many people simply fail to recognize how they are affected, especially in tasks requiring concentration, situation awareness, and complex coordination.

The body's circadian rhythm at first retains the characteristics of the point of departure following a journey across multiple time zones. The new environment soon forces new influences on these cycles, mainly the time of sunrise and onset of darkness. Endogenous circadian rhythms such as core temperature and other measures are relatively slow to adjust to this new context. It takes about one day for each time zone crossed for core temperature to adapt completely. Sleep is likely to be difficult for a few days, but **exogenous rhythms** such as activity, eating, and social contact during the day help to adjust the sleep–wake rhythm. Arousal state adapts more

Symptoms of Jet Lag

Poor sleep during the new night time

- Delayed sleep onset after eastward flight
- Early awakening after westward flight

Poor performance in mental and physical tasks

Negative subjective changes

- Increased fatigue and irritability
- Headaches and lapses in concentration
- Reduced vigour and motivation

Gastrointestinal disturbances

- Indigestion
- Frequency of defecation
- Consistency of stools
- Decreased enjoyment of meals

quickly than does body temperature to the new time zone. Until the whole range of biological rhythms adjust to the new local time and become resynchronized, athletes' performance may be below par (figure 4.3).

The severity of jet lag is affected by a number of factors besides individual differences. The greater the number of time zones traveled, the more difficult it is to cope with changes. A 2 hr phase shift may have marginal significance, but a 3 hr shift (e.g., British or Irish teams traveling to play opponents in Russia, or American athletes traveling coast to coast within the United States) will cause desynchronization to a substantial degree. In such cases the flight times—time of departure and time of arrival—may determine the severity of the symptoms of jet lag. Training times might be altered to take the direction of travel into account. Such an approach was shown to be successful in American football teams traveling across time zones within the United States and scheduled to play at different times of day (Jehue et al., 1993).

When journeys entail a 2 to 3 hr time-zone transition and a short stay (2 days), it may be feasible to stay on "home time." Such an approach is useful if the stay in the new time zone is 3 days or less and adjustment of circadian rhythms is not essential. This approach requires that the time of competition coincide with daytime on home time. If this is not the case, then adjustment of the body clock is required. A European team that is to compete in the morning in Japan or in the evening in the United States will require an adjustment of the body clock, because these timings would otherwise be too difficult to cope with.

Symptoms of jet lag recede after the first 2 or 3 days following arrival but may still be acute at particular times of day. There will be a window during the day when time of high arousal associated with the time zone departed from and the new local time overlap. This window may be predicted in advance and should be used for timing of training practices in the first few days at the destination.

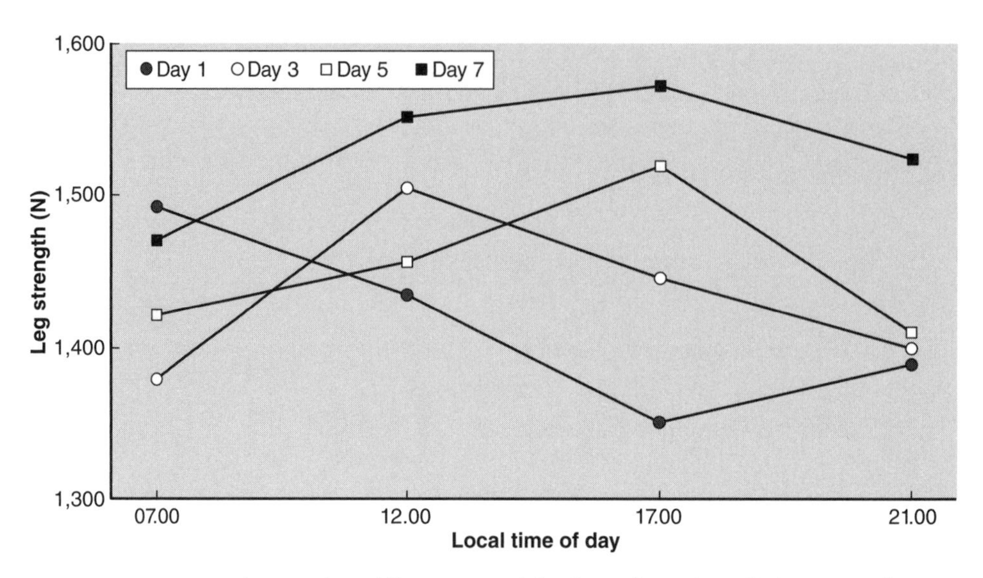

Figure 4.3 Leg muscle strength at different times of day for 7 days after a flight westward across five time zones. Note the curve is still adjusting between days 3 and 5.

Adapted from *Sport, Exercise and Environmental Physiology,* T. Reilly and J. Waterhouse, pg. 102, copyright 2001, with permission from Elsevier.

The direction of travel influences the severity of jet lag. Flying westward is easier to tolerate than is flying eastward. On flying westward, the first day is lengthened and the body's rhythms can extend in line with their natural free-wheeling period of about 25 hr and thus catch up. Traveling to Japan (9 hr in advance of British Summer Time) and Malaysia (7 hr in advance of British Summer Time) requires more than 9 and 7 days, respectively, for jet lag symptoms to disappear in some individuals. In contrast, readjustment is more rapid on returning to Britain from the east (Reilly, 2003). However, when time zone shifts approach near-maximal values (e.g., a 10-12 hr change) there may be little difference between eastward and westward travel and the body clock is likely to adjust as if the latter had occurred (Reilly et al., 2005).

Sleeping pills have been used by some traveling athletes to induce sleep while on board flight. Drugs such as benzodiazepines are effective in getting people to sleep but they do not guarantee a prolonged period asleep. They were ineffective in accelerating adjustment of the body clock in a group of British Olympic athletes traveling to the United States (Reilly et al., 2001). Besides, these drugs have not all been satisfactorily tested for subsequent residual effects on motor performances such as sport skills. They may in fact be counterproductive if administered at the incorrect time. Nonbenzodiazepine sedatives such as zopiclone and zolpidem have fewer side effects and minimal interference with normal sleep architecture (Lemmer, 2007). Melatonin is one substance that can act directly on the body clock as well as being a hypnotic, but the timing of administration is critical. Travelers between the United Kingdom and Australia, a journey that can elicit the most severe jet lag symptoms, were found to have no benefit from melatonin (Edwards et al., 2000). Melatonin administered in the few hours before the trough of body temperature will have a phase-advance effect whereas if administered in the hours after this trough will delay the circadian rhythm. Ingestion of melatonin at other times will have no chronobiotic effect but will help to induce drowsiness. Drugs do not provide an easy solution to preventing jet lag, and a behavioral approach can be more effective in alleviating symptoms and hastening adjustment (Reilly et al., 2005).

The timing of exposure to bright light is key in implementing a behavioral approach. Light demonstrates a phase-response curve, opposing the effects of melatonin (Waterhouse et al., 1998). Exposure to natural or artificial light before the trough in core temperature promotes a phase delay, whereas a phase advance is encouraged by light administered after this time, meaning "body clock time." Exposure to light at 10 p.m. in Los Angeles following a flight from London would promote a phase advance on the first night rather than the required phase delay, administration occurring after the trough in core temperature (Waterhouse et al., 2007). Where natural daylight cannot be exploited, artificial light from visors or light boxes can be effective for phase-shifting purposes (see figure 4.4); these commercially available devices have been used in treating seasonal affective disorder found among natives of northern latitudes during the winter seasons when the hours of daylight are limited. The malaise is not a common affliction among athletes.

The athlete should adjust as soon as possible to the local daytime and nighttime in the new environment. Focusing on the local time for disembarkation can help in planning the rest of the daily activity. Natural daylight inhibits melatonin and is the key signal that helps to readjust the body clock to the new environment. There may be other environmental factors to consider such as heat, humidity, or even altitude.

A phase delay of the circadian rhythm is required after traveling westward, and visitors may be allowed to retire to bed early in the evening. Early onset of sleep will be less likely after an eastward flight. In this case, a light training session on that evening will instill local clues into the rhythms. Exercise can hasten the adaptation to a new time zone, and a light training session on the afternoon of arriving in the United Kingdom after a flight has proved beneficial (Reilly, 1993). Training in the morning is not recommended after a long-haul, eastward flight because it exposes the individual to natural daylight and could delay the body clock rather than promote the phase adjustment required in this circumstance. This strategy of avoiding morning sessions until it was deemed appropriate was used by British Olympic athletes arriving in Australia for the Sydney Olympics in 2000.

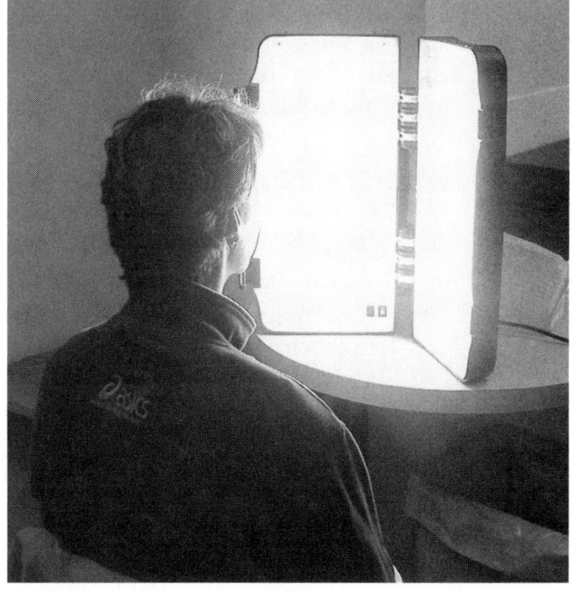

© Tom Reilly

Figure 4.4 A light box can be used to correct circadian rhythm disturbances.

Exercise should be light or moderate in intensity for the first few days in the new time zone, because training hard while muscle strength and other measures are impaired will not be effective (see figure 4.3). Skills requiring fine coordination are also likely to be impaired during the first few days, and this might lead to accidents or injuries if technical training sessions are conducted too strenuously. When a series of tournament engagements are scheduled, it is useful to have at least one friendly competition before the end of the first week in the overseas country. Naps should be avoided for the first few days because a long nap at the time the individual feels drowsy (presumably at the time he or she would have been asleep in the time zone just departed from) anchors the rhythms at their former phases and so delays the adaptations to the new time zone.

Some precautions are necessary during adjustment to the new time zone. Alcohol taken late in the evening is likely to disrupt sleep and so is not advised. Normal hydration levels may be reduced following the flight because of respiratory water loss in the dry cabin air, and so fluid intake should be increased. A diet recommended for commercial travelers in the United States entailed use of protein early in the day to promote alertness and carbohydrate in the evening to induce drowsiness. This practice is unlikely to gain acceptance among athletes, although they could benefit from avoiding large evening meals. The evening meal might include vegetables with a choice of chipped, roasted, or baked potatoes; pasta dishes; rice; and bread with sufficient fiber to reduce the risk of becoming constipated.

By preparing for time zone transitions and the disturbances they impose on the body's rhythms, the athlete can reduce the severity of jet lag symptoms. There has been little success in attempting to predict good and poor adaptors to long-haul flights. The fact that a person feels relatively unaffected on one occasion is no guarantee that she will do so again on the next visit. Regular travelers benefit from their experiences and develop personal strategies for coping with jet lag (Waterhouse et al., 2002). The disturbances in mental performance and cognitive functions have consequences not only for athletes but also for training and medical staff traveling with them, who are also likely to suffer from jet lag symptoms. The long periods of inactivity during the plane journey may lead to the pooling of blood in the legs and in susceptible people cause a deep-vein thrombosis. Moving around the plane periodically during the journey, say, every 2 hr, and doing light stretching exercises are recommended. Travelers should also drink about 15 to 20 ml extra fluid per hour, preferably fruit juice or water, to compensate for the loss of water from the upper respiratory tract attributable to inhaling dry cabin air (Reilly et al., 2007b). Without this extra fluid intake, the residual dehydration could persist into the early days in the new time zone.

Sleep Deprivation or Disruption

Sleep deprivation refers to an amount of sleep below what the individual is accustomed to. Effects may accumulate so that a "sleep debt" is experienced. The circumstances range from minor disruption to a total lack of sleep at nighttime (or rest time), as occurs in activities such as watch keeping, military maneuvers, or extreme sports. Sleep loss is associated with nocturnal shift work and is a by-product of altered circadian rhythms such as occur in crossing multiple meridians. Complex tasks are affected more easily by sleep loss than are gross motor functions. Effects of sleep loss are normally self-limiting, and sleeping pills are not necessarily useful for athletes.

Total Sleep Deprivation

Some years ago, Thomas and Reilly (1975) showed that it was possible to maintain continuous exercise at moderate intensity for at least 100 hr nonstop. Energy intake was provided to match the rate of energy expenditure (30.77 MJ/day) and delivered as a glucose syrup drink. Despite the consistency in muscular power output (which was controlled), the heart rate decreased over the first 2 days of the trial, suggesting a reduction in sympathetic drive. Lung function (indicated by vital capacity and forced expiratory volume in the first second of expiration) displayed a deteriorating trend over the 100 hr, superimposed on circadian periodicity. There was a significant trend in slowing of visual reaction time with each successive day without sleep. Errors in a signal detection test appeared after the first night of sleep loss and in mental tasks requiring short-term memory after the second night, although neither task demonstrated a significant circadian rhythm under these conditions. The observations highlighted the erratic nature of performance tasks in these circumstances and the suppression of some circadian rhythms in conditions that demand a constant level of muscular power output.

Where participants attempt to achieve entry into the *Guinness Book of World Records* for extreme endurance, the activity is sustained at a self-chosen intensity. When two teams playing five-a-side soccer for 91.8 hr were monitored, the work rate demonstrated a significant circadian rhythm each day and a decline from day to day (Reilly and Walsh, 1981). The rhythm in activity was in phase with that of body temperature, and in this instance the heart rate response showed cyclical variation corresponding to the physical activity (figure 4.5). Impairment in mental performance was evident after only one night; lapses in attention and delays in reaction time became more pronounced than changes in physical measures such as grip strength, which proved resistant to fatigue effects induced by lack of sleep.

When individuals are deprived of sleep over successive nights, bizarre behavioral episodes, illusions (visual, auditory, and olfactory), or hallucinations are often noted. The cause of disturbances in cognitive and perceptual processes was examined in a group of soccer players playing five-a-side games indoors for 72 hr (Reilly and George, 1983). Blood samples were obtained every 4 hr and mood states were monitored at

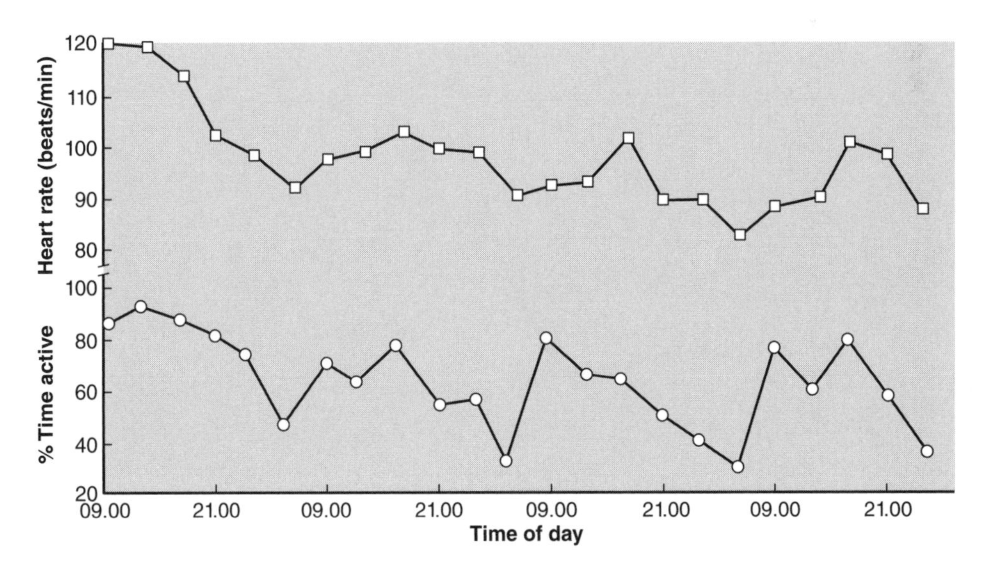

Figure 4.5 The mean heart rate and percentage of time active in indoor soccer players over 4 days without sleep. The measurements were made every 4 hr, heart rate 1 hr later than activity.

Adapted, by permission, from T. Reilly and Walsh, 1981, "Physiological, psychological and performance measures during an endurance record for 5-1-side soccer play," *British Journal of Sport Medicine* 15: 122-128.

the same time points in the 5 min rest allowed every 50 min. Unprepared reaction time was sampled by means of a portable device worn on a harness for administration of a test protocol while play was continued (see figure 4.6). The data suggested that β-phenylethylamine, a naturally occurring brain amine, plays a role in the cycles of unusual behavior and mood states occurring in these circumstances, because the concentrations of the free amine demonstrated a circadian rhythm superimposed on a progressive day-to-day increase. Despite the occasional episode of erratic behavior, grip strength remained relatively stable over the 4 days, allowing for the circadian rhythm that existed.

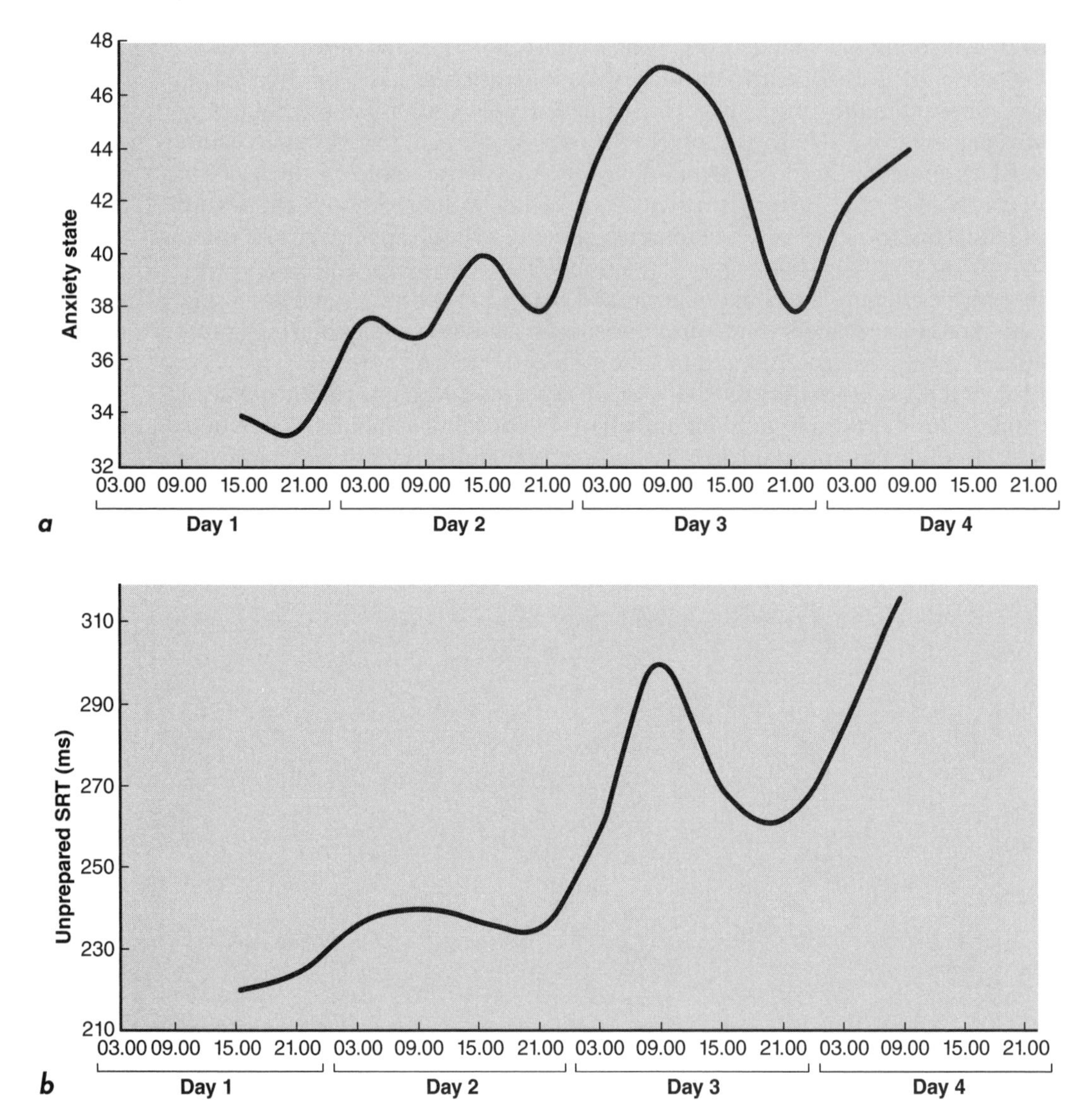

Figure 4.6 Changes in *(a)* anxiety state and *(b)* unprepared simple reaction time in five-a-side soccer players over 72 hr without sleep.

Reprinted, by permission, from T. Reilly and A. George, 1983, "Abstracts of the Society of Sports Science Conference: Urinary phenylethamine levels during three days of indoor soccer play," *Journal of Sport Sciences* 1: 70, Taylor & Francis Ltd, www.tandf.co.uk/journals.

Sleep restriction is commonly experienced by sailors and soldiers. How and colleagues (1994) used a battery of tests that covered cognitive and physical performance when studying naval seamen deprived of sleep for more than 72 hr. The more pronounced declines were observed in cognition, speed, and precision, whereas smaller effects were found in routine tests of physical measures. The changes became more evident after 36 hr: All performance measures displayed a diurnal rhythm and troughs coincided with the highest ratings for sleepiness.

A similar correlation between subjective states and skill performance was reported when military recruits were monitored while being kept awake for three successive nights (Froberg et al., 1975). An increase in self-rated fatigue coincided with a decline in accuracy of rifle shooting, both measures exhibiting circadian rhythmicity over the 3 days. The performance curves were in phase with circulating noradrenaline concentrations, which increased progressively with a peak each daytime. The increased concentrations of noradrenaline were thought to reflect an increased mental drive necessary to maintain performance in the face of sleep deprivation. The observations showed how circadian rhythms can persist alongside a progressive trend in fatigue under conditions of complete sleep deprivation. A similar picture is presented for anxiety and unprepared simple reaction time of subjects playing soccer indoors for 72 hr without any sleep (shown in figure 4.6).

The significant impact of sleep deprivation on psychomotor performance was confirmed in a meta-analyses of relevant studies. Koslowsky and Babkoff (1992) concluded that the longer the period without sleep, the greater was the effect on performance. Furthermore, decreases in speed were greater than decrements in accuracy. In a second meta-analysis, Pilcher and Huffcutt (1996) showed that mood measures were more sensitive than cognitive tasks, which were in turn more sensitive than motor tasks during sleep loss. Sport skills frequently incorporate decision making as well as physical components, errors in either of which are reflected in performance outcomes. Any deterioration in mood is also likely to affect performance when maximum effort and determination are required of the participant.

Chronic Sleep Loss

Observations on chronic sleep loss in realistic conditions have relied mainly on ultra-endurance races, long-distance sailing, and military operations. In these instances some sleep is allowed or is taken according to strategies for the competitive event or necessitated by weather conditions.

Smith and colleagues (1998) studied competitors in the Race Across America, a solo-bicycle race over 4,640 km in the United States, which takes 8 days. Over 3 years the average sleep taken by the winners was 2 hr per night. In a comparable Eco-Challenge event completed in 7 days and 2 hr, the sleep taken voluntarily by the winning team in 2002 averaged 2.4 hr/day. After experiencing extreme physical and cognitive fatigue the previous year, the victorious team had decided to go no more than 30 hr without sleep (Smith and Reilly, 2005). Although participants can complete these competitions over challenging terrain and difficult environmental conditions, the events exact a huge toll on their physical and mental resources.

Chronic sleep disturbances are anticipated by sailors in races across the great oceans and around the world. Bennet (1973) studied 19 solo sailors during a transatlantic race that took about 38 days. Most participants awoke at intervals to check weather and direction, one sailor making these checks every 30 min each 24 hr. Errors were common among the sailors, and hallucinations were reported by some.

The British sailor Ellen McArthur used the cluster-napping technique promoted by Stampi and colleagues (1990) to sail in the 40,000 km Vendee Globe Race alone. Nighttime sleep varied in duration in accordance with weather conditions but was supported by daytime naps. The strategy entails separating a long sleep into shorter units of 25 to 40 min each, during which quick checks are conducted on the boat, its navigation equipment, and the weather conditions while staying awake, and immediately resuming sleep once these chores are completed. McArthur's average nap lasted 36 min, and total sleep averaged 5.5 hr/day over the 94 days of the 2001 race, in which she finished second overall.

Military personnel have also been studied with a view to charting the effects of an arduous physical regimen while on restricted sleep rations. Rognum and colleagues (1986) considered that Norwegian soldiers were ineffective at the end of 4 days with only 2 hr of sleep each night. This conclusion was based on deteriorated performance over a 1 km assault course, a shooting test, and a 3 km run. A diet high in energy intake did not prevent the impairment.

In another study, 27 soldiers expended 21 MJ/day over 5 days on a combat course, taking less than 4 hr of sleep each day. The participants were divided into three equal groups according to energy intake; those on low intake had 7.6 MJ/day, a medium-intake group had 13.4 MJ/day, and the soldiers on high intake were given 17.6 MJ/day (Guezennec et al., 1994). The participants on the low-energy intake experienced an 8% decrease in maximal oxygen uptake ($\dot{V}O_2max$) and a 14% decline in anaerobic power by the end of the course, whereas the other two groups did not show a significant decrease in either function. It seems that a large energy imbalance leads to deterioration in both aerobic and anaerobic power production when activity is sustained over days and sleep is reduced. In this study there were insufficient intermediate observations to show transient decreases in performance while the soldiers were on combat maneuvers.

A decrease in $\dot{V}O_2max$ is not inevitable with sleep deprivation, and no change in $\dot{V}O_2$ is observed at work rates up to 80% $\dot{V}O_2max$ (Horne and Pettit, 1984). Maximal oxygen uptake is itself a robust function, but a difficulty facing researchers is to get subjects who are deprived of sleep to exercise at progressive work rates until voluntary exhaustion is reached. Criteria showing that $\dot{V}O_2max$ is actually attained include a plateau in $\dot{V}O_2$ before termination, a high blood lactate concentration, and a respiratory exchange ratio ($\dot{V}CO_2$:$\dot{V}O_2$) greater than 1.10. Some subjects have shown a small decline in $\dot{V}O_2max$ after incurring a sleep debt over two successive nights (Chen, 1991; Plyley et al., 1987) but other researchers (Martin and Gaddis, 1981) have found that maximal aerobic power can be retained, at least after one whole night's sleep loss. Disruptions to normal eating and drinking patterns and to the individual's motivational climate may contribute to a failure to sustain exercise on an incremental test to exhaustion—as is required to satisfy the standard criteria that a maximal physiological state was reached.

Changes in gene expression may provide insights into the consequences of sleep loss on energy processes. Genes expressed during wakefulness to regulate mitochondrial activity and glucose transport are likely to reflect increased energy needs. One gene for the enzyme arylsulfotransferase has shown stronger induction as a function of the length of sleep deprivation. This induction was suggested to reflect a homeostatic response to continuing central noradrenergic activity during loss of sleep (Cirelli,

2002). The neuradrenergic activity is thought to reflect increased central effort to maintain wakefulness.

Partial Sleep Deprivation

Most sports entail competition within a single day, and so the study of partially reduced sleep in the day or days prior to sports contests has more relevance than the study of total sleep deprivation or chronic sleep loss. Research designs have entailed substantially reduced sleep allowances, partly to ensure that all sleep stages are affected and partly also to safeguard against a type II experimental error. Those studies relevant to sport have included time trials or components of performance, whereas others have used laboratory-based measures that have more generic applications.

Performance of swimmers on restricted nightly sleep (2.5 hr of sleep per night) was studied by Sinnerton and Reilly (1992). Eight swimmers were tested in a 50 m pool on 4 consecutive days, morning (6:30 a.m.) and evening (5:30 p.m.), under conditions of normal sleep and under partial sleep deprivation. Grip and back strength, lung function (vital capacity, forced expiratory volume in 1 s), resting heart rate, and mood states were recorded, and the swimmers performed four trials at 50 m and one trial at 400 m. No decrements were observed with sleep deprivation in back or grip strength, lung function, or swim times, although these variables demonstrated an effect of time of day. Sleep loss affected mood states, increasing depression, tension, confusion, fatigue, and anger while decreasing vigor significantly. The data supported Horne's (1988) brain restitution theory of sleep, suggesting that the primary need for sleep is located in nerve cells rather than in other biological tissues.

Effects of partial sleep deprivation were investigated by Reilly and Deykin (1983) in a group of trained men (3 nights of sleep loss and a single night of subsequent recovery sleep), by means of a battery of psychomotor tests, measurements of physical working capacity, and subjective-state tests. To investigate the effects of exercise as an antidote to sleep loss, the investigators measured the participants' ability to carry out various performance tasks while running on a treadmill at 10 km/hr. Gross motor functions including muscle strength, lung power, and endurance running on a treadmill remained unaffected by three nights of severely restricted sleep. Decrements occurred in a range of psychomotor functions, the majority of which were evident after only one night of reduced sleep. Exercise promoted arousal after sleep loss, providing an obvious temporary counteraction to decreases in mental alertness. All functions monitored were restored to normal after a full night of recovery sleep, so that the effects of sleep deprivation were short-lasting.

Women seem to experience the same effects of sleep deprivation as do men. Reilly and Hales (1988) restricted the sleep of well-trained women to 2.5 hr per night for 3 nights. Baseline measures were obtained for 4 days as a control. Measurements were made each morning (7-9:30 a.m.) and evening (7-9:30 p.m.) for oral temperature, lung function, grip strength, anaerobic power output, limb steadiness and speed, and subjective sensations at rest and during exercise. Apart from hand steadiness, all measures showed diurnal variations in phase with the variation in oral temperature. Gross motor functions were less affected by sleep loss than were the tasks requiring fast reactions. A 5 min submaximal exercise bout at 60% $\dot{V}O_2max$ was effective in reducing the feeling of sleepiness, which was more pronounced in the morning than in the evening. The exercise was rated more difficult in the morning than in the evening, and the rating was increased with successive days of partial sleep deprivation.

It was concluded that the effects of sleep loss may be masked if time of day is not taken into consideration.

In single all-out efforts athletes may be able to overcome the adverse effects of sleep loss, yet they may be unable or unwilling to maintain a high level of performance in sustained exercise and in repeated exercise bouts such as those that occur in extended training sessions. Reilly and Piercy (1994) focused on weightlifting tasks, using typical weight training exercises such as maximal lifts and a psychophysical approach to assessing repeated submaximal efforts. The investigators found no significant effect of sleep loss on performance of maximal biceps curl, but a significant effect was noted on maximal bench press, leg press, and deadlift. Trend analysis indicated decreased performance in submaximal lifts for all four tasks; the deterioration was significant after the second night of sleep loss. These changes were evident in the perception of effort—whether rated for breathing, muscles, or general whole-body feeling—as indicated by the responses to biceps curl (figure 4.7 *a-d*) and deadlift (figure 4.7 *e-h*). It appears that submaximal lifting tasks are more affected by sleep loss than are maximal efforts, particularly for the first two nights of successive sleep restriction. The greatest impairments were found the later in the protocol that the lifts were performed, indicating that a cumulative fatigue effect accrued during the training sessions that was attributable to sleep loss.

The fact that muscle strength may be resistant to the effects of one night's sleep deprivation has been confirmed, whether loss was total (Meney et al., 1998) or partial (Bambaeichi et al., 2005). Meney and colleagues noted that body temperature did not decline as a result of no sleep, and isometric strength of back and leg muscles was retained. Bambaeichi and colleagues conducted measurements at 6 a.m. and 6 p.m. on female subjects using isokinetic dynamometry. Peak torque was about 5% higher in the evening compared with the morning for concentric actions of knee flexors at angular velocities of 1.05 and 3.14 rad/s. The variations were in phase with changes in rectal temperature but were unaffected by restriction of sleep to 2.5 hr overnight. These findings suggest that circadian variations in muscle performance are more robust than are the effects of sleep deprivation.

Performance has taken different guises in the various studies of partially deprived sleep. A taxonomy proposed by Reilly and Edwards (2007) is shown in table 4.1 (p. 94), suggesting how performance in certain types of activity might be affected. Such a classification itself entails broad generalization, because the effects of sleep loss can be mitigated by the challenge that the activity presents to the subject.

Ramadan

During the holy month of Ramadan, strict adherents to the Muslim faith avoid food and fluid ingestion during daylight hours. The focus on Ramadan rather than other forms of self-deprivation is because normal circadian rhythms are disturbed during the holy month. This fasting regimen applies to approximately 18% of the world's population. This practice displaces energy intake and hydration to the hours of darkness and partly reverses the normal circadian pattern of eating and drinking. The long period of abstinence during the day causes an increase in hunger and subjective fatigue, a reduction in energy levels, and a progressive dehydration compared with habitual diurnal experiences at other times of the year.

A negative energy balance is often experienced during Ramadan, but this deficit in energy intake is not a universal finding. Energy expenditure may be reduced because of a lowered daytime physical activity level, and there may be two separate meals between onset of darkness and returning to sleep. A delay in bedtime has consequences the following morning, because breakfast must be taken before dawn. The sleep–wake cycle may be further displaced if the daytime period of fasting is fractured by spells of sleeping in nonstrict adherence.

The performance consequences of Ramadan fasting were reviewed by Reilly and Waterhouse (2007). Effects were noted in increased incidence of accidents and adverse moods with reduced vigor and increased fatigue. Physical performance tends to show

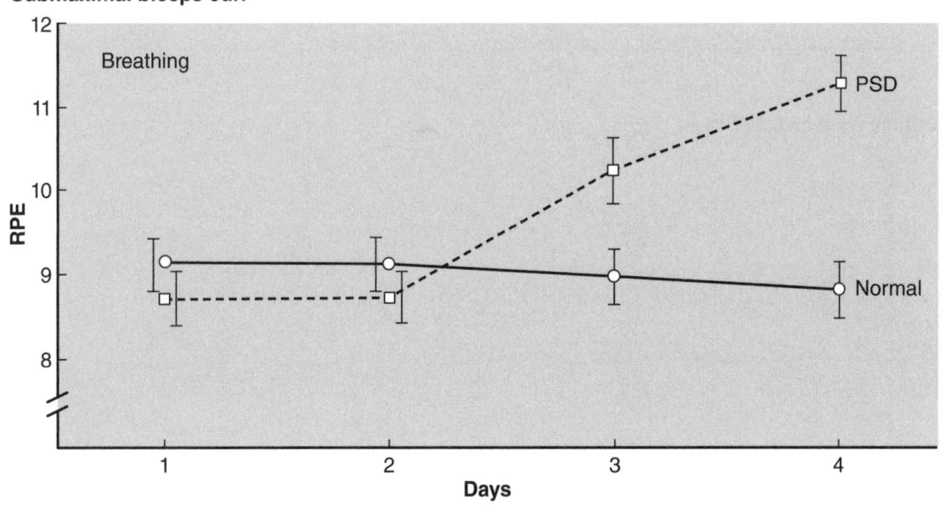

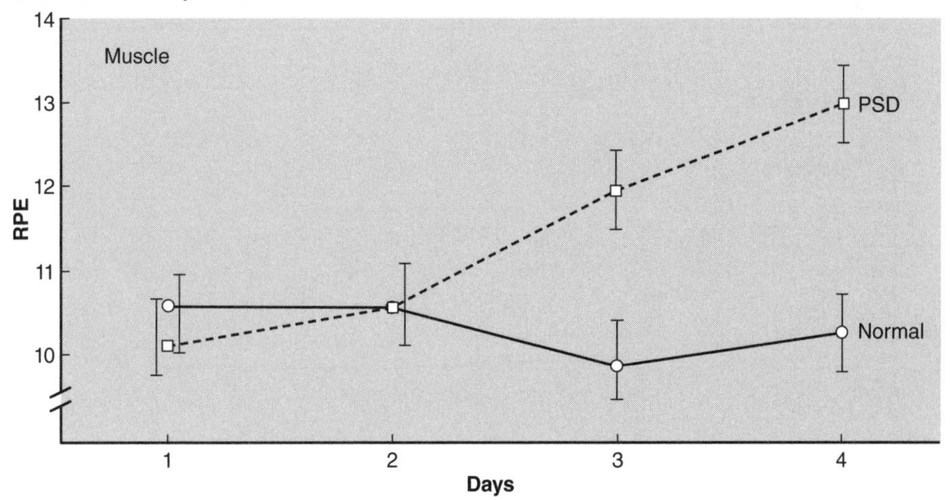

Figure 4.7 Perceived exertion (RPE) during sustained biceps curl *(a-d)* and deadlift *(e-h)*, rated for breathing, muscle, and general whole-body feeling. The CRS scale refers to category ratio. Day 1 is a baseline day after normal sleep, and PSD refers to partial sleep deprivation.

Reprinted, by permission, from T. Reilly and M. Piercy, 1994, "The effect of partial sleep deprivation on weight-lifting performance," *Ergonomics* 37(1): 107-115, Taylor & Francis Ltd, www.tandf.co.uk/journals.

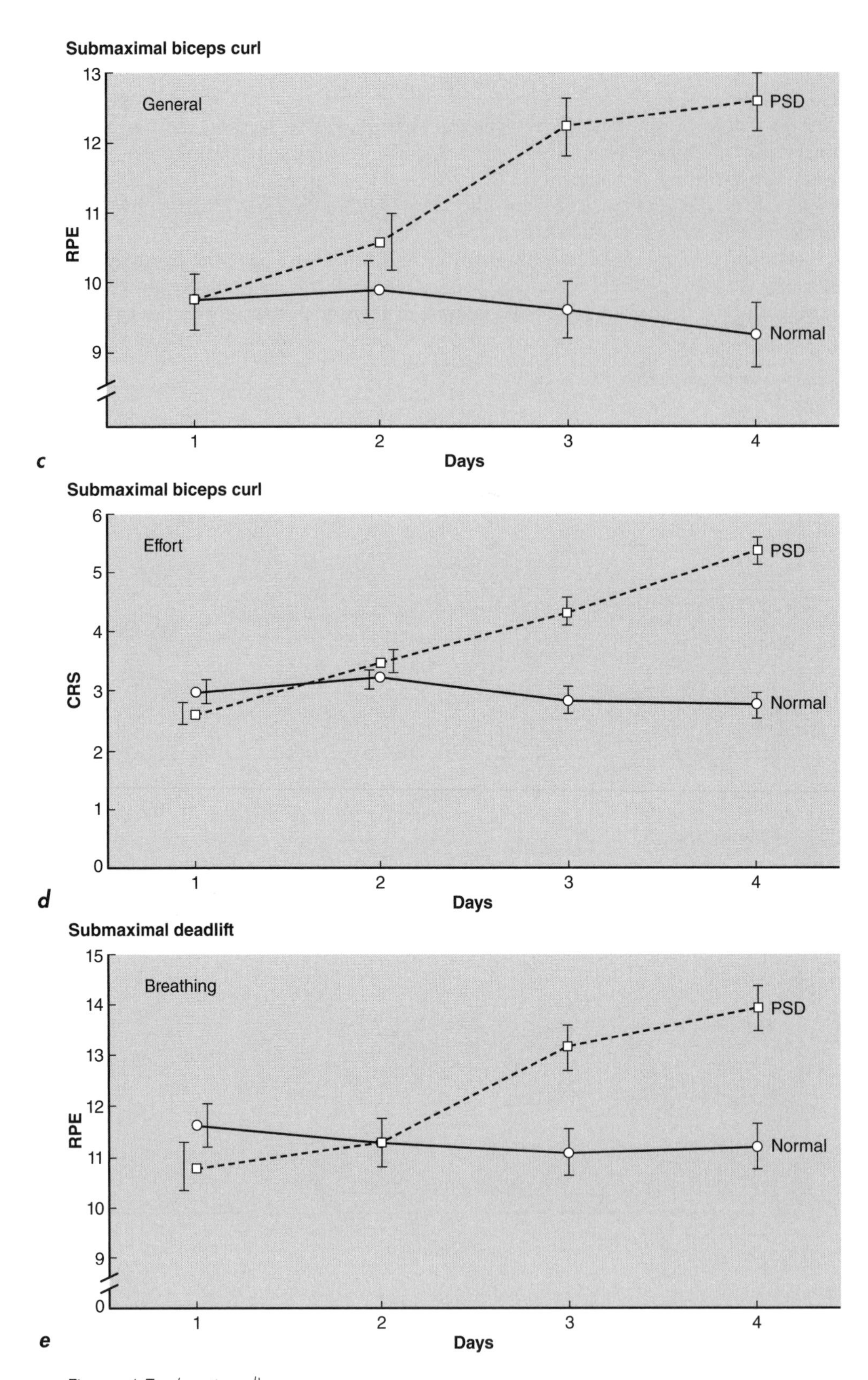

Figure 4.7 *(continued)*

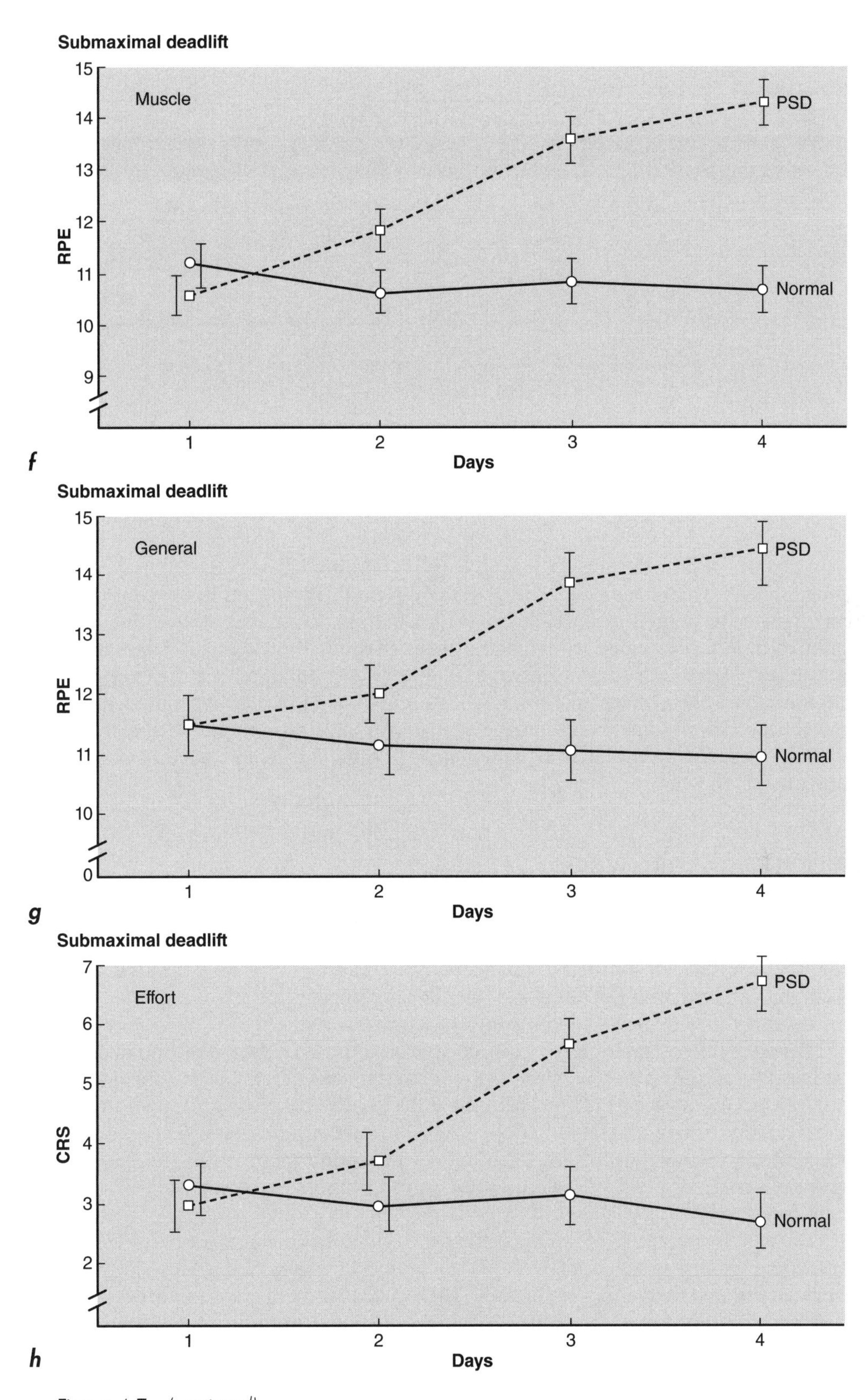

Figure 4.7 *(continued)*

Table 4.1 Taxonomy of Sports Affected by Sleep Loss

Characteristics	Sports	Effects
Low aerobic, high vigilance	Sailing, road cycling, aiming sports	Errors ↑
Moderate aerobic, high concentration	Field sports, team games, court games	Decision making ↓
High aerobic, gross skills	Running 3,000 m, swimming 400 m	Marginal
Mixed aerobic and anaerobic	Combat sports, swimming, middle-distance running	Power ↓
Anaerobic	Sprints, power events	Marginal
Multiple anaerobic efforts	Jumping events, weight training	Fatigue ↑

a progressive decline during the month. Exercise training is best practiced in the evening, preferably 2 hr or so after the first evening meal. There does not appear to be an increased health risk associated with fasting in this context, except for individuals on medication who neglect to test their prescribed drugs. Professional athletes training alongside non-Muslim teammates may be at a disadvantage unless their personal circumstances are taken into account. Reilly and Waterhouse (2007) concluded that the psychological adjustments during the month have certain similarities to the disturbances in circadian rhythms experienced in different circumstances, including nocturnal shift work.

Nocturnal Shift Work

Participation in nocturnal shift work disrupts circadian rhythms. The stress provided by working at night differs from that of traversing multiple meridians in that the environmental signals for biological timekeeping stay constant and the work–rest cycle stays out of phase with the alternations of day and night. This relative permanence means that the body never adapts fully to nocturnal work.

The difficulty of sleeping during the day is compounded by the distraction of noise and also social influences. Both the amount of sleep taken each day and the quality of sleep as indicated by EEG recordings are decreased in night workers who must sleep during the day (Akerstedt, 2006). The problems experienced by shift workers in adjusting to a nocturnal shift force many to abandon night work. The unsocial hours of work and circadian rhythm disturbances have been a concern to the workers' health and well-being (Reilly et al., 1997a).

There are many reports of impaired mental and physical performance of shift workers during the nighttime hours. Decreases in attention, increases in errors, decreases in vigor, progressive fatigue (Bohle and Tilley, 1993), and impaired performance in perceptual–motor tasks (Monk and Folkard, 1992) have been associated with failure of circadian rhythms to adapt. In contrast, rhythms adjust relatively quickly to a day-work routine and normal sleep patterns are quickly reestablished. Petrilli and colleagues (2005) showed that a tracking task that measured hand–eye coordination was sensitive to fatigue-related errors during shift work and so could be used to determine fitness for duty in the workplace.

Shift workers are presented with difficulties in organizing their domestic, athletic, and occupational commitments. Few workers on shift schedules compete in sport at a high level (Reilly et al., 1997b). Adoption of an optimal shift system would alleviate adverse effects of night work and promote a more active lifestyle. There is a wealth of evidence that a forward-rotating shift work program (morning shift, afternoon shift, night shift) facilitates adjustment to working at night, although such an option is not always accepted in industrial contexts (Fisher et al., 1993).

Overview and Summary

The sleep–wake cycle is the most discernible of human circadian functions, activity being associated with the hours of daylight and sleep with the hours of darkness. This daily recurrence is linked with pineal responses to the environment, its secretion of melatonin being promoted at dusk and inhibited on exposure to morning light. A myriad of other biological functions are knit into a common system of circadian rhythms, cycles in behavior, and biological functions that recur with a period of *circa diem* (about 24 hr).

Although the study of sleep itself is inherently attractive to researchers because of its fundamental nature, it is beset with methodological problems. Smith and Reilly (2005) outlined three features of research protocols required to define the effects of sleep deprivation on athletic performance with the desired level of accuracy. First, the experimental protocol should isolate the homeostatic from the circadian components because these frequently confound each other. Second, the protocol should include an externally valid competitive event to reduce motivational confounds and decrease the distortion associated with extrapolation to the real conditions. Third, the research protocol should effectively reduce the many confounding variables affecting sport performance, for example, home advantage, climate conditions, change in fitness, and individual circumstances.

There remains the problem of separating the circadian component from the homeostatic drive to sleep. This difficulty arises irrespective of whether sleep is lost in the early morning or because of a late bedtime. The two processes are compounded in travelers, especially those going westward on long-haul flights. The forced desynchrony procedures of Cajochen and colleagues (2002) or other experimental models may have value in the study of sleep and circadian effects on athletes. Such studies are methodologically and practically challenging.

The body's normal circadian rhythms have a profound impact on human activity and can influence performance. Effects become obvious when rhythms are disrupted. This disturbance applies to sleep deprivation, working across nocturnal shifts, and traveling across time zones. In the latter case the disruption gives rise to jet lag, and a knowledge of rhythms enables travelers to cope with travel issues.

Sport Ergonomics

PART II focuses on sport ergonomics. The first of the three chapters describes the ergonomics models and training methods used in the world of sport, the next is concerned with competitive stress and training loads, and the third covers sports equipment and surfaces. Thus part II provides a comprehensive account of the tools available to the ergonomist in analyzing and understanding the demands placed on the individual participant.

There is no one single ergonomics approach to analyzing task demands in competitive sport and training contexts. This diversity both in models adopted and alternative modes of training is covered in chapter 5. Neither is there necessarily a formulaic solution to problems that arise. A range of methods for analyzing movements, motions, and actions have been developed exclusively for application to sport. There are also generic models that can adequately describe the main approach to problem solving, such as the systems approach (for complex scenarios), anthropometric approach (for design), and workstation analysis (gymnasiums and playing areas). Models have been constructed for application to the training process itself.

Chapter 5 outlines various models, including fitting the task to the person, the systems approach, and workplace engineering design. Design principles are relevant in human applications and relate to the population accommodated when attempting to fit the person to the task. Training and overload are placed in context of the processes experienced. Peaking for performance and the differentiation between individual team approaches are elements of sport that contrast with conventional occupations.

There is a call also to analyze competitive performance so that participants benefit from the feedback provided. In contemporary sport, a range of analytical techniques are available to identify areas to be targeted for improvement. Just as task analysis is a cornerstone on which an occupational ergonomics project is developed, a formal analysis of competitive sport is essential to understanding the nature of its demands.

PART II

Methods of monitoring the performance of participants include work rate analysis, notation analysis, and biomechanical techniques that involve computer-aided methods. A range of physiological responses can be used to reflect the relative strain on participants. Performance also can be monitored from psychological angles that include the periods before and during competition. The contributions of each individual in a team sport must be placed in context to determine how the strengths and weaknesses of different athletes can be combined for the squad to become an effective unit.

Training programs take many different forms depending on objectives, the sport concerned, the experience of the individuals or groups, and individual demographics. Although there is no single pathway to success in sport, the level of achievement is influenced by implementation of systematic training that is suited to the athlete and is specific to the sport concerned. The program can combine different goals, for example, concomitantly training muscle strength and muscle power or training aerobic power alongside anaerobic capacity. The overload needed for physiological systems to adapt and performance level to improve can be induced by functional resistance, sport simulators, standard training apparatus, sport-specific equipment, or a range of ergometers. Optimizing the training stimulus entails striking a balance between training at an intensity too low to elicit physiological benefits or an intensity high enough to cause harmful overload. Creative alternatives such as deep-water running or complementary activities are available that provide training stimuli at a maintenance level without risking impact-related damage to soft tissues.

Chapter 6 considers a variety of methods for quantifying competitive stress. The relative loading on the individual can be quantified when observations on physiological responses are related to measured maximal capacities. The rich array of techniques range from measurement of muscular output such as forces produced or associated electrical activity to whole-body measures like oxygen uptake or core body temperature. Physical loading can be measured using biomechanical principles, and spinal shrinkage can serve as an index of loading on spinal structures.

Competitive stress also encompasses psychological issues, before and during competition as well as in its aftermath. Behavioral measures of anxiety are used on their own or allied to more established inventories or newly designed questionnaires designed to suit the circumstances of the sport. The level of effort associated with exercise is gauged by subjective scales, the most common of which was initially validated against physiological criteria. Such scales are helpful in gauging the competitive pace and evaluating the severity of training sessions as a whole.

Chapter 7 targets the participant's interface with equipment, clothing, and playing surface. Materials used in sports equipment are more advanced than in formative years of the sport, alloys and synthetic composites being nowadays more common. Manufacturers give continuing attention to improving equipment design, both for gaining a competitive edge in performance and for improving safety in use.

Participants in a number of sports rely on the performance of the equipment they use and the surfaces on which they play or compete. At an elite level the design may be dedicated to the individual athlete rather than the mass market of recreational users. Harmony between athlete and machine can be achieved by means of aerodynamic factors, an example being the benefit of wind-tunnel observations on the posture adopted by racing cyclists and the characteristics of their machines. Participants need to be versatile and capable of coping with different playing conditions and surfaces, top tennis players having to excel on clay, grass, and synthetic courts for all-around success.

© Tom Reilly

Ergonomics Models and Training Modes in Sport and Leisure

DEFINITIONS

critical path analysis—A method of analyzing events so that a track through a sequence of actions is determined.

deep-water running—Exercise, usually in a swimming pool, in which the feet are kept off the floor.

dynamical systems—Highly interconnected composition of numerous interacting parts capable of changing their state of organization at all times.

lactate threshold—The exercise intensity at which lactate concentrations in blood begin to increase.

notation analysis—A method of recording movements during sports and activities such as dance.

task analysis—A formal method of breaking an activity down into its key components.

work rate—Measurement that reflects the exercise intensity or power output of an activity; in games, expressed as distance covered per unit of time.

workstation analysis—A scheme for examining the layout of hardware, equipment, and facilities and for considering environmental factors, with the interface between the worker and the task as the priority.

five

ERGONOMICS principles are equally as applicable in sport, leisure, and recreation contexts as they are in occupational and military settings. Elite sport is performed at the highest level mostly when participants are in their third or fourth decades of life, but competitive sport can include younger and older participants. Differences between the sexes are reflected in separate competitions, and there are different weight classes for sports such as boxing where body mass would otherwise provide a major advantage. Special populations are acknowledged in disability sports, wheelchair athletics, and the Paralympics. There is diversity also in the characteristics of those who engage in active leisure and recreational activities with respect to age, sex, physical characteristics, and capabilities.

Each sport and each recreational activity has its own unique features and attractions. Consequently no single ergonomics model can be applied across all sports. Methods of **task analysis,** for example, that were designed for industrial operations have only limited relevance in sport and leisure. Nevertheless, modified approaches to ergonomics principles can be applied to sports to suit the circumstances of the sport concerned.

Any application of ergonomics must first consider three types of resources, namely hardware, software, and the human participant. Both hardware and software vary in nature and complexity between sports and level of participation. These resources interact with the environment, which can present many different risks and sources of stress. Hence the choice of analytical tools is determined according to a scheme that first identifies the key problems to be investigated and then derives solutions for them. Various generic approaches can be used, and these are covered in this chapter.

Fitting the Task to the Person

The focus in ergonomics is primarily on the human, whose characteristics, capabilities, and limitations are essential considerations. Relevant information is needed about the demands of the activity and the individual participant's capability to meet these demands. This direct matching of capabilities and demands is relatively straightforward in individual sports such as swimming, cycling, and running. It becomes more complicated when events are combined, as in triathlon competitions, and when skills are varied, as in the Olympic sports of modern pentathlon, heptathlon, and decathlon. The mapping of capabilities and demands is even more complicated in team events where more than one participant is involved and all team members must act in concert, such as in rowing eights, in sailing events such as the America's Cup, and in field games such as hockey or football. In these events the entire group of members must aim to be an effective competitive unit.

Competitive sport entails confrontation with a single opponent or a team, both sides behaving according to the regulations of play as administered by a referee or group of officials. Victory is assigned to the athlete who can run, swim, cycle, or row faster than opponents and to the team that outscores the opposition. Success is often accomplished despite imperfections in performance and without deficiencies in specific fitness components being exposed by the opponent. The higher the competition level, the more likely it is that such weaknesses will be exploited by other contestants or that they will result in injury. For example, hamstring injuries are usually incurred by sprinters in the weaker of the two lower limbs (see Reilly, 1981) and often when the sprinter is under pressure from another competitor.

The capabilities of athletes are assessed by means of fitness testing. There are now standard tests for measurement of aerobic power, anaerobic power and capacity, **lactate threshold,** peak muscle torque at fast and slow angular velocities, flexibility, and agility (Reilly, 1991). Such fitness tests are most relevant when they are specific to the sport in question (see figure 5.1). Their interpretation requires some knowledge of the sport, the phase of the season, and the emphasis on training at the time of each assessment.

a

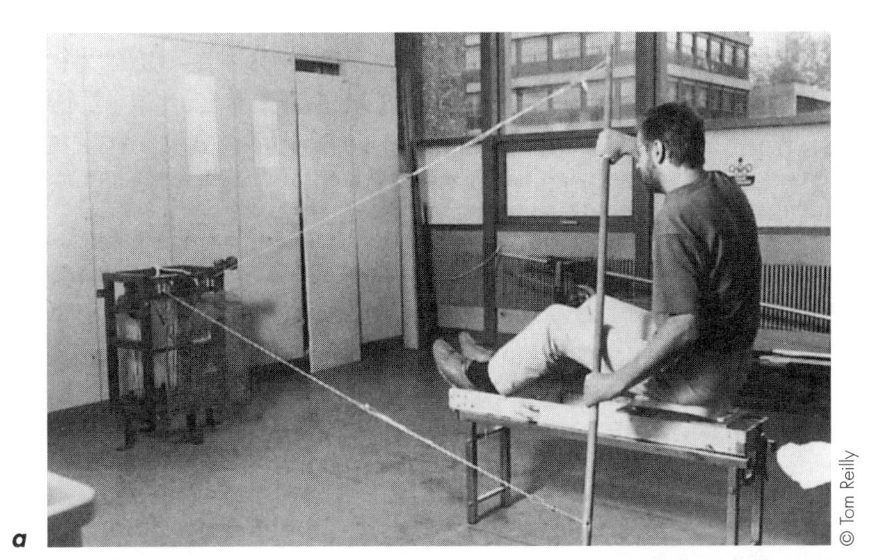

b

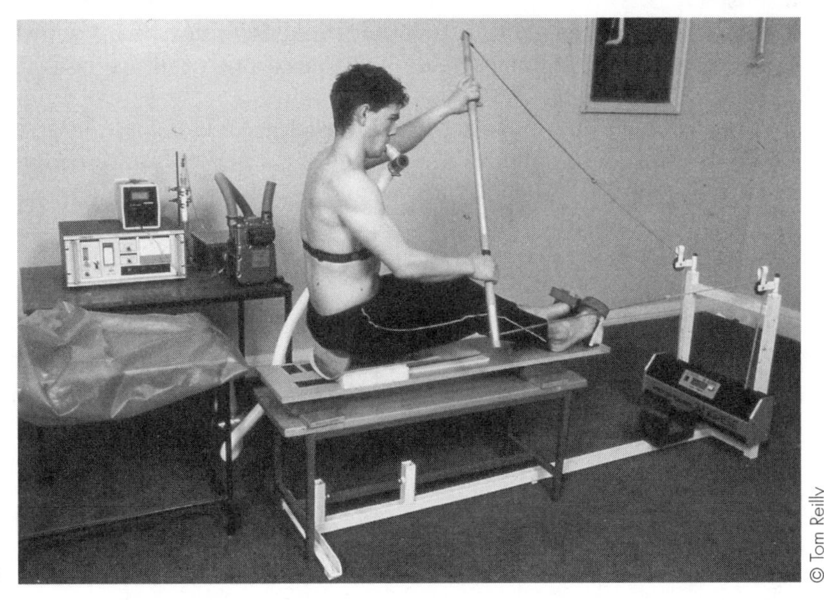

Figure 5.1 *(a)* Kayak ergometers designed on wind resistance principles and *(b)* adapted from a swim bench.

Knowledge of people's capabilities provides a rubric for matching participants to activities and roles. The practical difficulties in doing so differ between occupational and sports contexts. In the former case interpretations are often based on population norms and databases. The competitive basis of sport implies that participants, whether individuals or teams, strive to gain a winning edge over the opposition. It is not surprising, therefore, that methods of analyzing task demands vary between these two domains.

Task Analysis

The form and complexity of any task analysis will vary according to the problems to be faced. Task analysis techniques attempt to represent human performance in a particular scenario being investigated. The approach may be analytical with implications for training needs or evaluative with respect to workload, fatigue, comfort, or usability. Where interactions with machines or other people are involved, the tasks under review are broken down into discrete task steps for further consideration.

Stanton and colleagues (2005) reviewed methods of task analysis for use by human factors specialists in system design and evaluation. There can be some overlap between the various methods, as listed in table 5.1, and certain scenarios may call for more than a single approach to the problem being addressed. The ergonomist must be careful in choosing the method for application before embarking on data collection.

A clear definition of the tasks under examination and the collection of relevant data are prerequisites for task analysis. The necessary data may be obtained by conducting individual interviews or focus group meetings, administering questionnaires, and shadowing participants. Formal observations can generate quantitative evidence and objective information, but the time consumed in data handling should be considered. The data being collected should be compatible with the task analysis method chosen.

Hierarchical task analysis results in an exhaustive description of task activity. This type of task analysis was originally developed to examine cognitive tasks and has found widespread applications in military, civil aviation, driving, emergency services, and many other contexts. Commercial software packages for hierarchical task analysis are available. The approach entails a description of the activity concerned in terms of goals, subgoals, operations, and plans. The method is essentially descriptive but flexible and can be adopted in a variety of sport applications where the main goal can be placed at the top of a hierarchy.

Table 5.1 Selection of Recognized Task Analysis Techniques

Method	Key feature
Hierarchical task analysis	The overall goal and intermediate steps toward subgoals are set out.
Critical path analysis	The critical path toward accommodating parallel operations is determined.
Verbal protocol analysis	The operator thinks about the processes involved.
Charting techniques	The tasks under analysis are graphically described.

Critical path analysis is commonly used in project management, for example, to estimate the duration of a project for which some activities can be conducted in parallel. A given task cannot be started unless all preceding tasks that contribute to it are finished. The order in which tasks are conducted and their duration and dependency must be known to determine the critical path. The approach can be applied to human performance models as well as to research projects relevant to sport.

Verbal protocol analysis is used to produce a descriptive account of the cognitive and physical processes that a person uses to complete a task. A written transcript of behavior during this performance is generated based on the self-report of the operator. Verbal protocol analysis has been used to gain insight into cognitive aspects of complex behaviors. The method has relevance to many sports tasks, but the transcription and encoding of the data are time-consuming.

Charting techniques encompass a number of approaches to describe and represent graphically the activity being analyzed. These are essentially descriptive approaches to considering the processes involved. The techniques include event tree analysis and fault tree analysis, the latter being suited to identifying potential sources of error from interacting with the system in question. Treelike diagrams are used in event tree analysis to represent possible outcomes associated with different task steps. This method may be suitable for depicting task sequences and their possible outcomes and can be used to model team-based tasks. Similarly, fault tree diagrams define events in system failures and display possible causes in terms of hardware breakdown or human error.

Notation Analysis

Notation analysis refers to the detailed examination of sport performance using a coding system for recording actions and events. Originally developed as a manual method of detailing sequences of movements in dance and later in sports such as basketball, notation analysis has been developed into a powerful computer-based tool for analyzing team sports (Carling et al., 2005). It is now used as an effective means of providing athletes and their coaches with feedback on performance and patterns of play in team games (see figure 5.2). The information provided by notation analysis is a critical element in this communication loop.

When notation analysis is applied to team games, information is revealed about each action, the position on the field where it occurred, and the players involved. The outcome of each action is registered, and the sequence of actions is followed to establish its success or failure. In this way a pattern of play can be constructed and the effectiveness of different tactical plays evaluated. The analysis can be conducted by scrutinizing video recordings of matches and used to study the performance of the two competing sides.

Hughes (2003) described the four main purposes of notation as analysis of movement, tactical evaluation, technical evaluation, and statistical evaluation. Computerized systems can be used to provide immediate feedback to participants, develop a database for coaches, reveal areas requiring improvement, evaluate performance, and search video recordings. Because of the comprehensive use of computer-aided analysis within professional sport, many elite teams employ a match analyst on their support team. Match officials also benefit from feedback of their performance during ongoing professional development programs.

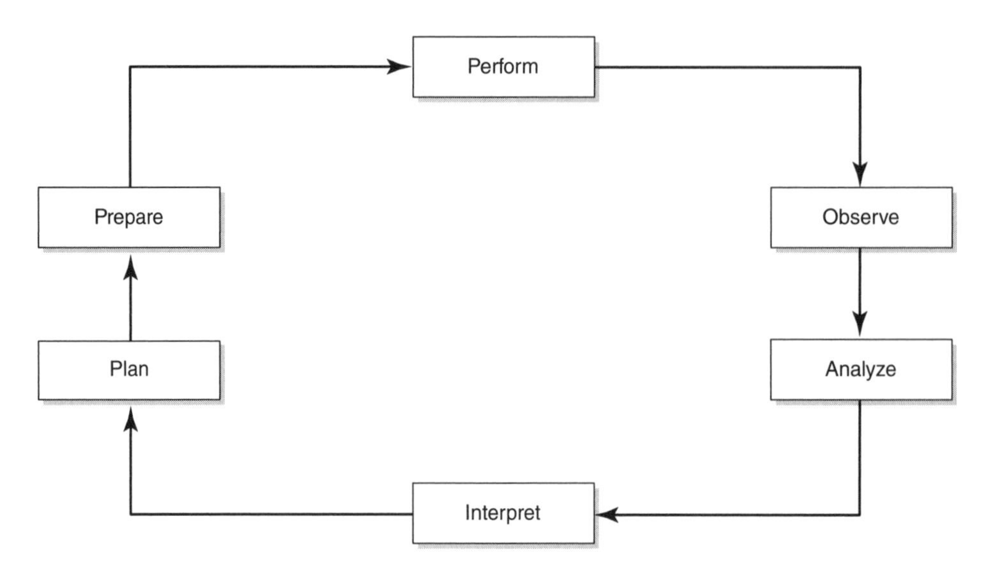

Figure 5.2 The feedback cycle implemented by means of notation analyses for coaches.

Moving **notation analysis** beyond being merely a descriptive tool has been a challenge for its proponents. To develop predictive models, research groups have had to use more sophisticated statistical and mathematical approaches. McGarry and Franks (1994) used stochastic processing to predict outcomes in competitive squash. Later, McGarry and colleagues (2002) offered examples of dynamic interactions in dyadic (squash) and team (soccer) sports as well as some predictions from a dynamic systems analysis for these types of sport contests. There has been tentative application of artificial intelligence through interactive video systems for teaching psychomotor skills in tennis (Rush et al., 1990) and later fuzzy logic and artificial neural networks. Artificial neural networks have been more widely used in performance analysis than have expert systems (Lees et al., 2003). The potential for multilayer neural networks to optimize decision making or prediction has not been realized as yet.

Figure 5.3 demonstrates how, subsequent to training, the Kohonen self-organizing map visualizes a series of sagittal plane joint angles as a single chain of nodes within an artificial neural network. Three neighboring values of the four curves correspond to a single location on the topological map, which carries a label of a complex movement pattern (Barton, 1999). This artificial neural network can help identify gait characteristics that are indicative of injury risk or assist in monitoring pregress during rehabilitation.

Memmert and Perl (2009) demonstrated how artificial networks could overcome the limitations of notation analysis that focuses on quantifying descriptive aspects of performance rather than on qualitative evaluations. These authors outlined a framework for analyzing types of individual development of creative performance based on neural networks. Neural networks were applied to field hockey and soccer to gain insight into the extraordinary creative behavior evident in experts in these games. Neural networks can be used as a tool for detecting influential structures in learning processes when behavior is complex.

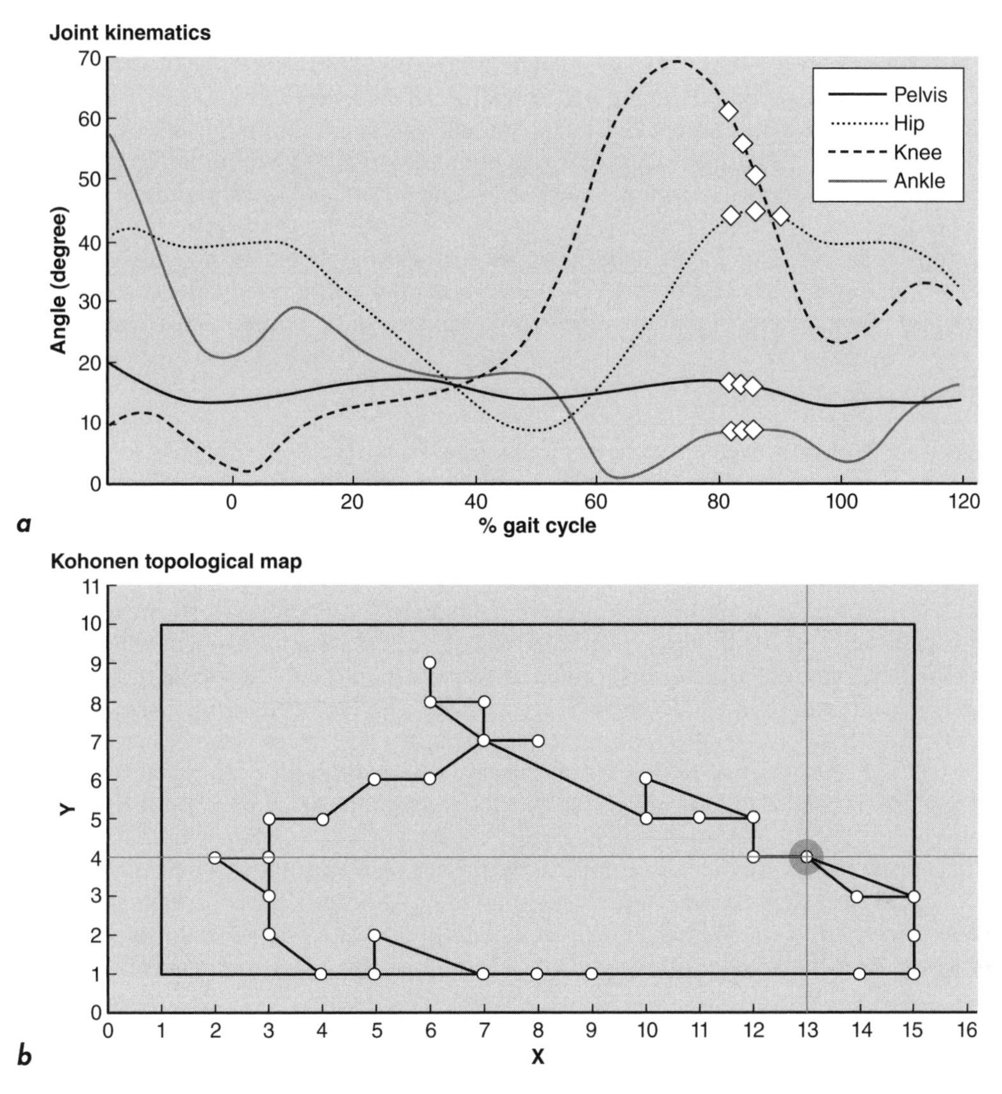

Figure 5.3 Subsequent to training, the Kohonen self-organizing map visualizes a series of sagittal plane joint angles *(a)* as a single chain of nodes *(b)*. Three neighboring values of the four curves (highlighted) correspond to a single location on the topological map, which carries a label of a complex movement pattern (Barton, 1999).

Adapted from *Gait and Posture*, Vol. 10, J.G. Barton, "Interpretation of gait data using Kohonen neural networks," pgs. 85-86, copyright 1999, with permission from Elsevier.

Motion Analysis

Analysis of human movement has a rich history dating back to the original cine-photographic technique used by pioneers in the field. Methods have progressed through strobe photography of sequences in skilled movements, video analysis with pause facilities to digital video recording, fast-film analysis, and synchronized cameras for analysis of motion in three dimensions. The technologies available are continually being updated, and the associated software has eliminated the drudgery of data extraction using the earlier methods (see Atha, 1984; Reilly and Lees, 2008).

Contemporary methods incorporate specialized transducers and automatic analyzers with built-in calibration systems. The systems have fast-frame and pause facilities that allow sport scientists to analyze movements and skills in their minutiae. Examples include studies of the golf swing, tennis serve, long-jump take-off, and pole vault. It is possible also to perform accessory monitoring—for example, using electromyography or force analysis—to supplement the information observed from motion analysis. Where cost is an issue for prospective users, video-based systems that allow qualitative analysis of motion provide a practical alternative to three-dimensional biomechanical analysis.

Movement Analysis

Analysis of gross movement behavior at a whole-body level is appropriate for field games. Movement profiles can be used to indicate overall **work rate**, and work rate indices can be used as measures of performance. The principle is that the work rate reflected in the amount of distance covered in a given time determines the energy that is expended.

The first validated application of movement analysis to indicate work rate was in association soccer (Reilly and Thomas, 1976). The overall distance covered represents a global measure of work rate that can be broken down into the discrete actions of an individual player for a whole game. The actions or activities are classified according to type, intensity (or quality), duration (or distance), and frequency. The activity is juxtaposed on a timeline so that the average exercise-to-rest ratios can be calculated. These ratios are used in physiological studies to represent the demands of the sport and also in conditioning elements of the players' training programs. These work rate profiles can be complemented by monitoring of physiological responses where possible.

In the early applications of movement analysis to professional soccer, activities were coded according to intensity of movements, the main categories being walking, jogging, cruising, and sprinting, while other game-related activities such as moving backward or sideways and playing the ball were investigated. The observer used a learned map of pitch markings in conjunction with visual cues around the pitch boundaries and spoke into a tape recorder. The method of monitoring activity was checked for reliability, objectivity, and validity (Reilly and Thomas, 1976) and is still considered to be the most appropriate way of monitoring one player per game.

An alternative approach to data collection is to set the activity profile alongside a timeline. This method permits establishment of fatigue profiles and exercise-to-rest ratios, which are useful both in designing training drills and interpreting physiological stresses. This approach is straightforward now that video systems are linked with computerized methods of handling the observations.

Work rate analysis is now used by most professional teams in soccer, rugby, field hockey, and other field games. The more substantial systems incorporate multiple cameras, usually three or four distributed along the stand over each side of the pitch, whose records are later synchronized for computer-aided analysis. Detailed feedback on activities is provided to aid interpretation and evaluation of performance, including its high-intensity components. Contemporary software enables real-time data analysis.

Methods of movement analysis have helped to highlight persistent features of performance. These features include the influence of positional role on work rate

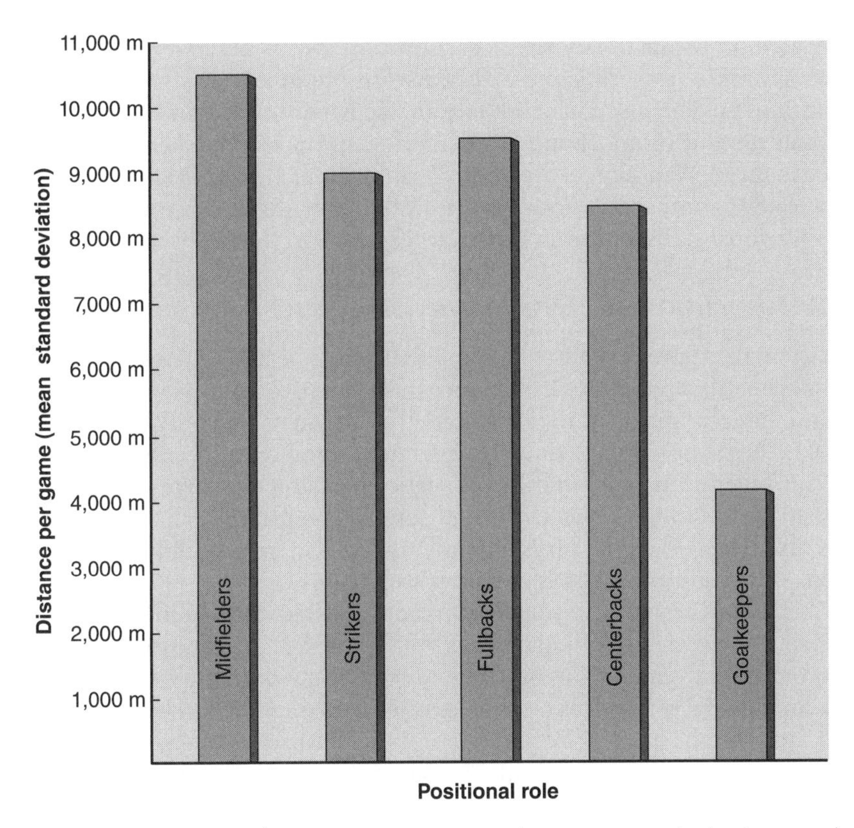

Figure 5.4 Distance covered in a soccer game according to positional role determined using motion analysis. The highest work rates are associated with playing in midfield positions.

characteristics (see figure 5.4), the occurrence of fatigue as muscle glycogen stores near depletion toward the end of a game (Saltin, 1973), and the transient experience of fatigue attributable to successive bouts of high-intensity exercise (Mohr et al., 2003). The adverse effects of high ambient temperatures on work rate characteristics have also been demonstrated (Ekblom, 1986) as have the influences of playing style and fitness status. In particular, the correlation between aerobic fitness and total distance covered in competitive games has implications for the training programs to be used.

The techniques associated with work rate analysis are equally applied to match officials as to players. Reilly and Gregson (2006) considered referees and assistant referees among special populations, concluding that the relative physiological strain on the referees equaled that on players. Both groups showed evidence of fatigue occurring before the end of play and exhibited unorthodox movement patterns that elevated energy expenditure over normal locomotion. The high standards of physical fitness and decision making expected from match officials have consequences for the training programs and nutritional practices they adopt.

The physiological responses of games players to match play show that a combination of demands are imposed on participants during competition. The critical phases of play for an individual player call for anaerobic efforts, but these are superimposed on a background of largely aerobic submaximal activities. The intermittent and acyclical nature of activity during competition means that it is difficult to model game-related protocols in laboratory experiments. It is likely that field studies with a greater specificity to each game will be used more in future investigations of the physiology of field games. The work rate and activity profiles can be used to design appropriate training protocols to optimize fitness and ensure that performance during play is enhanced. Although ergonomic considerations have a place in a systematic preparation for competition, performance ultimately depends on the quality with which individual skills and team tactics are executed in the face of stern opponents.

Signal Detection and Eye-Movement Tracking

The behavioral responses of the visual apparatus have been of interest in ergonomics since the classical observations in vigilance tasks that subjects missed periodic signals even if the subjects appeared to be focused on the screens on which the signals were displayed. These observations fit with signal detection theory, the changes in performance attributed to two parameters, a fatigue effect and a change in the sensitivity for recognizing a stimulus. The theory had particular application in long-lasting tasks such as monitoring radar displays, identifying faults in an assembly line, sailing, or working under conditions of sleep deprivation.

The behavioral responses of sensory organs can yield relevant information about performance in visual recognition tasks, especially when there is an emphasis on speed. Monitoring eye movements can help identify the behavior of the visual apparatus of experts compared with novices or their less able adversaries. Such monitoring also is used to establish the existence of fatigue and the occurrence of errors. Two cameras are engaged, a scene camera to locate the participant in his or her environment and an eye camera to produce a close-up image of a force platform on which the subject stands in order to record the whole-body responses in conjunction with the data on eye movements.

Within applied environments, most techniques for recording eye movements are video based. These systems illuminate the eye with infrared light, which is parallel to the axis of the camera viewing the eye. The incident light reflects both off the retina, which makes the pupil seem bright, and off the cornea on the front of the eye surface. Processing of these reflections enables point of gaze to be calculated.

Eye-tracking techniques have been applied to work and sport settings. Wooding and colleagues (2002) used a public exhibit of a novel eye-tracking system to obtain data for eye movements of a large population. Ball and colleagues (2003) measured on-line attentional processing by tracking eye movements during inspection and selection tasks. The technique can be used in a wide variety of sports when incidents are presented typically on video screens. A penalty kick in soccer or a serve in lawn tennis can be scrutinized to identify the visual cues that the goalkeeper or the receiver uses in anticipating ball flight.

Generic Models

In many instances the ergonomist needs to have an overall perspective on the system being analyzed. The features observed may be broad rather than specific and may entail some simplification if inferences are to be drawn from the model. Generic models refer to how the entity concerned is pieced together into a coherent whole. These models have some use at the early stages of an ergonomics project.

Systems Approach

Many ergonomics problems, especially those linked to technology, are addressed by use of systems analysis. A system may be defined as an assembly of functional units with a common overall purpose and forming a connected whole. The concept can be applied to represent a sport organization, a team of games players, a water skier in tow, or a biological system. Knowledge of overall objectives is implied in adoption of the systems approach as are the inputs and outputs, discrepancies between which are relevant in error analysis. In assessing overall performance, ergonomists examine various aspects of system behavior.

The ergonomist is mainly concerned with how the human harmonizes with other elements in the system. In human–machine systems, the concern may be with which functions are allocated to the human and which are done mechanically or automatically (see table 5.2). In operational systems, the issue may be the compatibility between controls, displays, and human sensory mechanisms. In a sport context, controls may consist of pedals or handlebars of a bicycle or the joystick and steering wheel of a power boat. A systems approach also can be applied to the management of a ski resort by means of computer assistance to avoid congestion and reduce accidents on the mountain slopes.

In human–machine sports, both the machine and its operator must behave in harmony for the benefit of the system as a whole. Machine characteristics can be compared with human patterns when tasks are being allocated to humans or to mechanical devices. This harmony is essential for avoiding injuries and optimizing performance by using machine characteristics to best effects. Whatever system is concerned, the most important human functions relate to information input, information processing, decision making, and action or response. In serial operations that require continuous control, some form of feedback must be provided to the operator to permit the correction of errors. This essential part of any system enhances performance and reduces accidents.

Although the original allocation of tasks to machines and humans was based on the work of Fitts (1951), the scheme may be inadequate for contemporary complex systems. Fitts listed those tasks at which the human is better alongside those in which machines are more capable of performing. It is now clear that other criteria are relevant, and allocation decisions should not be made solely on performance issues. Designers should discuss function allocation methods with potential users in the context of the entire design process. This would broaden the set of concerns, widen the scope of methods to be used, and help designers create supportive tools and techniques. This method could be incorporated in both sociotechnical and the more recent macroergonomic approaches to system design. These approaches embrace subsystems dealing with personnel, technology, and organizational structure in parallel with the macroergonomic emphasis of the external environment (Waterson et al., 2002). These approaches apply equally to plant safety and to various aspects of sport stadium or training facility designs.

The systems approach has stimulated the study of human movements and sport skills in fine detail. Insights into the processes of skill acquisition, motor learning, and proprioception have followed from inspired applications of control theory to these contexts. Use of biofeedback—the presentation of biological signals to the person generating them—has been helpful in a variety of sport contexts such as the

Table 5.2 Human Versus Machine Characteristics

The human excels in	The machine excels in
Detecting certain forms of very low energy levels	Monitoring (both men and machines)
Showing sensitivity to an extremely wide variety of stimuli	Performing routine, repetitive, or very precise operations
Perceiving patterns and making generalizations about them	Responding very quickly to control signals
Detecting signals in high noise levels	Exerting great force, smoothly and with precision
Storing large amounts of information for long periods and recalling relevant tasks at appropriate moments	Storing and recalling large amounts of information in short time periods
Exercising judgment when events cannot be completely defined	Performing complex and rapid computation with high accuracy
Improvising and adopting flexible procedures	Showing sensitivity to stimuli beyond the range of human sensitivity (infrared, radio waves)
Reacting to unexpected, low-probability events	Doing many different things at one time
Applying originality in solving problems (i.e., finding alternative solutions)	Displaying deductive processes
Profiting from experience and altering course of action	Showing insensitivity to extraneous factors
Performing fine manipulation, especially where misalignment appears unexpectedly	Repeating operations very rapidly, continuously, and precisely the same way over a long period
Continuing to perform even when overloaded	Operating in environments that are hostile to humans or beyond human tolerance
Reasoning inductively	Reasoning inductively

regulation of training loads and the control of precompetition anxiety levels. This form of feedback has also aided neuromuscular training during rehabilitation and correction of faults in skills training.

Nonlinear **dynamical systems** are highly interconnected compositions of many interacting parts that are capable of changing their state of organization constantly. Examples of such subsystems include weather, communities, and sport contests. Dynamical systems theory is an interdisciplinary framework utilized to examine coordination processes in physical, biological, and social systems. Many sport scientists have resorted to this theory in an attempt to explain phenomena in sport.

It has been used in examination of skills to establish similarities between localized subsystems and the global system. It was deemed by Davids and colleagues (2005) to have potential for modeling coordination processes in team games, with implications for coaching behavior. Because of the complexity of dynamical systems theory and the huge numbers of microcomponents within dynamical movement systems, this approach is more a conceptual tool than an ergonomics technique that can be applied immediately.

Anthropometric Approach

The human population is characterized by diversity in a range of anthropometric measures. These differences apply to size, shape, proportions, breadths, cross-sectional areas, and circumferences. It is not surprising, therefore, that anthropometric databases are valid only for the population, gender, or ethnic group from which the observations were derived.

Despite these constraints, ergonomists often adopt an anthropometric approach to solving the problems that confront them, either in helping design new products or evaluating existing setups. Examples include the matching of sports clothing and equipment to individual users, the design of seating for multiple use in sport stadiums, and the choice of entrance and exit sizes in enclosed spaces. Anthropometric criteria also apply when designing workspaces such as the arrangement of seats, layout of control displays, and internal environment of sports cars or sailing boats.

Designers may consider adjustable ranges to increase the number of people to be accommodated by their product or artifact. Percentile values may be chosen when finalizing doors, safety exits, and escape hatches. The lower percentiles are relevant in accommodating young and old people in terms of the minimum forces to be applied in activating alarms. Design for the average person is an unwanted compromise because few people fit average values when combinations of anthropometric measures are considered. Sport participants may have unique anthropometric features that distinguish them from the general population from which databases are drawn.

The concept of size applied to clothing and footwear allows participants the opportunity to select an item that best fits the contours of their body, segment, or limb. Sport shoe manufacturers now recognize the specific design requirements of children, women, and different ethnic groups. Running shoes are available with built-in antipronation features to accommodate those athletes who pronate the foot excessively on landing, thereby protecting them against injury. Astute location of cushioning and flexibility features in shoes for training and for racing increases the ergonomic properties of sport footwear. The stud configurations in soccer shoes have been reevaluated with a view to avoid overloading sensitive areas of the foot such as the metatarsal heads. Individual profiling of the insoles of the shoe has also been used to minimize discomfort caused by poor distribution of pressure associated with stud placement.

In ski boots, release features are important considerations complementing fit. International standards in setting ski bindings are aimed at determining the forces needed to release the ski in the event of a fall. The settings are based on anthropometric variables related to body size.

Sports equipment may be sized to the individual elite performer in track and field athletics, golf, tennis, cycling, and football. The same principle applies to machines used in high-performance sport. It is considered good practice to design the workspace within powered vehicles around the individual characteristics of racing drivers and to design bobsleds and skeletons to suit winter-sports competitors. In the former case, the need for heat-protective clothing and a helmet poses additional design requirements.

Headgear is provided in generic form, with individual choice of the most appropriate fit. Helmets are worn for protection in many sports ranging from cycling and tobogganing to amateur boxing and ice hockey. Their effectiveness stems from cushioning compressive forces to the head and reducing its acceleration when hit. The design of helmet varies with the sport concerned, special helmets with ear holes being used by hang gliders to allow the participant to sense air flow and hence speed of movement. Supplementary protection in contact sports is offered by mouth guards; when constructed over an accurate model of the wearer's teeth, these devices lower the intracranial pressure resulting from a blow to the chin. In many sports the problem of protection is of such a scale that a helmet alone cannot be fully preventive, and additional safety strategies need to be adopted.

It may be necessary to evaluate how clothing fitted to human structure affects function, particularly when the purpose is protection. Mobility should not be unduly restricted when padded sportswear is used for protective purposes. The same applies when gloves are worn in that the safe operation of controls should not be impaired by any reduction in tactile sensation. Padded gloves are used by baseball catchers and hockey goalkeepers to avoid injury on impact with the ball. In contrast, tight-fitting gloves are worn by cyclists and rowers to avoid blisters when training.

Workstation Analysis

The underlying principle of the **workstation analysis** approach depicted in figure 5.5 is that the design process is concentrated first on the human and then progresses outward. The tasks involved and the interface with equipment, tools, and machinery are then evaluated in turn. The workplace is constructed or reconstructed around these factors before projections are made for the complete environment.

In analyzing the workstation, the ergonomist uses a safety checklist to ensure that key aspects of the environment are given attention. This approach is important in outdoor activities and adventure sports, where the safety of others is a priority. It is also relevant in consideration of the overall layout in a fitness center or indoor arena. When these facilities are designed de novo, ergonomics is an assistant technology to architecture, engineering, and interior design. In this case application of ergonomics principles can save embarrassment of mandatory redesign when mistakes appear obvious after construction. For this reason professional sports clubs consult both experts and prospective users when planning training facilities.

The involvement of users in such projects is described as "participatory ergonomics." This approach is adopted to involve workers in decisions about work, promote a safety culture, and empower the workforce. The philosophy is an acknowledgment that support from people at all levels within an organization is needed to improve performance and reduce injuries. The approach has been used in attempts to reduce musculoskeletal injuries in an industrial context (Brown et al., 2001). It is applicable also in sport contexts, notably in managing change in human resources. Participatory ergonomics is not an alternative to an expert-driven intervention, and many sports team managers may engage their players in

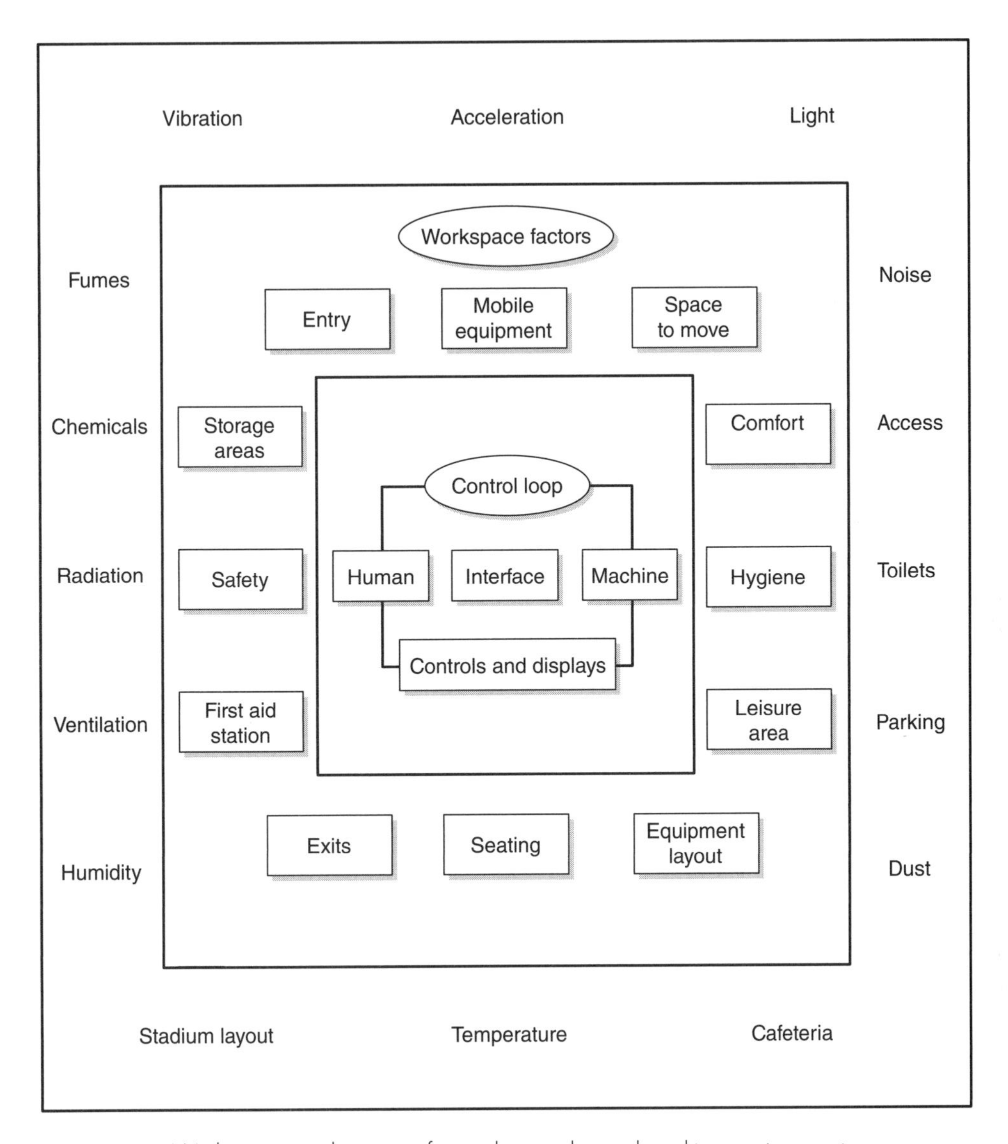

Figure 5.5 Workstation analysis map for evaluating the total working environment.

debate over practical issues before basing their decisions largely on their own summary evaluations.

Feedback from users can be obtained in more traditional ways, such as providing a means for lodging complaints or making suggestions. There may be certain patterns in accidents and errors that become apparent only when incident statistics are reviewed. A checklist approach is not exhaustive, and specific measurements of the workplace may be needed, such as ambient temperature, dust and pollutant concentrations, noise levels, or illumination levels.

The eyes are delicate sensing mechanisms, and so lighting is an important aspect in workstation analysis. Indoor sport facilities require artificial illumination that must conform to the standards set by the sport governing body. In ball games, the faster the action, the more essential it is to have good lighting. Glare, whether emanating from polished floors or adjacent glass paneling, can affect performers adversely. Players may need a contrasting background against which to judge the flight of a shuttlecock or a squash ball. Attention to these features is often neglected when sport arenas are restructured for specific tournaments. This neglect may extend to the design of floodlit outdoor areas, particularly for recreational uses.

The Training Component

The purpose of training is to prepare the participant for the tasks in hand, which usually entails improving individual capacity. A variety of ways are available to achieve this end.

Fitting the Person to the Task

Training programs embrace the concept of fitting the human to the task. Ideally, training regimens are designed with specific goals in mind and can be evaluated in terms of reaching specific physiological targets. In practice, training goals are multifaceted and programs vary in a progressive fashion throughout the annual cycle of competition. Furthermore, performance itself is mostly multivariate and requires an interplay of different physiological systems for optimal fine-tuning.

Circuit weight training, for example, can be used for aerobic training as well as musculoskeletal training. Plyometric training focuses on stretch-shortening cycles of muscle action as used in bounding, hopping, or drop-jumping regimens. A corollary of this form of training is referred to as *delayed-onset muscle soreness,* in which intramuscular structures sustain microdamage. A prerequisite to plyometric training is a foundation of strength training. During the days in which delayed-onset muscle soreness is experienced, exposure to deep-water running can help maintain fitness and alleviate muscle soreness (Dowzer et al., 1998) but cannot substitute for a progressive program of preliminary conditioning.

Sporting events such as triathlon call upon a combination of skills and specific fitness measures. Such events require attention to how the training components complement each other rather than cause interference. Cross-training programs have particular health benefits by virtue of their variety, which maintains motivation and reduces the dropout rate normally associated with a population of new exercise participants.

The down side of physical training is when adaptation does not accrue and breakdown occurs. Overtraining is the condition in which underperformance is experienced despite continued or even increased training. A vicious cycle of more training produces lower performance and chronic fatigue. The phenomenon represents the classical ergonomics model of task demands outstripping human capability to cope with the training load.

Technology has been used extensively to improve the design of training equipment and produce supplementary training aids. These may be designed for training in specific sports or may have more widespread applications. Some of these artifacts and alternative modes of training are now considered. In examining training approaches, athletes and trainers should consider how a piece of equipment can be used, understand why it is used in a prescribed way, and consider its advantages over other equipment or apparatus.

Alternative Training Models

Not all training methods are conventional approaches to improving strength, power, and endurance. There is a continued search for novel and creative means of enhancing capacity. A few of those that have scientific support are now described.

Drop-Jumping

Drop-jumping (or depth jumping) uses the participant's body weight and gravity to exert force against the ground. The participant steps out from a box, drops to the ground, and immediately drives the body upward as quickly as possible. The eccentric part of the action where the lowering of the body is controlled is known as the amortization phase, before the body is directed vertically. Learning to coordinate the whole movement into a smooth performance is essential for this exercise to be fully effective. The aim in depth jumping is to enhance the training stimulus for power production. Its advantage is that it isolates the stretch-shortening cycle in muscle action by inducing a stretch in the active muscles prior to their concentric action against gravity.

Depth jumping has been adopted with success by high jumpers, triple jumpers, and sprinters. It is relevant in sports where the lower-limb muscles generate high power output in fast, explosive actions. It was originally prescribed by Verhoshanski (1969), who recommended a box height of 0.8 m for achieving maximum speed and 1.1 m for developing maximal dynamic strength. He recommended no more than 40 jumps in a single workout. In later studies, a box height of 20 to 40 cm was thought to be sufficient (Boocock et al., 1990).

Although drop-jumping is recognized as effective in enhancing muscle power, a disadvantage of drop-jumping is the delayed-onset muscle soreness caused by the eccentric, or stretch, component of the activity. This soreness attains a peak some 48 to 72 hr after exercise. Biological markers of muscle microtrauma include elevation of creatine kinase and myoglobin concentration in blood. Athletes habituate to this form of training by means of the so-called repeated bouts effect. There is also a transient loss of maximum force production when the soreness is induced.

Pendulum Training

The pendulum device provides a method of training that was developed in Eastern Europe. Its main advantage is that it activates the stretch-shortening cycle without the accompanying delayed-onset muscle soreness. The participant is seated in a device like a child's swing and pushes off toward a wall, absorbs the contact with the wall, and immediately pushes off again. The desired muscle action of a stretch followed by a shortening of the leg muscles is produced while body weight is supported on the seat (see figure 5.6).

For a series of experimental studies, Fowler and colleagues (1997) built a force platform on a laboratory wall to record the forces generated when subjects trained using the pendulum method. The system was found to be effective as a training method and reduced both the acute load on the skeleton and the degree of transient muscle damage. A disadvantage of this system is that its use is limited to one person at a time. In addition, this system is not widely available, and it is used more for laboratory work than as a training tool.

Figure 5.6 Drop-jumping is performed from a box *(a)*, whereas exercise using the pendulum *(b)* is performed from a seated position.

Adapted, by permission, from T. Reilly, 2007, *The science of training: Soccer* (London: Routledge), fig. 4.3, pg. 57, and fig 4.4, pg 59.

Complex Training

Different forms of training at high intensities can be combined into a single session. This integration is referred to as complex training. It might include jumping actions with overload in formal arrangements such as 3 or 4 sets of 6 repetitions. These are performed alongside repetitive bounding, for example for 30 s, and exercises with loose weights as free-standing resistance. The regimen includes concentric as well as eccentric actions and exploits the force–velocity characteristic of muscle, most of the exercises being performed toward the faster end of the curve.

The aim of complex training is to introduce different training stimuli in a single integrated exercise session. An advantage of this type of training is the variety that a combined session entails. Although complex training has gained in popularity in the training of games players, its effectiveness has not been examined thoroughly. A potential disadvantage is that the training stimulus falls short of specific effects given the low frequency of repetitions of particular exercises.

Isokinetics

Isokinetics describes the form of exercise permitted by machinery with the inbuilt facility to adapt resistance to the force exerted. Normally when weights are lifted through a range of movement, the maximum load is limited to that sustainable by the muscles involved at the weakest point in the range. Consequently, other points within the range undergo submaximal training stimuli. With isokinetic machines the angular velocity is preset, and a speed regulator in the apparatus allows the resistance to adapt to the force applied. In this way, the greater the effort exerted, the greater the resistance, and maximal effort can be performed throughout the complete range of movement. Where comparisons have been made, training programs using isokinetic machines have proved superior to isometric and typical progressive resistance programs, with high speeds producing best results. These results may in part reflect the fact that the training programs are mostly evaluated using isokinetic equipment.

The aim in using isokinetic machines for training is to isolate a fixed angular velocity at which to operate. This allows the user to select low angular velocities for strength work and faster velocities for speed work. This form of training is most effective when used in training to correct specific muscle weaknesses or during rehabilitation programs to restore muscle function to preinjury values.

Modern isokinetic equipment (see figure 5.7) allows eccentric as well as concentric muscle actions. Typically, the top angular velocities available on the equipment are

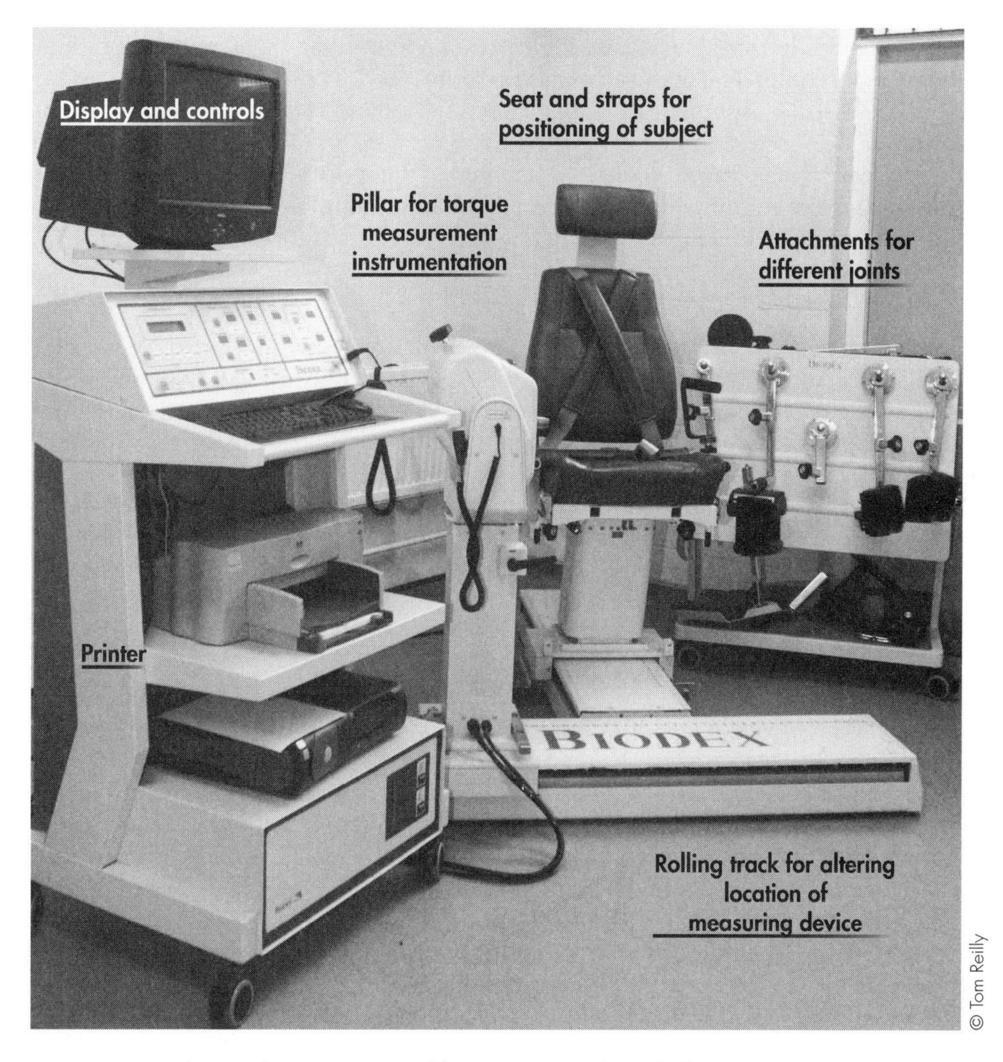

Figure 5.7 Isokinetic dynamometer used for assessment of muscle function.

higher under concentric than eccentric modes of action. Nevertheless, the angular velocities that are possible are well below the maximal velocities achieved in playing actions such as kicking a ball or serving in tennis. A related limitation is that the linear motions may lack specificity for some sports.

Training at high velocities is likely to assist slow movements, whereas training using slow movements is likely to assist only slow movements. The velocity-specific adaptations are linked to the pattern of motor unit recruitment. Improvements in muscle strength are to be expected from training at slow angular velocities, because of the recruitment of a large population of motor units. Such actions close to maximal efforts can induce muscle hypertrophy, provided repetitions are sufficient in number and the program of training is sustained for some months. To avoid muscle hypertrophy and at the same time improve strength (attributable to neuromotor factors), no more than 3 sets of 6 to 8 repetitions are recommended (Reilly, 2007).

Isokinetic facilities are expensive and are not normally available for team training. The area where they are of major benefit is in training muscle strength during rehabilitation. In this instance, isokinetic dynamometry can be allied with physical therapy in a comprehensive progressive program.

A limitation of isokinetic exercise is that it may interfere with the natural pattern of acceleration used in competitive actions. Furthermore, movements are linear and so do not correspond to musculoskeletal function in the sport. Nevertheless, assessment using isokinetic equipment is very effective in identifying deficiencies at individual joints. The appropriate muscle groups can then be isolated for remedial training.

Multistation Equipment

A series of weight training exercises are organized for sequential performance in a circle. The participants rotate in the circle as they progress through the training session. The aim of organizing exercise as a circuit is that different muscles are used at each workstation, so local muscular fatigue is avoided. In theory, this method is ideal for team training provided the number of participants does not exceed the number of workstations available. In practice, group organization invariably presents some problems given interindividual differences. A fixed load at a given station may not be suitable for all participants, but altering the loads delays the workout and allows unwanted recovery. A homogeneous group, a well-organized routine, and repetition of the circuit or even supplementary training are necessary to achieve objectives.

An advantage of multistation exercise machines is that they overcome the organizational problems of circuit training and the injury risks of weight training using traditional resistance modes. The machine illustrated in figure 5.8 was one of the first to be designed on ergonomic principles to provide the training stimulus requirements for strength, power, and local and general muscle endurance. Resistance is alternately supplied by body weight, weighted stacks, and isokinetic machines. Muscle groups are altered from station to station; the muscles worked include abdominals, leg, shoulder, arm, and back. Each station is adjustable to accommodate people of different body size and different exercises. Physiological studies have shown that the training stimulus to the circulatory system is significantly greater than that induced by conventional circuit training routines (Reilly and Thomas, 1978). However, because the delay in altering loads at any one station is minimal, the circuit of 12 stations can be repeated to perform two or more sets in a single training session.

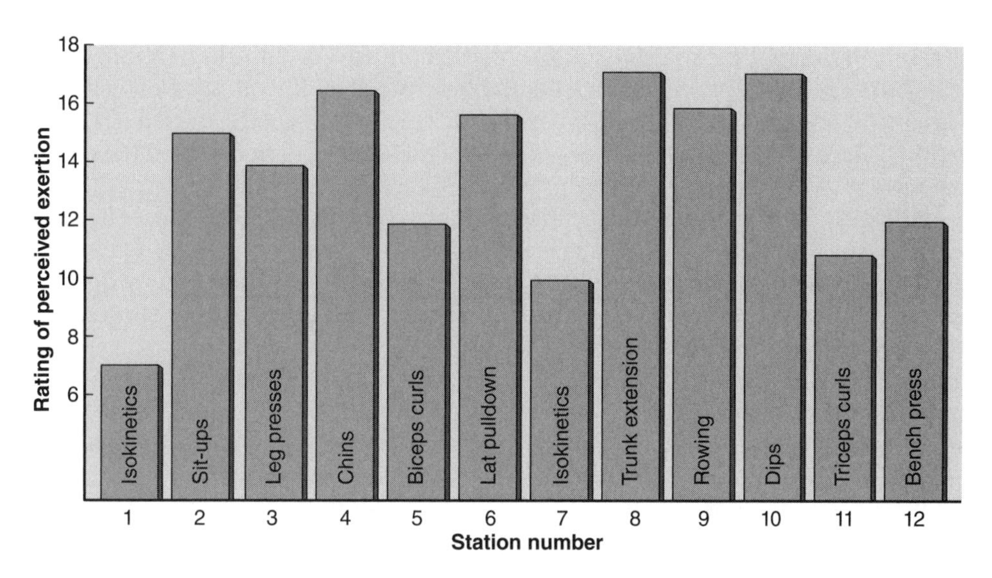

Figure 5.8 The original multistation equipment validated by Reilly and Thomas (1978) is accommodated within a single setup with interconnected stations.

Adapted from *Applied Ergonomics*, Vol. 9, T. Reilly and V. Thomas, "Multi-station equipment for physical training: Design and validation of a prototype," pgs. 201-206, Copyright 1978, with permission from Elsevier.

Portable stainless steel systems allow professional teams to integrate conditioning work into their daily training. Multiple stations are built into an overall frame into which up to 20 individual places can fit. This innovative development allows multiple users to exercise simultaneously. The exercises may include chin-ups, press-ups, squats, and core stability work and plyometric bounds, leaps, or hops. The device is a pragmatic means of controlling team conditioning during training sessions.

Stationary equipment for resistance training is safer than free weights, which carry the risk of accidents. The arrangement at each station can be changed quickly to accommodate different physiques and capabilities. This type of equipment is available at most fitness centers and sport training complexes. The design of such systems is constantly being improved by manufacturers to gain a competitive commercial edge. The mode of resistance—hydraulic brakes, air brakes, or loaded weights—varies according to cost and design preference. A disadvantage is that without attention to the design of training sessions, these systems do not guarantee physiological adaptations.

Technological Training Aids

A variety of small devices have found use in training contexts. Typically they act as stimulation modes for engaging the muscles to be trained. The aim is to engage muscles under conditions of resistance to muscle action.

Mechanical Devices

Various devices have been developed to facilitate strength and power training. In some instances commercial claims have not been supported by laboratory studies.

In particular, claims for securing muscle hypertrophy in 1 to 2 weeks have not been corroborated. Furthermore, devices that incorporate electrical stimulation may be relevant for rehabilitation but lack specificity for a majority of sports. The main reason for the failure of these products is that their designers have not understood the training process. Some devices that have proven merit are now described.

Pulley systems incorporate both concentric and eccentric actions for back and thigh muscles. The best devices regulate the magnitude of the eccentric phase by reference to performance in the concentric actions. Alternatively, simpler elastic or bungee devices are used for speed training. The advantage in their use is that the path of motion (or locomotion) can mimic the actions of the sport.

Bungee systems are incorporated alongside instrumental platforms to optimize training of vertical jumping. The devices enable muscles to be trained at controlled velocities of movement. Infrared systems that measure jump height, flight and ground contact time, and power output are available to monitor performance and provide feedback on jumping performance.

Exercise machines with the arc of motion dictated by a cam device have been in use for some decades. The cam design allows the resistance to be increased at parts of the joint's range of motion where force is decreased. In this way the resistance accommodates to the force exerted throughout the range of movement, conforming to the force–angle curves for the joint in question.

Vibration Plates

Vibration platforms are used for strength training by track-and-field athletes, by some of the top European soccer clubs, and by professional Rugby Union players as part of their conditioning program. These platforms have also been installed in training facilities of sport centers and commercial gymnasiums. There have been some positive results, but it seems that these apply to specific vibration frequencies. Devices like the Galileo Sport machine (Novotec, Germany) have a tilting platform that delivers oscillatory movements to the body at frequencies from 0 to 30 Hz around a horizontal axle. The participant is largely passive, merely standing on the platform while it vibrates at a preset frequency.

For many years ergonomists have recognized the potential adverse effects of vibrations delivered to the human body by handheld tools such as pneumatic drills and chain saws. Recently, exercise devices that include whole-body vibrations have been promoted as training aids in public sports centers. Vibration of soft tissue is a natural phenomenon. When a person walks or runs, the tissues in the lower limbs vibrate at their natural frequencies with the shock impact of landing as the heel strikes the ground on each foot strike. When exercise devices are used that deliver whole-body vibrations, the frequency of stimulation should be tuned to the resonant frequency of the participant to induce training effects. Results of studies have been mixed, negative results being attributed to too low amplitude, incorrect frequencies, or excessively long durations (Cardinale and Wakeling, 2005).

Benefits to performance have generally been observed with sinusoidal vibrations at frequencies of 26 Hz and amplitudes of 4 to 6 mm. Acute applications of whole-body vibration for 5 min at 26 Hz and 10 mm amplitude were reported to shift the force–velocity curve of well-trained subjects to the right (Bosco et al., 1999). Other positive effects have been observed on skeletal health and on flexibility (Cardinale and Wakeling, 2005). Acute enhancement of flexibility and muscle power has been attributed to stimulation of the muscle spindles and recruitment of additional motor units via activation of multiple nerve synapses (Cochrane and Stannard, 2005).

Issurin (2005) distinguished whole-body vibration from vibratory stimulation of local tissues combined with strength training exercises and static stretching. In these instances vibratory stimulation is superimposed on muscle contraction or stretch. The balance of research evidence has indicated positive effects with short-term use but suppressive effects on muscle force when stimulation was prolonged (6-30 min). Greater effects were observed in dynamic than in isometric muscle actions and were most pronounced in fast movements. The main benefit of this method may be in physical therapy for individual muscles and when resistance training methods are completed on the vibration platform. The main disadvantage is that the vibration sessions are ineffective and may even cause harm unless the appropriate frequency and amplitude are used.

Functional Overload

Various forms of natural resistance may be provided to overload the active muscles. These include running uphill, on sand dunes, or ankle-deep in water. Traditionally, soccer coaches used stadium terrace steps in the preseason training of their players, whereas athletics coaches used sand hills for their groups. The aim of functional overload training is to increase resistance to motion beyond what is normally experienced. An advantage is that the extra resistance is provided during actions that are directly relevant to the sport.

Activities related to sport are also set up with overload in mind. Jumping with ankle weights or jackets weighted with lead is an example. Graham-Smith and colleagues (2001) examined the use of an additional load of 10% body weight in the form of a weighted vest during a typical plyometric training session of 5 sets of 10 repetitions of vertical jumping with 3 min recovery between sets. The unique load distribution around the shoulders, lumbar region, and waist led the investigators to conclude that device was safe for resistance training. Improvement in performance after using such devices can be pronounced: Bosco (1985) reported an average increase in vertical jump performance of 10 cm after wearing a vest (11% of body weight) for only 3 weeks.

Running harnesses can be used to create resistance against the athlete attempting to accelerate from a standing start. The amount of resistance is controlled subjectively by a partner or trainer (see figure 5.9). Attachment to a sled or similar load by means of an abdominal belt offers an alternative means of resistance training.

Figure 5.9 Harness running for resistance training.

For optimal effects, the normal vigorous running action should not be modified unduly. Parachutes have been advocated to increase air resistance, but their efficacy has not been seriously addressed.

Functional resistance to motion may slow the athlete too much for the resultant training effects to be of benefit in competitive performance. Murray and colleagues (2005) showed that speed was generally reduced over 10 to 20 m as soccer players towed weights ranging from 0% to 30% body mass, using a waist harness and a 5 m rope. Loads up to 10% body mass are favored, and the procedure seems to improve peak anaerobic power rather than maximal running velocity.

Medicine balls are used to improve sport-specific skills such as throw-ins. Throwing distance can be increased with a dedicated strength training program for pullover strength and trunk flexion (Togari and Asami, 1972). These balls are also used for one-hand exercises as in the case of the goalkeeper, for whom the skill of throwing the ball long distances is directly relevant. Medicine balls are used in drills for javelin throwers, shot-putters, and cricket bowlers. Both forms of functional resistance provided by medicine balls and harnesses should be used alongside practice of the skill itself for optimal effects.

Deep-Water Running

Deep-water running can introduce novelty into the training program. It is performed in a deep hydrotherapy pool or in the deep end of a swimming pool. The participant tries to simulate the normal running action used on land while wearing a buoyancy vest to assist flotation. Given biomechanical differences between running in water and on land, a definite attempt must be made to push the hips forward to maintain good posture. This training modality is used to prevent injury, promote recovery from strenuous exercise, and provide supplementary training for cardiovascular fitness (see table 5.3).

Deep-water running is deemed safe and is suitable for a range of sport and recreation populations. It can be used by recreational and veteran games players as well as by professional players. Its uses, listed in table 5.3, emphasize that it can serve different purposes for different groups.

In deep-water running, impact is avoided and the risk of injury to the lower limbs eliminated because the feet do not touch the floor of the pool. The buoyancy provided by deep-water running decreases the compressive forces on the spine that are evident during running on land. Dowzer and colleagues (1998) reported that while participants exercised at 80% of mode-specific $\dot{V}O_2$peak, there was reduced axial loading on the vertebral column during deep-water running compared with treadmill running. Running in deep water enables participants to reduce impact loading while maintaining training intensity.

Deep-water running also accelerates recovery after competitive games or strenuous training. In one investigation, the recovery of muscle strength after stretch-shortening exercise designed to induce muscle soreness was accelerated by deep-water running compared with treadmill running (Reilly et al., 2001). It was concluded that deep-water running temporarily relieves soreness while enhancing recovery. The temporary relief of muscle soreness allows formal training to continue when training on land would be uncomfortable.

The physiological responses to exercise in water and in air differ largely because of the hydrostatic effect of water on the body in deep-water running. The changes in blood compartments, cardiovascular responses, and pulmonary and renal function

Table 5.3 Uses and Benefits of Deep-Water Running

Population	Purpose	Benefit
Injured sport participants	Rehabilitation	Prevents detraining Accelerates rehabilitation
Games players	Recovery from delayed-onset muscle soreness	Accelerates recovery from matches Promotes pain-free exercise Maintains flexibility
Athletes	Complementary training	Avoids overtraining effects Maintains central training stimulus
Untrained people	Aerobic training Strength training	Avoids injury resulting from the initiation of land-based training Increases shoulder strength
Physically debilitated people	To allow movement	Prevents injury from falls Increases subjective sense of comfort and security
Overweight people	Aerobic training	Aids weight reduction by increasing energy expenditure Reduces load bearing on the joints Allows exercise to be performed without embarrassment

that occur during this form of training were reviewed by Reilly and colleagues (2003). Heart rate is reduced by reflex action immediately on immersion. Both stroke volume and cardiac output increase during immersion in water: An increase in blood volume largely offsets the cardiac decelerating reflex at rest. At submaximal exercise intensities, blood lactate responses to exercise during deep-water running are elevated in comparison to treadmill running at a given oxygen uptake ($\dot{V}O_2$). Although $\dot{V}O_2$, minute ventilation ($\dot{V}E$), and heart rate are decreased during maximal exercise in the water, deep-water running nevertheless provides an adequate stimulus for cardiovascular training. Responses to training programs have confirmed the efficacy of deep-water running, although positive responses are most evident when measured in a water-based test. Aerobic performance is maintained with deep-water running for up to 6 weeks in trained endurance athletes; sedentary people benefit more than athletes in improving maximal oxygen uptake. There is some limited evidence of improvement in anaerobic measures and in upper-body strength in people engaging in deep-water running attributable to the arm actions being performed against the resistance of the water (Reilly et al., 2003).

A disadvantage of deep-water training is that it cannot fully replace conventional training programs. Furthermore, if the exercise intensity is too light, the training stimulus is insufficient to maintain fitness levels.

Ergometers

Ergometers are devices for calculating muscular power output. The load against which the person exercises is set, and the mechanical work done to overcome this resistance is measured. The amount of mechanical work done per unit of time indicates the power output, and this function is calculated in watts. The higher the exercise intensity, the greater the power output. An advantage of using ergometers is that the training load can be quantified precisely.

The ergometer is a fundamental apparatus in a sport physiology laboratory because it precisely indicates exercise intensity. Ergometers used for experimental studies include cycles and motor-driven treadmills. In cycle ergometry, the work rate is regulated by altering the pedaling speed or the loading. The exercise intensity is controlled on the treadmill by varying the belt speed and increasing the gradient, but calculating the power output is quite complex. For exercise purposes, the value of power is not necessary.

Although ergometry was first used for scientific purposes, ergometers have found favor for health-related exercise. The devices as modified for use in gymnasiums are able to grade exercise intensity but cannot measure power output. The ergometers used in fitness centers range from inexpensive resistance systems to relatively sophisticated computer-controlled devices that allow the user to preset the dimensions of the exercise session.

The cycle ergometer is the most basic of the fitness devices. It supports body weight, so the lower limbs are not subject to the repetitive-impact loading that occurs in running. Exercise can be performed at a high intensity or sustained for a long duration at a lower intensity. Sessions can be set up as continuous exercise for a fixed duration. As fitness improves, the intensity and duration can be increased progressively. Alternatively, the intensity can be varied systematically to correspond with interval training, for example, 60 s at a high intensity followed by 120 s at a low intensity performed for 12 repetitions.

Cycle ergometry is an ideal form of exercise for maintaining aerobic fitness and restoring muscle strength during rehabilitation from lower-limb injury. It has value as part of a warm-up regimen for elevating the metabolic rate and core body temperature. Cycle ergometry can also be used in recovery training for games players, for example, between competitive matches scheduled close together.

Rowing ergometers have also found use as training aids. The muscular power output is displayed in watts or, for performance purposes, is indicated on a digital display as equivalent distance covered in meters. This feedback is valuable to the performer in gauging the session and monitoring improvement with repeated use.

An advantage of rowing exercise is that major muscle groups are engaged. The power is generated by the extension of the quadriceps, complemented by activity in the trunk muscles and in the upper body. The starting posture for the next effort is resumed in a controlled manner. A disadvantage is that unless the exercise is conducted in a smooth and coordinated manner, low back pain can occur when ergometers are used by novices.

A well-equipped fitness center generally has a number of treadmills available to users. Programmable treadmills allow the participant to set the target load and exercise duration. In some setups the treadmill speed is controlled by the heart-rate response of the participant. In an alternative treadmill design, the belt is propelled by the person who runs on it, who thus dictates the exercise intensity by his or her own effort. This type of treadmill is completely safe because the belt stops immediately when the participant halts.

Professional soccer clubs tend to have a treadmill available so that injured players can gradually return to running. Treadmills also can be used by players who have been prescribed supplementary endurance exercise. In such cases, running outdoors, in parkland or forest paths, may be preferable although there may be professional reasons for completing the activity within the club's premises.

Stair steps and ladder mills have been designed as modes of exercise for health promotion. Care is needed when the step height approaches 45 cm in case the lead leg slips or the strain through the patellar tendon becomes excessive. The legs should be used alternately in stepping down; otherwise delayed-onset muscle soreness is experienced unilaterally. An advantage of this form of exercise is the novelty introduced to encourage compliance among those engaged in health-related fitness programs.

The work done in stepping up onto a bench or set of stairs may be calculated once the body mass and the vertical distance are known. Completing repetitive step-ups in a given period of time enables the power output to be calculated in kilograms per minute and converted to watts. Because the rate of stepping can be controlled, this activity was used in fitness assessments such as the Harvard Step Test. This test involved stepping onto a bench 20 in. (50.8 cm) high at a rate of 30 steps per minute for 5 min. The pulse rate was recorded for 30 s at 1 min, 2 min, and 3 min postexercise and a fitness score calculated. The test is hardly now used as a measure of fitness in athletes. As a mode of exercise for fitness training, step-ups may be of most value for recreational players or incorporated as a single station in circuit training. Benches are still used by exercisers, who bound over them or drop from them onto the ground when doing plyometric exercises.

Table 5.4 Energy Requirements (kJ/min) of Selected Sports and Recreation Activities

Light	Moderate	Heavy	Very heavy
Archery (13-24)	Aerobic dance (21)	American football (30-43)	Cross-country running (16-17 km) (66-67)
Billiards (11)	Badminton (26)	Basketball (38-46)	Cross-country skiing (41-78)
Bowls (17)	Baseball (20-27)	Boxing (38-60)	Cycling (>21 km/hr) (46-84)
Fencing (21)	Cricket (21-33)	Circuit training (33-44)	Rowing (59)
Golf (20)	Gymnastics (10-50)	Handball (46)	Professional soccer (50-69)
Table tennis (15-22)	Horse riding (13-42)	Hockey (36)	Squash (42-76)
	Volleyball (24-27)	Lawn tennis (29-46)	
	Waterskiing (29)	Rugby (33-60)	
		Step aerobics (36-42)	

Values in parentheses are kilojoules per minute. They are based on typical gross energy costs collected from various sources for male subjects. These values may underestimate the energy expended in top-flight competition in some cases.

These different ergometers permit the continuation of training indoors when weather conditions are unfavorable. A disadvantage is that the athlete forsakes natural conditions in which competition takes place.

Complementary Sports

Many components of fitness for sport and many skills are relevant to particular games. It is not surprising therefore that many games players have a good all-around athletic ability and can participate with reasonable competence in a range of sports. Invariably they have to abandon these other interests to concentrate on their own sport if they are to realize their aspirations in their prioritized sport.

The value of a specific sport as a training stimulus is reflected in its average energy expenditure. Even so, sports that are physiologically demanding may be unsuitable if there is a risk of injury. For this reason, cross-country skiing and downhill skiing as well as contact sports are not appropriate recreational activities for athletes specializing in other individual sports or team games.

Sports such as volleyball include jumping and coordination activities without a necessity for physical contact. These games can be used, for example, by football players for low-intensity activity the day following a hard training session. Alternatively, the principles of play can be modified with the ball being played with the feet or the head. In this way, athletes can practice game in an unobtrusive manner and in a relaxed circumstance.

Of the sports listed in table 5.4, golf is one that is suitable for professional games players on days off. The energy expenditure is light so the physiological strain is negligible. Nevertheless, the participant may take 3 hr or more to complete 18 holes of play, so a round of golf can constitute activity suitable for recovery on the day following the more strenuous training or competition days.

Overview and Summary

Various ergonomics models have been adapted from industrial settings to fit sport and leisure contexts. These largely fall into a framework of fitting the task to the person. Many top athletic coaches gain inspiration from business models and conversely business managers recognize the uncompromising attitudes of sport performers toward their tasks. A variety of methods are available for analysis of these tasks, the method of choice depending on the sport (or occupation) concerned. Where environmental factors are concerned, these are monitored and modified as far as possible to enhance human safety and comfort. Design principles accommodate anthropometric criteria as far as is manageable creatively. In competitive sport the challenges posed may be uncompromising, and participants undertake training programs so they can meet the demands of the activities. Ingenious equipment designs have enhanced performance capabilities and the achievement of excellence. The use of training devices is based on the principle of lifting the person to a higher level of performance.

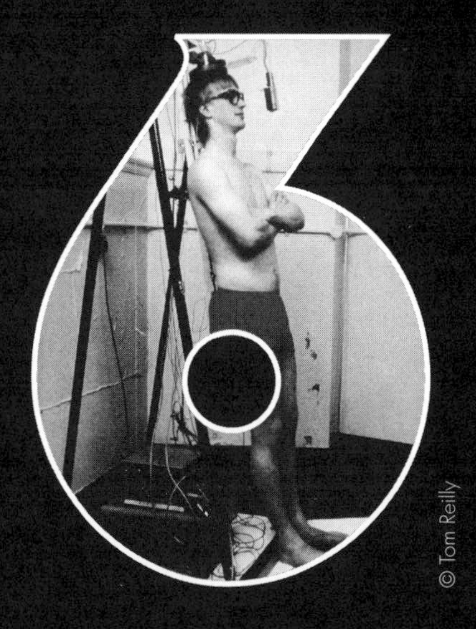

© Tom Reilly

Competitive and Training Stress in Sport

DEFINITIONS

aerobic interval training—Moderate- to high-intensity exercise, discontinuous in nature, used to improve oxygen transport and utilization.

circuit weight training—Resistance training in which the exercises are laid out in a circle of workstations.

drop-jumping—A form of exercise in which the athlete jumps vertically, having started to move by lowering the body from a box or raised platform; also known as depth-jumping.

hydration status—The reference of body water stores to their normal physiological range.

oxygen cost—The energy consumed during activity expressed in units of oxygen and allowing for resting metabolic rate.

radio telemetry—Transmission of signals in aerial waves from the body without requiring the subject to be tethered to equipment.

rating of perceived exertion—A scale for quantifying the subjective effort during physical activity.

respiratory gas analysis—Calculation of oxygen consumed from measurement of the O_2 and CO_2 in the air the individual breathes out and the ventilation rate.

shrinkage—Loss of height within the spinal column due to axial loading on intervertebral discs.

weight training—Use of external loads as resistance when training, usually for strength and power.

six

THE competitive nature of sport dictates that individual participants will engage in activity in a highly stressful environment. The prospect of doing so may induce anxiety prior to the contest, a certain level of physiological arousal being necessary for optimal performance. Training for competitive sport also places demands on psychological and physiological resources. Those who aspire to compete regularly in sport must be able to tolerate high training loads and must have the motivation to comply with training requirements.

The severity of competitive demands can be gauged by monitoring physiological responses to activities. Where methods are invasive and likely to interrupt the activity being studied, investigators sometimes examine simulations in model contests instead of the real events. Indirect measures may have validity as surrogates, such as the use of work rate profiles determined from video recordings to replace **respiratory gas analysis.** Investigators also gain insights by monitoring training intensities that provide more access to volunteer subjects. Ergonomics investigations entail a balance between ecological validity on one hand and adequate control of the study conditions on the other.

Psychological strain can manifest as an alteration in physiological state, such as an increase in endocrine secretion or activity in neurotransmitter pathways. A variety of measures are available to reflect these responses, although many may be too obtrusive to apply in precompetitive or competitive sport contexts. This difficulty has prompted the use of questionnaire-based and qualitative measures to explore aspects of psychological stress and emotional strain among participants in sport and recreation.

In this chapter, various methods are presented for investigating the relative loading on participants in different kinds of sport. Although the main focus is on psychological approaches, physical and physiological loading criteria are also considered. The more appropriate method of assessing the strain on participants must fit the context, and frequently a combination of methods is adopted.

Physiological Loading

Physiological loading is associated with engagement in physical activities. Relative loading is the percentage of maximal capacity that is required to carry out the task. Ergonomists express the load in terms of internal responses, some of which are now considered in detail.

Oxygen Consumption

The severity of physical activity is reflected in the energy expended in accomplishing the tasks concerned. Scientists measure energy in kilojoules, the conventional units being kilocalories. The amount of oxygen consumed reflects the energy cost of activity: If the respiratory exchange rate is known (the amount of oxygen consumed expressed as a ratio of the carbon dioxide produced), the energy expenditure can be calculated.

The **oxygen cost** of exercise is expressed as an absolute value (L/min) or in relative terms, either $ml \cdot kg^{-1} \cdot min^{-1}$ or as a percentage of $\dot{V}O_2max$. The $\dot{V}O_2$ cost of activity indicates the metabolic load on the individual, the proportional use of $\dot{V}O_2max$ indicating the relative loading. In prolonged endurance events, the athlete who can sustain activity at a high percentage of $\dot{V}O_2max$ likely has the edge over opponents.

The fuel source used as substrate for muscle varies with the intensity and the duration of exercise. As exercise intensity increases, oxygen consumption increases as does the respiratory exchange ratio, indicating a preferential utilization of carbohydrate as a fuel source. Fat is the preferred source of fuel when the exercise intensity is low and when exercise is prolonged. The increase in use of fat as a source of energy spares glycogen stores, which otherwise would be depleted rapidly.

Oxygen consumption ($\dot{V}O_2$) has traditionally been measured using respiratory gas analysis after collecting expired air in Douglas bags or meteorological balloons. The minute ventilation ($\dot{V}E$) is calculated from the amount of air expired, and the O_2 and CO_2 content is analyzed. The $\dot{V}O_2$ is calculated using standard procedures (Cooke, 2003). Because the use of Douglas bags tends to hamper activity, portable respirometers such as the Kofranyi–Michaelis system gained use in the early ergonomics work of Durnin and Passmore (1970). A limitation of these systems was that only oxygen was analyzed, the respiratory exchange ratio ($\dot{V}CO_2$:$\dot{V}O_2$) being assumed or estimated. Furthermore, activity needed to be sufficiently long in duration for a steady state to be achieved.

The advent of metabolic analyzers that are lightweight and wearable as a backpack during sport provided more precise information about physiological demands of these activities. This type of equipment operates by means of short-range radio telemetry. The data for $\dot{V}O_2$ and $\dot{V}E$ can be downloaded postevent and matched to other measures such as heart rate responses. Even a lightweight backpack can be intrusive in games, so the use of portable respirometers has been limited to comparison of games drills (e.g., Kawakami et al., 1992) or to individual sports. The equipment has been used in orienteering to quantify the additional energy costs associated with running on uneven terrain (Creagh et al., 1998). A portable radio telemetry system was used in an experimental investigation by Hulton and colleagues (2009) to establish the added physiological demands of solo running with the ball in Gaelic football compared with normal running. Where simulations of game actions are examined under laboratory conditions, on-line gas analysis can be used. Reilly and Ball (1984) used metabolic gas analysis to study the additional energy cost of dribbling a soccer ball. In a similar study, Reilly and Bowen (1984) showed how the extra cost of moving backward and sideways increased in proportion to the speed of locomotion.

Heart Rate

Heart rate increases during exercise because an increased cardiac output is needed to supply oxygenated blood to the active muscles. The stroke volume also increases although the relative increases in both of these functions are not equal. The elevation in heart rate is determined by the exercise intensity, and on this reasoning the heart rate during exercise has been proposed as a surrogate of energy expenditure. To use heart rate (HR) to estimate oxygen consumption, the HR–$\dot{V}O_2$ relationship, which is unique to each person, must first be obtained in laboratory conditions. For locomotor sports, the regression line relating HR to $\dot{V}O_2$ should be established during treadmill running. The error involved in using this method of estimating energy expenditure is small (Bangsbo, 1994). Allowing for any imperfections in such extrapolations from laboratory to field conditions, investigators find the heart rate to be a useful indicator of overall physiological strain during play.

Table 6.1 Mean Values for Heart Rate (beats/min) During Soccer

Subjects	Heart rate	Match-play situation
Czech players	160	Model 10 min game
Czech players	165	Model 10 min match
English League	157	Training matches
Japanese players	161	Friendly matches (90 min)
Scottish players	169	Friendly match (90 min)
Danish players	167	Competitive game (90 min)
University players	161	Competitive game (90 min)

Traditionally, long-range **radio telemetry** has been used to monitor heart rate data during friendly matches or simulated competitions. In recent years, the use of short-range radio telemetry has been adopted on a widespread basis. Observations generally confirm that the circulatory strain during games play is relatively high and does not fluctuate greatly during a game (see table 6.1). The variability increases in the second half of soccer play at university level because the players take more rest periods (Florida-James and Reilly, 1995). Rhode and Espersen (1988) reported that the heart rate was about 77% of the heart rate range (maximal minus resting heart rate) for 66% of the playing time. For the larger part of the remaining time the heart rate was above this level, reflecting the periodic high-intensity elements of the game.

The heart rate during exercise can be influenced by emotional and thermal stimuli, factors that can lead to overestimation of the metabolic cost. Although the error has been considered small, heart rate is easy to record and provides a good representation of the overall physiological strain. This approach is valid for arm exercise as well as whole-body activities, the heart rate being higher for a given oxygen cost of arm work compared with leg exercise.

Heart rate monitoring has proven useful in a variety of field settings. Many athletes and cyclists use lightweight monitors to pace themselves during training and during competition. Heart rate monitoring has proved useful in professional soccer in identifying the physiological stimuli provided in training by adopting small-sided game drills (Sassi et al.,2005). Responses can also be used to regulate rather than just monitor training intensity, heart rate being kept within a light-exercise zone (120-125 beats/min) on a recovery day and in a high-intensity zone (170-180 beats/min) during **aerobic interval training** (see figure 6.1).

Blood Lactate

The severity of exercise is also indicated by blood lactate concentrations. This measure has been used after all-out efforts in swimming and running and also during stages of intermittent exercise. Progressively higher lactate concentrations have been observed in soccer matches from the fourth to the top division in the Swedish league (Ekblom, 1986). Gerisch and colleagues (1988) demonstrated that higher blood lactate concentrations are associated with player-to-player marking roles compared

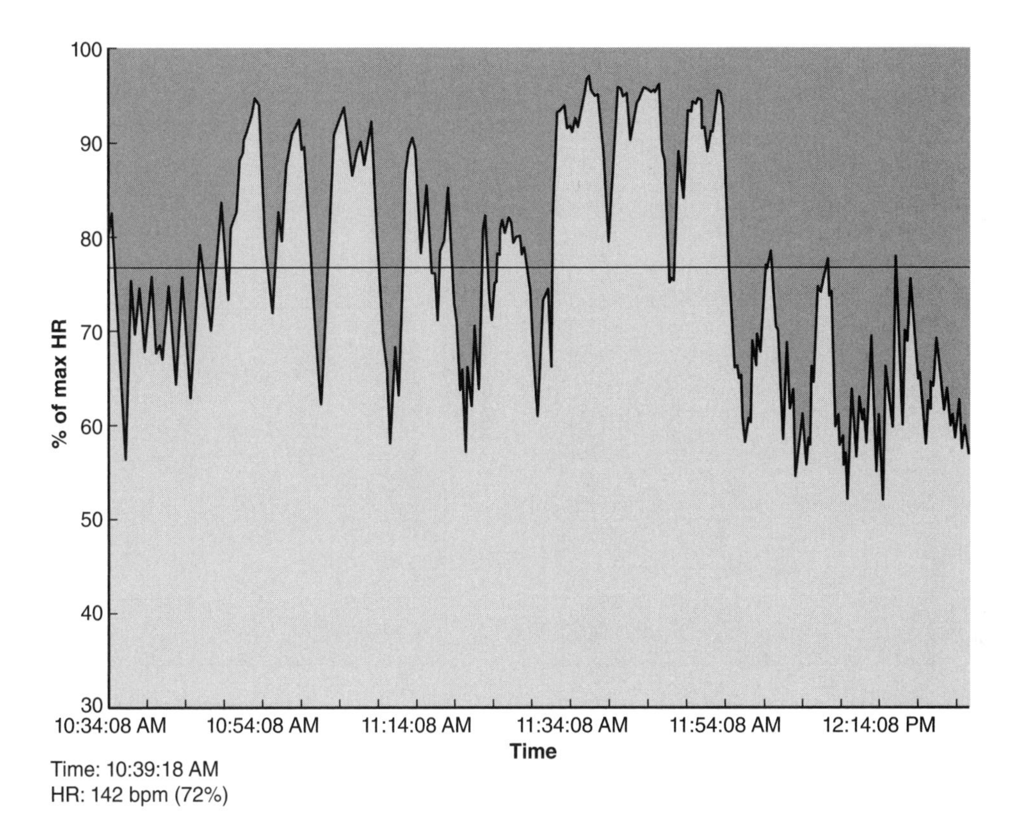

Figure 6.1 Heart rate (HR) responses at different training zones during intermittent-exercise training.

with a zone-coverage responsibility. Ekblom (1986) claimed that peak values greater than 12 mmol/L were frequently measured at high levels of soccer play. Activity could not be sustained continuously under such conditions, which reflects the intermittent consequences of anaerobic metabolism during competition. Although most studies of blood lactate concentration have shown values of 4 to 6 mmol/L during soccer play, such measures are determined by the activity in the 5 min prior to obtaining the blood samples. Higher values are generally noted when observations are made at halftime compared with the end of the match, because work rate tends to be lowered late in the game.

Blood lactate levels are elevated as a consequence of dribbling the ball in hockey and in soccer, for example, the increased concentrations are disproportionate at the higher speeds. In the study by Reilly and Ball (1984), the lactate inflection threshold was estimated to occur at 10.7 km/hr for dribbling but not until 11.7 km/hr in normal running (figure 6.2). This finding indicates that the metabolic strain of fast dribbling will be underestimated unless the additional anaerobic loading is considered.

It is now relatively straightforward to monitor blood lactate concentration. A capillary sample is obtained from a fingertip or an earlobe; a toe capillary is used in the case of a rower because the feet remain stationary during exercise while the arms and the head are inaccessible during exercise. Portable analyzers require only a small sample, and results are obtained quickly after obtaining the sample. Care should be taken in drawing the sample and in its timing. The concentration in the

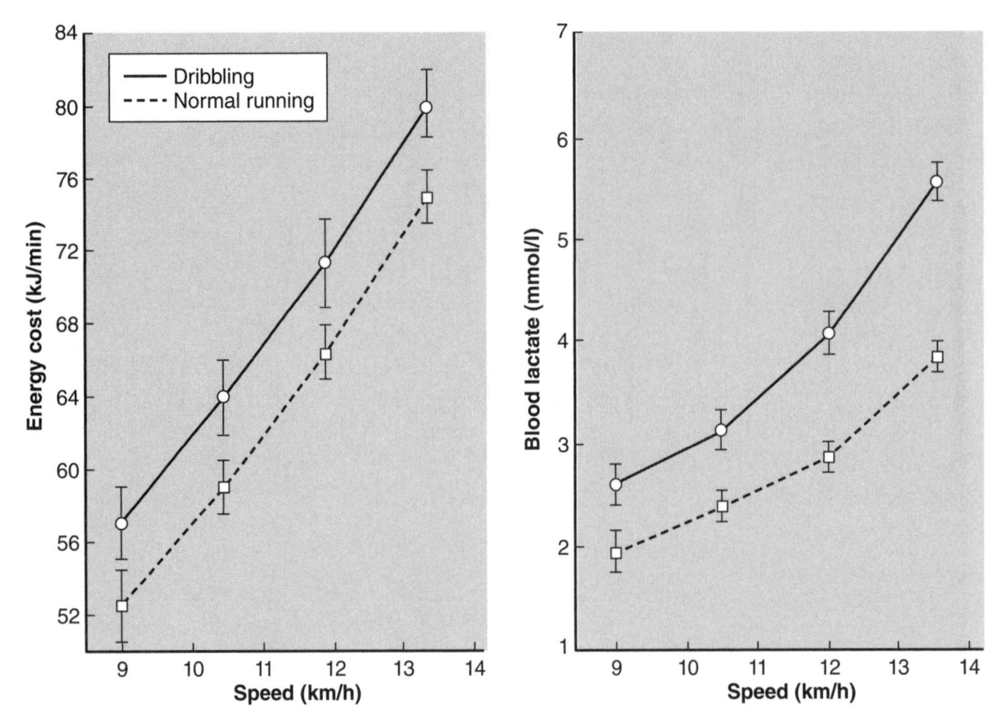

Figure 6.2 Physiological responses to running at different speeds are higher when dribbling a ball than in normal running.

Reprinted with permission from *Research Quarterly for Exercise and Sport*, Vol. 55, pgs. 269 and 270, Copyright 1984 by the American Alliance for Health, Physical Education, Recreation and Dance, 1900 Association Drive, Reston, VA 20191.

blood represents the balance between the production of lactate in the active muscles and its removal. Because lactate is a result of the anaerobic breakdown of glycogen, it indicates the contribution of anaerobic metabolism to exercise.

Core Temperature

Muscle actions generate internal heat because only about 20% to 25% of the energy used in each muscle contraction contributes to mechanical work production. The body acts as a heat sink during the first 5 to 10 min of exercise. Physiological mechanisms for losing heat include activity of the sweat glands and a redistribution of blood to the skin for cooling. Without these means of losing heat to the external environment, body temperature would increase within an hour or so of intense exercise to a level that proves fatal.

The increase in core body temperature is dependent on the relative exercise intensity, reflected in percentage $\dot{V}O_2max$. In uncompensated heat load, the elevation in core temperature limits performance, lowering the exercise intensity or completely stopping exercise. In some cases central factors operate to prevent damage from hyperthermia. When the exercise intensity is within the compensatable zone whereby the difference between heat gain and heat loss can be tolerated, core temperature is an index of metabolic loading and also indicates the thermal strain on the individual.

Various sites are used to measure the temperature within the body's core. Esophageal temperature is preferred by environmental physiologists in laboratory settings. Alternatives include tympanic and rectal temperature, the latter being traditionally chosen by exercise physiologists. Rectal temperature lags behind brain temperature when the internal temperature is changing rapidly. In recent years gut (intestinal) temperature has been used in training and sport competition contexts, the participant ingesting a temperature-sensitive pill some hours before the contest. The receiver is worn as a pack around the midriff, and the data are subsequently downloaded to a computer for analysis.

Hydration Status

Water is essential for life and constitutes approximately 73% of the body's lean tissue content. Body water is finely regulated to balance fluid intake and fluid losses **(hydration status).** The regulatory mechanisms are not immediate or precise, and thirst can be satisfied before water balance is restored after large losses of fluid such as can occur during exercise in the heat.

Athletes participating in weight categories often voluntarily dehydrate in preparation for the weigh-in prior to the contest. This practice applies to boxers, wrestlers, horse-racing jockeys, and others. The athletes then try to regain euhydration by the start of the contest but many fail in the attempt. Consequently, they experience negative mood states and impaired performance during competition.

Fluid loss affects performance, depending on the degree of dehydration incurred and the nature of the activity. Effects are more apparent in sustained activities, where the exercising muscles need to be continuously supplied with oxygen carried in the circulatory system, than in short-term events that depend on anaerobic metabolism. A fluid loss of 2% body mass has been sufficient in most studies to demonstrate a decrease in performance. Fluid is lost from all body water pools, including plasma and intramuscular and interstitial fluid.

Thirst is sensed when the water deficit reaches about 1% of body weight. This signal is caused by a change in cellular osmolality and dryness in the mucous membrane of the mouth and throat. Thirst may be satisfied before the fluid lost is fully replaced, indicating that reliance on this mechanism is insufficient in the short term. Urine osmolality is probably the best practical physiological marker for monitoring the hydration status of athletes who train in the heat, although urine conductivity is also easily measured and the color of the urine is a more visible indicator. The correlation of other measures with urine osmolality suggests that specific gravity is an acceptable alternative (table 6.2). The urine becomes more concentrated as the body attempts to retain water by secreting antidiuretic hormone from the posterior part of the pituitary gland.

Sweat contains electrolytes, urea, and lactic acid, the concentrations varying according to the site on the body, whether sweating was actively or passively induced, and the person's fitness level. People vary greatly both in the volume of sweat lost and its sodium concentration. Sweat is hypotonic relative to plasma, reflecting the body's conservation of its sodium content by the secretion of aldosterone. The primary need in fluid replacement when exercising in the heat is for water, both to reduce body temperature and to partially restore the water lost in sweat. Absorption of the fluid ingested is increased if some sodium is included in the drink. The energy content of sports drinks is relevant in endurance events where there is a likelihood that carbohydrate stores in the muscles and liver will become too low. In hot conditions, it is

Table 6.2 Correlations With Osmolality*

Measures of hydration status	r
Specific gravity (urinometer)	.86
Specific gravity (reagent strip)	.78
Color (1-8 scale chart)	.63
Conductivity (continuous scale)	.76

*Freezing point depression; N = 183.

good practice to start the competition well hydrated and take small amounts of fluid as occasion allows during the contest. A quantity of 150 ml every 10 to 15 minutes is still inadequate to compensate for the losses that can occur so that in hot conditions a water deficit is inevitable. This deficit must be countered by replacing fluid once exercise or training is terminated. Merely calculating the sweat lost is not enough, because some allowance should be made for the production of urine. It is recommended that an extra 50 ml of fluid be added to the 100 ml implied by a loss of body mass of 100 g.

In sports that have competitive weight classes, participants traditionally have attempted to lose weight by dehydrating in order to compete in the lowest possible category. Boxers and wrestlers in particular lose weight before their events to keep within the prescribed limit, and jockeys also must operate according to the loads outlined for their mounts. The practice is recognized as dangerous, especially if the forthcoming contest is held in hot conditions and severe levels of dehydration have been induced before the formal weigh-in. The use of dehydration as a strategy for making the desired weight has been roundly condemned and in some cases is replaced by a more systematic program of weight control based on nutritional advice. When a weight loss of 5.2% body weight was incurred over a 1-week period by a group of amateur boxers using a combination of food and fluid restriction, performance was reduced (Hall and Lane, 2001). The decrease in performance was accompanied by negative moods that included increased anger, fatigue, and tension and reduced vigor.

Spinal Loading

The low back is a common location of musculoskeletal problems for athletes as well as the general workforce, especially those engaged in lifting and lowering weights. There are various ways of indicating the biomechanical loading on the spine. Among these is the measurement of shrinkage, which registers the loss of height among the intervertebral discs caused by axial compression. Applications of spinal shrinkage are described in the sections that follow.

Theoretical Background

During everyday activities, the intervertebral discs experience axial loading and lose height as they are compressed. When the osmotic pressure of the discal tissues is exceeded by the compressive load, fluid is expelled (Tyrrell et al., 1985). The expulsion of fluid is followed by changes in the dynamic response characteristics of the intervertebral joint complex, and with time there is a reduced resistance to failure under static, dynamic, or vibratory load. In a degenerated disc, deformation under load is more rapid than normal (Kazarian, 1975). Thus, the cumulative effects of static and dynamic loading are significant contributors to back symptoms and back injury.

The spinal column constitutes about 40% of total body length, and approximately 33% of total spinal column length is occupied by intervertebral discs. The intervertebral discs respond to loading and unloading forces by altering their size. The loss of disc height with compressive loading has been attributed to elimination of fluid from the nucleus pulposus (Kazarian, 1975). Water is removed from the disc when the sums of the imbibition pressure of the protein polysaccharide complex of the nucleus pulposus and the osmotic gradient across the disc membrane are exceeded. Recovery ensues when the compressive forces are withdrawn or when the spine is distracted. Creep in the disc and subsequent recovery are not solely matters of fluid exchange because extension and contraction of the fibers of the annulus fibrosus are also implicated (Koeller et al., 1984).

Measurement of changes in stature can reflect shrinkage in spinal length and consequently in aggregate disc height. Variations in stature attributable to compression of appendicular structures are negligible compared with changes within the spinal column. Compression of soft tissue in the soles of the feet reaches equilibrium quickly when bearing body weight and is unaffected by experimental spinal loading. Exact reproduction of the standing posture in successive measurements, while allowing for differences in postural contour between individuals, is an essential requirement for measuring fine changes in stature (see figure 6.3).

The amount of **shrinkage** is related to the magnitude of the compressive load on the spine: Consequently, shrinkage has been used as an index of spinal loading (Corlett et al., 1987). Changes in total body length have been used to examine the effects of physical regimens that load the spine, the assumption being that the greater the load, the greater also the risk of back problems. The technique has been used also to evaluate maneuvers for unloading the spine, such as traction, gravity inversion, and the so-called Fowler position (see figure 6.4), in which the subject lies supine, with feet supported on a chair and hips at about 45° of flexion (Boocock et al., 1988).

An apparatus has been developed that allows intra-individual variation in spinal configuration to be accommodated and intra-individual postures to be reproduced accurately under relaxed conditions (Corlett et al., 1987). Relaxation is achieved by inclining the subject backward by up to 15°. Design features control the position and contour of the spinal curves, position of head and limbs, head angle in the sagittal plane, and weight distribution between heels and forefoot. The phase of the respiratory cycle is also controlled, after subjects undergo a brief training session to accustom them to the apparatus.

Weight Training

Static shoulder loads using rucksack and barbells were examined by Tyrrell and colleagues (1985). Observations were made at 2 min intervals during 20 min of experimental loading, measurements taking about 2 min. Shrinkage of 5.45 mm was incurred with the 10 kg rucksack and 5.14 mm with the barbell. Shrinkage increased with increased barbell loading to 7.11 mm (20 kg), 9.42 mm (30 kg), and 11.2 mm (40 kg). There was a linear relationship between shrinkage and external load (see table 6.3). There

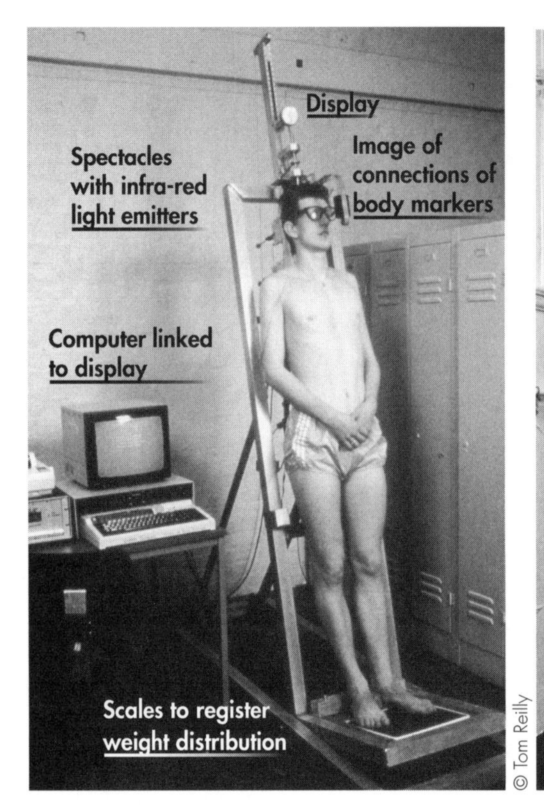

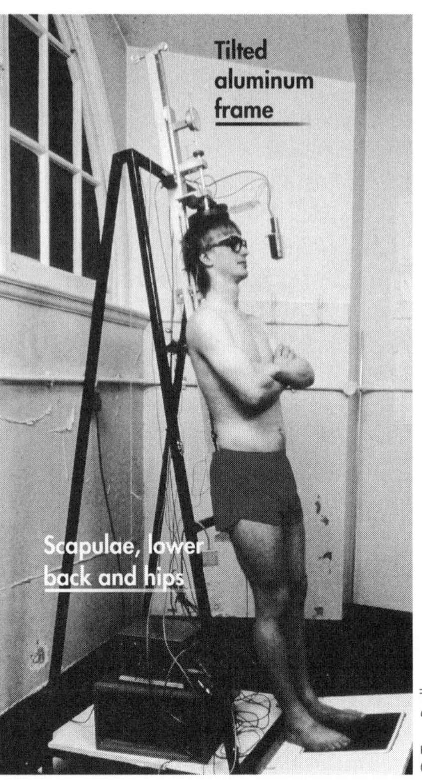

Figure 6.3 Subject in position for measurement of spinal shrinkage. A reference stature is measured, and change from this value is attributed to the intervention being studied.

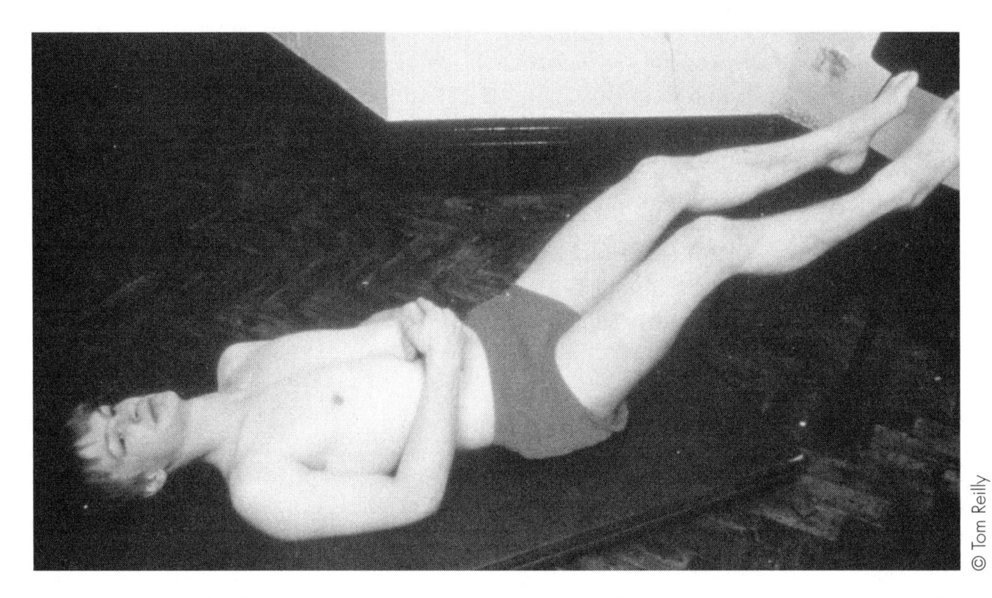

Figure 6.4 The Fowler position, which is used as a means of recovering from spinal loading. By altering joint angles at the hip, this position removes loading from the spine, which then recovers.

Table 6.3 Mean (±SD) Loss in Height Attributable to Experimental Static Loading of the Spine

Static load	Height loss, mm
2.5 kg rucksack	3.87 ± 1.98
10 kg rucksack	5.45 ± 2.12
10 kg barbell	5.14 ± 1.99
20 kg barbell	7.11 ± 3.18
30 kg barbell	9.42 ± 3.57
40 kg barbell	11.2 ± 4.60

is no clear indication of a threshold for back injury that can occur as a result of faults in technique or handling loads that are too heavy.

Circuit weight training, as used to stress the oxygen transport system rather than for muscle conditioning, has been examined. Nine males rotating around nine exercise stations for 25 min were found to lose on average 5.49 mm in stature (Leatt et al., 1986). The weights varied from 14 to 32 kg for the different exercises. A comparison of results with those of Tyrrell and colleagues (1985), who found shrinkage of 7.11 mm with 20 kg dynamic loading over 20 min, suggests that the strain on the spine may have been eased by some of the exercises in the circuit.

Similar responses occur in women as in men. Ten female subjects repeating a sequence of eight **weight training** exercises for 20 min were examined by Wilby and colleagues (1987). The regimen was performed immediately after rising from a night's sleep and again at 10 p.m. A greater loss of height was observed in the morning than in the evening, mean values being 5.4 and 4.33 mm, respectively. This difference was attributed to the diurnal variation in stature. The rate of change in stature varies throughout the day, being greatest in the morning, whereas disc imbibition and recovery are rapid in the early hours of sleep. Diurnal peak-to-trough variations in stature of 15.4 mm and 19.3 mm have been shown for females (Wilby et al., 1987) and males (Reilly et al., 1984), respectively.

Althoff and colleagues (1992) undertook a series of measurements on subjects using shoulder loads up to 30 kg. Spinal shrinkage was proportional to the load applied on the spine. The large intersubject variability was attributed to individual differences in cross-sectional area of the disc. Subjects with small disc dimensions incurred a greater spinal shrinkage than those with larger discs.

The effects of different load-carriage conditions on spinal shrinkage were examined by Reilly and Peden (1989). On three separate days, six females performed (a) 10 min stepping onto a bench 30 cm high without load, (b) the same activity with 10 kg and 15 kg carried on the back, and (c) the same task but with a load of 15 kg on the back. Mean shrinkage at the end of the unloaded condition was 1.1 mm. There was a significant increase in shrinkage at 10 min in both conditions with the 15 kg load (front bag 2.79 mm; back bag 2.78 mm). It was suggested that tasks such as repeated bench stepping performed with external load greater than 16% body mass can increase the risk of back injury.

In a relaxed standing posture after weight training, the participant regains height in an amount that is proportional to loading. About 75% of the losses incurred in static loading were regained within 10 min in the study by Tyrrell and colleagues (1985). The height regained in Fowler's position in a similar period exceeded the height lost during loading.

There is a question whether preventive measures have any influence on the load on the spine. Weightlifting belts are available commercially and are marketed with the aim of preventing back injuries while lifting heavy weights. It is believed that they do so by supporting and stabilizing the spine. These belts may also affect intra-abdominal pressure, the mechanism widely held responsible for reducing spinal compressive forces. It is common for U.S. workers engaged in manual materials handling to wear belts for protective reasons, although the evidence for reducing back injury is not compelling (NIOSH, 1994). However, Bourne and Reilly (1991) examined the effect of a weightlifting belt on spinal shrinkage in subjects performing a circuit weight training session and reported that wearing the belt induced less spinal shrinkage and caused significantly less discomfort compared with lifting without a belt. The observations suggested there are potential benefits in wearing a weightlifting belt and supported the hypothesis that the belt helps stabilize the trunk.

The protective effects of wearing a belt during weightlifting were further investigated by Reilly and Davies (1995). They examined the efficacy of a weightlifter's belt in attenuating spinal shrinkage during multiple repetitions of the deadlift. A further aim was to examine the relationship between shrinkage and the estimated cross-sectional area of the lumbar discs. The subjects performed 8 sets of 20 repetitions of the deadlift with 10 kg on an Olympic bar. This was done on two separate occasions, once while wearing a belt and once without the belt. The cross-sectional area of the L3-L4, L4-L5, and L5-S1 discs was estimated using the anthropometric procedure of Columbini and colleagues (1989). Shrinkage without the belt was 4.08 ± 1.28 mm compared with 2.08 ± 0.05 mm with the belt: Corresponding values for perceived exertion were 16.2 ± 1.6 and 13.4 ± 1.3. It was concluded that the weightlifter's belt was effective in reducing spinal loading during multiple repetitions of the deadlift. The magnitude of shrinkage incurred was related to both body mass and lumbar disc area. The decrease in shrinkage associated with wearing a belt was significantly related to body mass but not to estimated lumbar disc area.

Running and Jumping

Spinal loading is associated with activities where ground reaction forces exceed normal experiences. Running and jumping in particular induce high impact forces. In cases where the major leg muscles go through stretch-shortening cycles of activity, the impacts are partly absorbed by the skeleton. The consequence is an increase in spinal shrinkage, accentuated by external factors such as shoes worn and surfaces used.

Running Exercises

Running, particularly on road surfaces, induces repetitive loading of the spine. Shrinkage in experienced and novice runners exercising on a treadmill at 12.2 km/hr for 30 min was examined by Leatt and colleagues (1986). Loss of stature amounted to 2.35 mm for the experienced group and 3.26 mm for the other runners. The experienced runners did a further 19 km at 14.6 km/hr and lost another 7.79 mm on average. The duration of the run was an important factor in the total spinal shrinkage incurred.

Effects of running continuously at 10 km/hr for 40 min have been compared with those attributable to alternating a fast and slow pace regularly over the same time and covering the same overall distance. The pace fluctuated between jogging at 7 km/hr and sprinting at 21 km/hr. No significant difference was found between the intermittent and the continuous running in terms of spinal shrinkage, once the distance and duration of exercise were matched (Reilly et al., 1988).

The influence of running intensity on shrinkage was examined by Garbutt and colleagues (1989). Five male runners did three 30 min runs at 70%, 85%, and 100% of their competitive marathon pace. In the first 15 min, mean losses of stature were 4.25, 3.37, and 3.97 mm for the intensities increasing from half-marathon to 10K race pace. The influence of the quality of the running shoe on shrinkage may be important but has yet to be investigated.

Garbutt and colleagues (1990) examined the effect of three running speeds on two groups of runners, one with chronic low back pain. The two groups of seven male marathon runners exercised at 70%, 85%, and 100% of their marathon race pace for 30 min on separate occasions. Before and after exercise, the subjects were seated for 20 min with the lumbar spine supported. Stature was measured before preexercise sitting, before running, after 15 min of running, after 30 min of running, and after a period of sitting postexercise. There were no significant differences in the responses to the three running regimens between the groups. Shrinkage was significantly greater in the first 15 min, being 3.26 (±2.78) mm compared with 2.12 (±1.61) mm for the second half of the run. The faster the running speed, the greater the resultant shrinkage. The 70%, 85%, and 100% conditions caused 3.37 (±2.38), 5.10 (±1.90), and 7.69 (±3.69) mm of shrinkage, respectively. These observations suggest that in this group of runners who were able to maintain their training despite continuing back pain, low back pain was independent of the shrinkage induced by running.

Exercise in water has been promoted as a means of training without incurring high impact loading. Dowzer and colleagues (1998) used spinal shrinkage to evaluate the benefits of deep-water running in reducing impact loading on the skeleton. Running in deep water caused significantly lower shrinkage (2.92 mm) than running in shallow water when the feet touch the swimming pool floor (5.51 mm) or running on a treadmill (4.59 mm). These values were incurred over 30 min, but the higher rates of shrinkage were found in the first 15 min in all the conditions. The results support the use of exercising periodically in water as an alternative to running on hard surfaces.

Jumping Exercises

Jumping and bounding exercises have been increasingly implemented in training regimens to develop leg power. Landing from such exercises induces high impact forces, which the human body must absorb. The intervertebral disc is the principal shock absorber of the spine responsible for dissipating these high forces. Shrinkage measurements have been used to study the spinal loading resulting from such jumping and bounding exercises. A regimen of 10 sets of five standing broad jumps with 15 s recovery between each set, lasting on average 6.7 min, was found to cause a mean loss in stature of 1.7 mm (Boocock et al., 1988).

To assess the potential of unloading the spine preexercise, thereby increasing the discs' functional ability to absorb compressive loading, a 10 min period of gravity inversion was undertaken prior to the same exercise period. Inverting the subject at 50° was found to increase stature more than a 90° inclination, which in turn was superior to the Fowler position (Leatt et al., 1985). The unloading period caused a mean increase in stature of 2.7 mm, and the resulting exercise period when performed immediately after inversion induced twice the magnitude of shrinkage, 3.5 mm. It was concluded that the benefits gained by spinal unloading preexercise are short lasting.

Similarly, **drop-jumping** exercises, whereby athletes drop from a predetermined height and perform a rebound jump immediately on landing, have been found to induce shrinkage. Five sets of five drop-jumps from a height of 1 m, followed by rebounding over a hurdle 0.5 m high, caused a mean loss in stature of 1.74 mm (Boocock et al., 1990). On this occasion, postexercise unloading was investigated with a 20 m gravity inversion period directly following the exercise session. This inversion period caused an increase in stature of 5.18 mm compared with 0.76 mm from a standing period of similar duration. Stature was maintained for a further 40 min in which subjects stood. During the 40 min following the inversion, there was a rapid loss in stature of 4.07 mm. For the session involving standing postexercise, this same period caused only a little alteration in stature—0.04 mm. It was noted that 30 min into this 40 min recovery period there was no significant difference for stature alterations between the two experimental conditions. It was again concluded that the effects of unloading are only short lived. In some people a rapid regain in stature on unloading can adversely affect the dynamic response characteristics of the spine. If major exertion is undertaken immediately after the rest period is ended, a brief warm-up may be advisable.

Practitioners have advocated adding external resistance to stretch-shortening exercise to maximize the training stimulus. Fowler and colleagues (1994) compared the effects of drop-jumping with an 8.5 kg load added in a weighted vest. Shrinkage of 0.62 mm in unloaded drop-jumping was increased to 2.14 mm when the weighted vest was added. Furthermore, the rate of force loading rose from 20,742 N/s when the weighted vest was used. The results reflected the greater physical stress of loaded drop-jumping compared with the same exercise without external load.

Plyometric exercises such as drop-jumping give rise to high impact forces and therefore a high spinal loading (Boocock et al., 1990). Because this form of plyometric exercise is potentially injurious to the back, alternative modes of exercise that reduce this risk but provide the same stretch-shortening stimulus for muscle training are desirable. The pendulum swing provides such an alternative, whereby the athlete is positioned seated in a swing and directly in front of a vertical rebound surface. The athlete swings forward and backward on the pendulum, rebounding against this vertical surface. Fowler and colleagues (1994) showed that the device offers a significant training stimulus. Later the same group showed that the pendulum swing reduces the loading of the spine compared with drop-jumping exercises (Fowler et al., 1995). Given the lower shrinkage results and lower peak forces, it appears that the pendulum exercises pose a lower injury potential to the back than do drop-jumps performed from a typical height of 28 cm.

Other Applications to Physical Activity

Many types of physical activity, whether in recreational or occupational contexts, have a high prevalence of back pain. Common among these is loading on the spine, irrespective of whether imposed by manual handling, weightlifting and carrying, twisting, or working too long in an inappropriate posture.

Golf is a recreational activity in which players carry their golf clubs around the course. Wallace and Reilly (1993) simulated an 18-hole round of golf in a laboratory study. Three conditions investigated were walking the course without playing, walking and playing (without bag), and walking and playing carrying an 8 kg golf bag. The walking condition caused a smaller spinal shrinkage (3.58 mm) than did playing (4.98 mm) and playing combined with carrying the golf bag (5.82 mm). It was suggested that the high incidence of low back pain in golf players may be associated not only with compressive loading but also with high shear forces produced during the golf swing.

Spinal loading is implicated in back injury in cricket. Reilly and Chana (1994) used spinal shrinkage to identify specific consequences for the spine of fast bowling. Bowling every 30 s for 30 min caused shrinkage of 2.30 mm compared with 0.29 mm when a run-up without a delivery was used. The delivery rather than the run-up was found to be the main cause of spinal shrinkage in cricket bowling. A gravity inversion regimen preexercise was found to have a likely protective role in such practice conditions.

Field invasive games such as hockey make unique physiological and physical demands on players. Playing and dribbling the ball are usually executed in a position of spinal flexion. Evidence of the physical strain on the spine during field hockey was provided by Cannon and James (1984), who reported that over a 4-year period 7.6% of patients referred to a clinic for athletes suffering from back pain were hockey players. Reilly and Seaton (1990) observed an average shrinkage rate of 0.4 mm/min in players dribbling a hockey ball in a laboratory simulation, a value greater than previously reported for other activities. The investigators concluded that the peculiar postural requirements of the game caused physiological strain (indicated by oxygen consumption and heart rate) and spinal loading in excess of orthodox locomotion. Later, Reilly and Temple (1993) demonstrated that an enhanced crouched position when dribbling accentuated the subjective and physical strain on the spine. Their observations suggested that the strength of the back muscles may have a protective function in such conditions.

Spinal shrinkage has been measured in occupational as well as sports contexts. In view of the responsiveness of spinal shrinkage to load carrying, the technique has been used to evaluate new mail-bag designs for postal deliveries. Parsons and colleagues (1994) compared three new designs with the existing pouch mail-bag in laboratory-based and field trials. The investigators based their assessments on spinal shrinkage combined with biomechanical, physiological, and perceptual (subjective) responses. The combination of techniques was useful in interpreting the overall results and in highlighting the particular benefits of the individual designs.

Many current guidelines for lifting in industrial work are tailored to static and sagittally symmetric postures, yet the majority of tasks associated with manual materials handling have asymmetric components. There is evidence that low back disorders are related to lateral bending, axial twisting, and awkward postures (Marras et al., 1993). Au and colleagues (2001) analyzed the spinal shrinkage attributable to repetitive exertions confined to each of the three separate axes (twist, lateral bend, flexion). The experiment was performed twice with small technique modifications in the twisting task (and thus two data collections were performed). Subjects performed each task for 20 min at 10 repetitions per minute, where stadiometer measurements of standing height were taken prior to and immediately following the 20 min exertion. The twisting task demonstrated significant spinal shrinkage (1.81 and 3.2 mm in the two experiments) but no clear effect emerged for the other two tasks. These data suggest that repetitive torsional motions impose a larger cumulative loading on the spine than do controlled lateral or flexion motion of tasks of a similar moment.

Musculoskeletal effects of aging can influence responses to compressive loading on the spine and its resultant shrinkage. Reilly and Freeman (2006) applied precision stadiometry to assess spinal shrinkage in a comparison of two age groups (18-25 and 47-60 years) completing a regimen of circuit weight training (2 sets of 12 exercises). The two groups showed a similar pattern of spinal shrinkage, loss in stature being greater for the first set compared with the second set. Subjects gained height when placed in the formal recovery posture, but responses were inconsistent during warm-up, cool-down, and active recovery. Irrespective of age, the spine was less responsive to loading as the duration of exercise increased. The authors concluded that, provided loading is related to individual capability, healthy older athletes are not necessarily compromised by their age in lifting weights.

Physical Loading

The notion of physical load implies the incurrence of a physical challenge to the individual. The load can be quantified via biomechanical or physiological measures, which range from the external forces produced to internal biological responses that are provoked.

Ground Reaction Forces

Ground reaction forces require the subject to move over a force platform, the forces in three planes being monitored by specialized piezoelectric sensors. Force platforms are commonly installed in biomechanics and gait laboratories. They are also used in national institutes of sport, embedded in runways for sprinters, jumpers, and gymnasts, for example.

The ground reaction force trace provides useful information not only on the magnitude of the vertical force but also its time course. This trace indicates the reaction force applied to the body on contact with the ground. The running action entails two spikes in the track, one for forefoot landing and one when the heel comes to ground. Higher forces are recorded in training activities such as drop-jumping and landing on the first two phases of the triple jump.

When the foot contacts the ground, the ground reaction force of a runner may exceed 2.5 times body weight. This force is increased when the surface is hard. Sport shoes have built-in cushioning components to reduce the effect of these forces, although the cushioning properties may not last long.

Electromyography

The engagement of specific muscles in sports can be quantified using electromyography (EMG). It is more common to use surface electrodes than the indwelling varieties given the more invasive procedures associated with the latter. Electromyography has been used to analyze skiing turns, running actions, and swimming techniques (see Clarys et al., 1988). It has also been used to compare the drop-kick and the punt kick used in Rugby Union and Gaelic football and by soccer goalkeepers (McCrudden and Reilly, 1993).

Increase in the amplitude of the EMG is associated with increased intensity of exercise. The temporal sequence of individual muscles is an important part of skilled actions. Therefore, information from EMG profiles can help trainers locate and correct technical faults in sport actions.

In their description of EMG applied to sport ergonomics, Clarys and colleagues (1988) referred to uses of both on-line and telemetric EMG systems. They used EMG to evaluate dryland swim-training devices by comparing their EMG curves to those observed during real swimming. The investigators also explored how different ski materials influenced muscle activity. They claimed that EMG offers practical applications in sport and exercise but is only one measurement tool among others.

There is a possible relationship between the increased tension in a muscle and the amplitude of the EMG signal recorded. The amplitude does not directly match the buildup of tension in an isometric contraction. When many muscles cross the same joint or cross multiple joints, the relationship of force to amplitude is more questionable. Isometric muscle action generates less muscle activity than concentric actions when working against equal forces. With fatigue there is a loss of the high-frequency component of the EMG and a decreased median frequency in the signal.

Electromyography has traditionally been used as a tool in industrial ergonomics. The technique has particular value in identifying muscles unduly loaded because of faulty working postures. It may also help in identifying causes of fatigue in such cases.

Work Rate Profiles

Work rate profiles can be used to describe the intensity of activity during games. The various means of obtaining such player profiles were outlined when movement analysis was considered in chapter 5. Alternative approaches have included manual methods, filming individual players, or placing 6 to 8 cameras in strategic positions so that all individuals can be studied after the game. The multicamera systems are used by many of the top soccer teams in the major European leagues, providing detailed feedback for the players and coaches.

The severity of play can be indicated by the total distance covered by the player, energy expenditure being related to the overall distance. It is also possible to focus on the high-intensity efforts, because these bouts represent the most vigorous activities of the game. The durations of recovery can be gauged, and first-half and second-half performances can be compared to establish the extent of fatigue. The work–rest ratios provide a model for planning and modifying the training program.

In more recent years, global positioning systems (GPS) have found use in both training and competitive contexts. The first such system was designed in the late 1970s by the U.S. Department of Defense, and the technology has proved to be reasonably accurate in calculating the position of a person at any one time. The cost and size of the monitoring device, worn as a sling around the athlete's back, make it affordable to professional sports clubs. Differential GPS improves the accuracy of measurement, and the training device can incorporate an accelerometer to record impacts as well as time and location. This approach is useful in contact sports where the frequency and magnitude of hits on the opponent can be recorded. Current systems are not precise in measuring accelerations, given the frequency, typically 1 Hz, with which the location coordinates are monitored.

Psychological Loading

Psychological load implies a gamut of nonphysical sources of challenge to the active person. As an entity it ranges from cognitive and mental functions to attention and emotional measures. The criterion used depends on factors such as the type of task and the objectives of the activity. It is influenced also by the circumstances in which the activity is conducted.

Precompetition

Practically all athletes become nervous before competition, the cocktail of emotions including anxiety yet anticipation, apprehension yet confidence, uncertainty yet excitement. Without motivation and arousal, the athlete will not likely attain peak performance. There is no formula for reaching the ideal mental state before competitive sport; each athlete finds individual ways of coping. Various means have been used to quantify precompetition anxiety, and some of them are now considered.

Various inventories have been borrowed from clinical psychology and applied to monitor anxiety in sport. Among these is the questionnaire designed by Spielberger and colleagues (1970), in which anxiety is divided into A-state (or transient) and A-trait, which is more enduring. Sanderson and Reilly (1983) observed 38 women before and after competing in the English women's cross-country championship and 26 men competing in a 10K international road race. Anxiety state was correlated with performance in the women and A-state was related to performance in the men. There were clear postrace reductions in A-state for the high finishers in the women's race and for the whole group in the men's race. Elevated A-state levels postcompetition suggest that coaches should direct specific attention to athletes recognized as poor losers.

Martens and Simon (1976) developed an anxiety inventory for application to sport. They claimed that it improved the predictability of anxiety states and traits in athletes. Their inventory has been modified by others in subsequent years for work in experimental as well as field circumstances.

Another approach to monitoring mental states in a variety of sport contexts is to use the mood adjective checklist of McNair and colleagues (1992). The Profile of Mood States (POMS) presents assessments of tension, depression, anger, vigor, fatigue, and confusion. Elite athletes' values tend to be lower than a normative sample on the subscales of depression, tension, anger, fatigue, and confusion and higher than a normative sample for vigor. A composite measure of mood can also be calculated by summing the five negative mood scores, subtracting the score for vigor, and adding 100 as a constant. When the profile of athletes is plotted, it is shaped like an iceberg, with vigor above the surface. This trend is reversed in the overtrained state, suggesting that mood states are sensitive to this phenomenon.

Behavioral and psychological measures have been used as indices of emotional strain in industrial as well as sport contexts. Behavioral measures include hand tremor, observation techniques, and subjective scales. Physiological measures have included electrical skin conductance, the basis of the lie detection test, and analysis of stress hormones in blood and in urine. The hormones most commonly used are adrenaline, noradrenaline, and cortisol. Heart rate is useful as an index of emotional tension when there is no physical component to the elevation above the resting value. Soccer goalkeepers have been found to have higher heart rates prematch than outfield players, and players have been shown to have higher heart rates before playing in front of a critical audience at their home ground compared with playing away (Reilly, 1979). People can habituate to stress: Reilly and colleagues (1985) reported reductions in heart rate of 15 beats/min when subjects completed a second ride on a high-acceleration track on a leisure playground.

During Competition

Psychological stress can be manifested in many ways during sports competition. It may be reflected in a wrong choice of tactics in track and field and in cycle races or a failure to implement the prerace strategy. Stress during games can lead to unforced errors, and their frequency can be examined and feedback put in place for remedial action. Inferences can also be drawn about the immediate mood state of participants from their body language or nonverbal behavior. Contestants in contact sports can be intimidated by distractive behavior of opponents. They can also be influenced by the noise from spectators, particularly if it is in the form of criticism or jeering.

The heart rate and hormonal responses are invalid measures to use during competition because both measures have large physiological components. Similarly, simulations of competition provide no real test of how athletes will cope, although simulations can be useful as preparation.

Helson and Bultynck (2004) considered the frequency of decision making among referees and assistant referees of soccer games. Top referees made 137 overall decisions on average in a match, manifested as an intervention in play. After estimating invisible decisions, the investigators concluded that referees make three to four decisions each minute. Choices are made against a background where noise from an excitable crowd and a home advantage can influence the result.

Perceived Exertion

The subjective response to exercise, in the form of the **rating of perceived exertion,** gives a good idea of the strain experienced by the participant. Borg (1982) designed a scale that ranged from 6 to 20, anchored at 7 to correspond to "very, very light" and 19 corresponding with "very, very hard." Because heart rate mirrors the subjective strain experienced, originally it was thought that the ratings and heart rate would correspond. The rating is influenced by the nature of the work—whether it is intermittent or continuous, arm or leg exercise, isometric or isotonic.

The category-ratio scale (CR-10) has replaced the original design for assessing perceived exertion. This scale can be applied to arms, legs, breathing, and whole-body sensations. In more recent years it has been used as a tool for monitoring the total load in training, referred to as the *session RPE* (Impellizzeri et al., 2006).

Overview and Summary

Stress is a holistic concept with misleading negative connotations. It is associated with reactions to uncertainty, risk, and challenging circumstances. It is a necessary corollary of competing in sport. The concept of stress has many facets, which need to be placed in context.

Whether stress is positive or negative depends on individual reactions to circumstances. An athlete can be motivated by the severity of the challenge ahead, excited rather than daunted by it, and this can raise his or her level of arousal for the encounter. Alternatively, anticipation of the event can engender anxiety and a loss of focus. Stress can be considered alongside other concepts such as activation, arousal, and anxiety when these are viewed in broad terms.

Stress is identified as the environmental agent that causes a strain response in the organism. Ergonomists seek the most appropriate methods to quantify strain. Such measures imply that the magnitude of the strain response can be quantified, whether the source is emotional, physical, or physiological. Methods can then be devised to reduce strain and instill coping strategies to improve stress tolerance.

© Tom Reilly

Sports Equipment and Playing Surfaces

DEFINITIONS

aerodynamic javelin—Field implement that was designed based on principles of air flow to optimize performance.

mathematical modeling—A formal means of representing a real-life system using quantitative assumptions.

protective equipment—Devices, products, or clothing intended to safeguard the user against harm.

sweet spot—The point on a racket whereby the player feels no shock following impact.

synthetic surfaces—Playing areas where the top layer is made of nonnatural materials such as artificial pitches.

seven

SPORTS equipment embraces a gamut of devices that are used in training and competitive contexts and range in sophistication from simple implements to computer-controlled assistive artifacts. Equipment is used to enhance human capabilities, and skill in using implements is an essential feature of sport. Impressive skills in using implements are seen in pole vault; some track-and-field activities; racket sports such as badminton, squash, and tennis; and activities such as golf and field hockey. Whether the equipment takes the form of projectiles or extensions of human limbs, its design must suit the user for performance to be safe and effective. The user also has to adapt to different designs when the projectiles are balls, as in football, tennis, or golf.

Sports also embrace different forms of locomotion over set distances or on designated pitches whether on foot or by means of specific mechanisms. Attention has been given to providing a competitive edge by the design of specific sport shoes, skis, skates, cycles, or motor-propelled vehicles. Comfort and ease of use become relevant criteria in the design of these products. Creativity has also been evident in the choice of equipment for water and aerial sports and recreation, including the optimization of yachting keels, the aerodynamics of rowing shells, and ergonomic factors in human-powered flight.

Whether the activity is on land, at sea, or in air, the interface with the environment is a relevant ergonomic concern. In this respect, sport shoes and clothing and winter sports equipment such as skis, skeleton, and skates must have protective as well as performance-enhancing functions. Criteria for clothing design will vary with the sport depending on whether contact is involved, the likely environmental conditions, and the risk and performance elements of training and competing. In view of the huge commercial interests in sports apparel, fashion is a regular factor in altering designs of shirts, tracksuits, and shoes. In particular, apparel and other items with team logos have a huge appeal for the fans and stimulate a correspondingly huge market in the leisurewear industry.

Given developments in materials science and design technology, playing surfaces are an important part of sport ergonomics. Improvements in earth sciences have facilitated developments in grass pitches, and underground heating systems have given a large degree of freedom from the vagaries of poor weather. **Synthetic surfaces** as replacement for natural grass have enhanced recreational use of tennis courts, soccer pitches, and a range of indoor and outdoor games. Nevertheless, many sports are still held on surfaces that are not tuned to the characteristics of the human body, for example, running on asphalt or cobbled stone or playing basketball on hard concrete surfaces. The emphasis in planning such facilities is often financial cost and long-term economic value as opposed to the protection of users against injury.

Technology is used extensively to develop equipment for officiating at sport contests, to enhance training methods, to improve the presentation of information to both live audiences and television watchers, and to allow precise measurements of human capacities under controlled laboratory conditions. Electronic timing devices have overcome the difficulties inherent in manual timekeeping of races and spotting false starts in running and swimming, for example. Video technology has helped in referencing decisions such as line calls in tennis and tries in the rugby football codes. Communication devices for officials and their assistants have enabled referees to make informed decisions about incidents that might otherwise be missed. Electronic registration of contact in amateur boxing has helped officials in the scoring of bouts.

The physiological and cognitive burden on referees in field games and the human factors associated with these roles are ongoing issues that await resolution (Reilly and Gregson, 2006).

Developments in materials science have played a major role in the designs of contemporary sports equipment. The production of sports equipment covers a large industry in which competitors invest a great deal of money to gain a market edge. Clever design and careful choice of material are important in the development of new products. The most advanced materials are used for recreational and elite sports equipment, often the result of their development for the aircraft and other industries. The diversity of application includes athletic equipment, bicycles, boats, canoes, fishing rods, golf clubs, rackets, skis, sports shoes, and surf boards.

Practically all sporting goods are based on fiber composite materials. Material with one set of properties (such as a polymer) is mixed with a different material with another set of properties (such as glass or carbon fibers) to yield improved equipment properties. Characteristics important in sports equipment are listed in table 7.1, although no single material can be expected to possess all of the desired properties. Advanced aluminum and titanium alloys cover a range of specifications, combining desirable properties in a way not possible with a single solid material (Easterling, 1993).

Many sports activities do not require equipment in competitive contexts, yet participants use complex hardware in their training and preparation to compete. Runners, swimmers, and soccer players, for example, find novel modes of resistance to motion to improve their strength and power. Similarly, sport scientists have developed elaborate systems for measuring biophysical and perceptual-motion characteristics so that fitness can be assessed and adaptations to training regimens monitored. Fitness assessment techniques have become increasingly refined with respect to their sport specificity. Correspondingly, laboratory methods in the sport sciences have progressed in sophistication and at the same time spill over into training contexts with the use of ergometers and virtual-reality setups for training as well as assessment. This overlap helps the transfer of knowledge from theory to practice.

Mathematical modeling can help to improve the oar blade in rowing and sculling. Caplan and Gardner (2007) modeled the relative movements between the oar blades and the water during the drawing phase of the stroke and determined the lift and drag forces generated by this complex interaction. The curvature of the blade was

Table 7.1 Important Properties of Sports Equipment

Properties	Description	Example
Lightness	Low weight per unit volume	Golf club
Stiffness	Material's flexibility or resistance to bending or stretching	Fiberglass pole vault
Toughness	Material's resistance to cracking even when exposed to sudden impacts	Bobsled
Dampening behavior	Ability of the material to absorb rather than transfer impacting stresses to the human body	Tennis racket
Fatigue resistance	Material's resistance to cyclic stress	Racing bicycle

found to affect performance, the competitive curved blade increasing boat velocity by 1.1%. The investigators concluded that the difference between blades was a significant factor in enhancing rowing performance, because races are often won with lower margins of victory.

Performance in rowing is influenced both by the design of the blade and the length of the oar. Larger and hydrodynamically more efficient blade shapes need to be rowed with shorter oars that allow rowers to improve the propelling forces without increasing the forces at the handles. Designs have evolved since new blades were introduced in 1991, replacing the traditional Macon blade by having an asymmetric shape and an increased surface area. The so-called fat blade introduced in 2004 again increased the blade surface area, requiring an overall reduction in the length of the sculling oars to about 2.80 m. These developments were reviewed by Nolte (2009) who used biomechanical models in quantifying the main forces affecting rowing performance. Manufacturers were encouraged to continue their development of blades, and coaches were advised to experiment further with shorter oars to promote continuing improvements in the performance of rowers.

It is evident that sports equipment encompasses a broad range of artifacts used in leisure, training, competition, and laboratory contexts. These areas are the focus of this chapter. Requirements in the competitive arena broadly dictate the trends and developments in the other spheres. Sports products differ from their counterparts in industry in that for sports competition to be possible and performances to be comparable on a worldwide basis, some degree of standardization is mandatory. Standards are set by national sport governing bodies, which invariably must abide by the standards fixed by international ruling bodies. These standards largely determine the constraints on designers of new equipment. Designers must operate within the limits of the specifications set by the ruling body, irrespective of whether the governing body hinders or encourages novel developments in equipment. Designers should also consider the implications of any change in product design on safety as well as performance in both training and competition.

Sports Implements

The search for better and safer sports equipment has fostered improved technology in its design. Design innovations, such as the use of fiberglass materials in sports, aid performance but sometimes require modifications (or clarification) of existing competition rules before improvement can be realized. Often athletes must alter their game skills in order to use new equipment. The original **aerodynamic javelin** enhanced performance but required a lower angle of release than the traditional missile for optimal effect. Because design features were later altered for safety purposes when throwing distances were large enough to risk damage to others in the arena, the throwing technique needed further adjustments.

Injuries can be prevented by appropriate design in that physical strain is avoided if the equipment matches individual needs and characteristics. Compatibility may require a range of fabrication within the margins of applicable standards. Age, sex, and skill become relevant indices governing design as well. When introduced, aluminum tennis rackets were believed to cause elbow injuries if players did not alter their techniques. The later steel and aluminum rackets reduced injuries compared with wood (Sanderson, 1981).

Racket properties may affect the development of lateral epicondylitis (tennis elbow). The shock and vibration transmitted to the arm of a tennis player are influenced by the location at which the ball makes impact with the head of the tennis racket as well as by the stiffness of the racket and the grip force. Hennig (2007) showed that novice players incur increased loads attributable to hitting the ball lower on the racket head compared with more experienced players. Lack of good coaching advice may therefore be a factor in tennis elbow among beginners.

The physical properties of the equipment, ease of operations, anthropometry, and subjective responses of users are consistently pertinent to performance. A reliable anthropometric method for determining the correct handle size for tennis was presented by Nirschl (1977). The formula can easily be corrected for application to other racket sports such as squash and badminton to account for differential impact forces, forearm strength, and implement mass across these sports.

The aim of racket designers is to improve the dynamic characteristics of the equipment, make the ball easier to control, and dampen potentially injurious vibrations set up by the stroke. For optimal effects, contact with the ball is best made at the **sweet spot** of the racket. This is defined as the point on the strings at which the player feels no shock following impact and at which there is a minimum of vibration when impact occurs. This latter part is known as the nodal region. The sweet spot is influenced by the damping characteristics of the racket, attachment of damping materials to the grip, and stiffness characteristics. The size of the sweet spot is therefore variable depending on the racket's construction. Contemporary rackets for tennis, badminton, squash, and racketball are no longer made from the traditional wooden and later aluminum constructions but are made of composite materials. These materials consist of multiple fiber-reinforced layers around a soft inner core with injected polyurethane foam or a honeycomb structure (Easterling, 1993).

Golf continues to attract the attention of designers in view of the large market for golf equipment at both highly competitive and recreational levels. This attention is reflected in the 28% increase in the average drive distance of U.S. tour golfers over the last 25 years (see figure 7.1). The increases prior to 1993 were in large part attributable to improved conditioning and training of golfers. The subsequent increases accompanied innovations in club design, and the impressive improvements between 2001 and 2003 were linked to introduction of more solid balls. In their review of the recent developments in golf science, Wallace and colleagues (2008) explained how a better understanding of the factors affecting the velocity of the club face helped to improve the construction of the shaft and the club head. Between 1993 and 2000 titanium-based alloys were used in hollow oversized drivers with a resultant increase in drive distance. The development was facilitated by the replacement of cast titanium–based alloys with forged metastable alloys coupled with control of the size and position of the joints between the face, crown, and sole. The outcome is that design characteristics approach the limits for coefficient of restitution set by the ruling bodies—the USGA and the Royal and Ancient Golf Club. Consequently the average drive distance has leveled off in the last 3 to 4 years.

Alterations in the design of golf clubs have not been restricted to drivers, putters also being periodically considered for redesign to enhance performance. Limb tremor induced by anxiety (and known as "the yips") can be detrimental to the skill of putting, a crucial aspect of performance whose success determines the score at each hole. The most radical change was the introduction of a lengthened shaft used

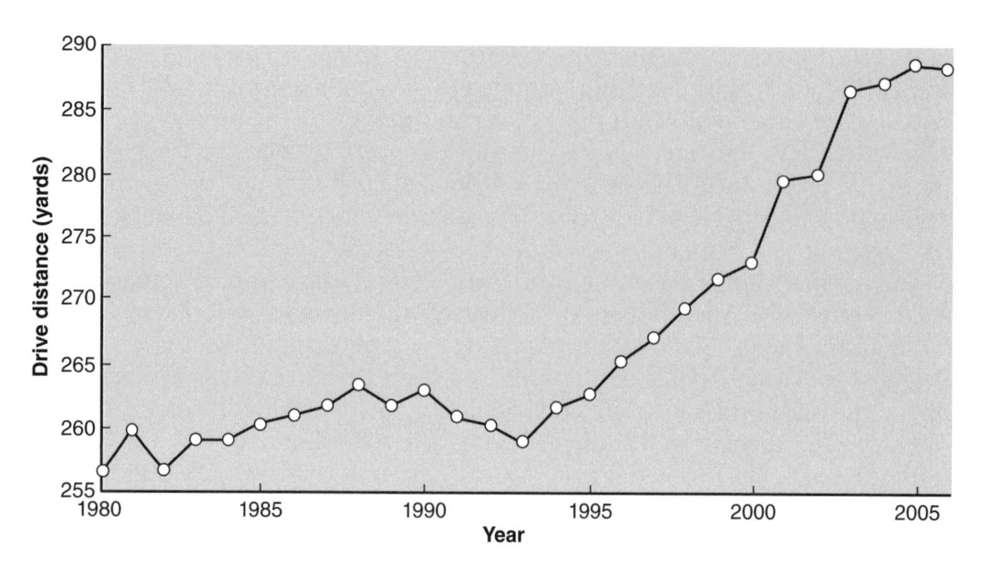

Figure 7.1 Increases in average drive distances in USPGA golfers for each year since 1980. Each point represents approximately 32,000 measurements.

This figure was published in *Science and sports: Bridging the gap*, T. Reilly, pg. 100, copyright Shaker Publishing 2008.

by some golfers to improve their grip on the handle of the shaft and enhance their stability when executing this skill. Its use necessitated a change in posture as well as alteration of putting technique.

The availability of cutting-edge analytical facilities has contributed to studies of the behavior of different designs of golf balls. The interactions of golf balls with normal and inclined plates (grooved and ungrooved) have been quantified and related to material properties and ball construction. Ranges of material properties for generating ball spin have been identified, spin generation being important in approaches to the green. Wallace and colleagues (2008) concluded that with the greater understanding of models of ball behavior, it is possible to design golf balls appropriate to the swing speeds of individual competitors.

Angling is a major recreational activity pursued worldwide. Composite materials have replaced the traditional bamboo in fishing rods. The ideal fishing rod must have resilience, a material property that incorporates elasticity, strength, and toughness and the ability to reduce vibration in the rod's shape when casting (see figure 7.2). Contemporary rods also contain line-runner rings with coated aluminum oxide on their inside to minimize resistance to the running of the line. The subjective feel of the rod is aided by incorporating layers of fiberglass weaves and carbon fiber in its backbone. The result is a finely tuned, lightweight instrument (Easterling, 1993).

Improvements attributable to advances in materials science are evident in the progress of results for the pole vault. Performance was improved immediately when the first lightweight aluminum pole replaced the traditional bamboo in the 1960s. Two decades later the introduction of polymer (epoxy) poles with reinforced carbon fiber material heralded further advances in the event. The new poles were lighter and stronger than their predecessors, possessing the right stiffness and springiness for top vaulters to maximize their performances. However, new gymnastics techniques are required to exploit the properties of the fiberglass poles. The performer is obliged to

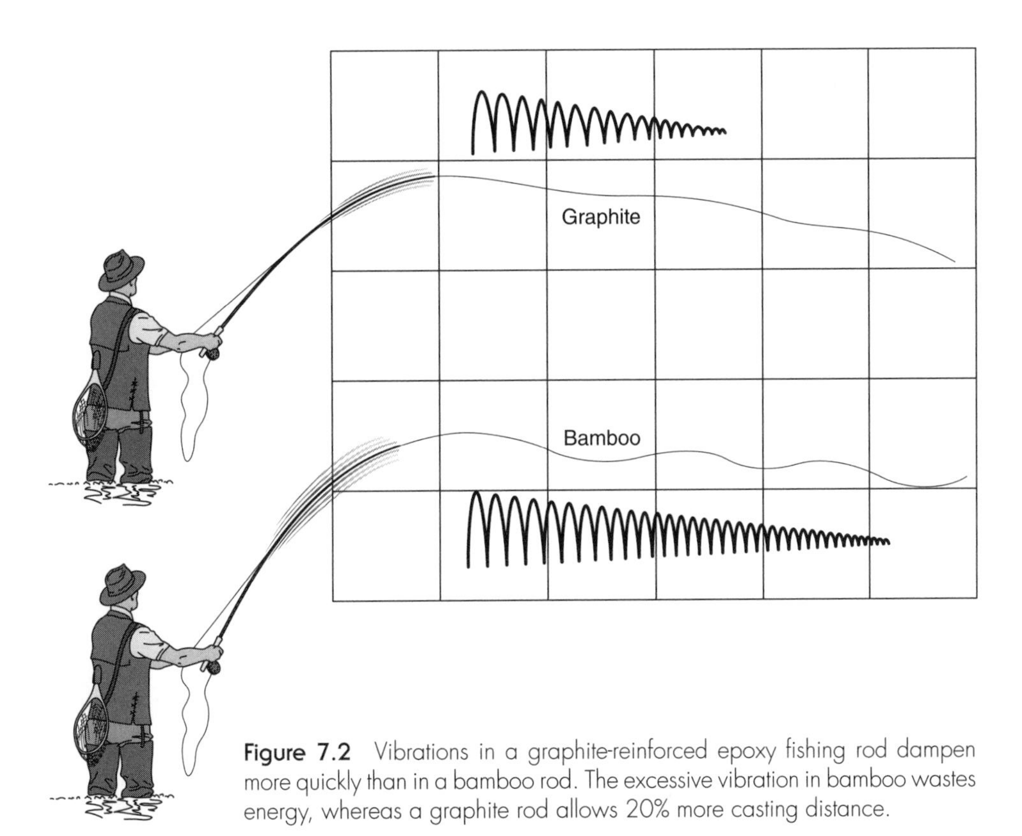

Figure 7.2 Vibrations in a graphite-reinforced epoxy fishing rod dampen more quickly than in a bamboo rod. The excessive vibration in bamboo wastes energy, whereas a graphite rod allows 20% more casting distance.

hang back more on the ascent as the pole is bent and virtually achieve a handstand to push off the pole as it straightens near the top of the athlete's trajectory.

The archer's performance is facilitated by technological features incorporated in the design of the bow, so that lack of perfection at top level now lies mainly with the human factor. The modern bow has two slot-in limbs that insert into a magnesium handle section and stabilizers that help minimize vibration and turning of the bow. Muscle strength is needed to draw the bow fully, and holding it steady requires muscle endurance; the bow is held drawn for up to 10 s while the target is being sighted. Deflection of the arrow tip by 0.02 mm at source would cause the arrow to miss the target 90 m away, illustrating the importance of hand steadiness. Before releasing the arrow, the archer pulls it toward and through the clicker—a blade on the side of the bow that aids in assessing appropriate draw length. The archer reacts to the sound of the clicker hitting the side of the bow handle by releasing the string, which in effect shoots the arrow. Archers are coached to react to the clicker by allowing the muscles to relax. Electromyography has been used in such circumstances to examine the smoothness of the loosening action (Reilly and Halliday, 1985) and the degree of arm tremor affecting performance.

Sport Surfaces

The surface on which sport is played must be considered for both training and competition purposes. The compliance of the surface should suit the characteristics of the human body if safety and performance are to be optimal. Very often, financial cost is one of the most important design criteria for indoor facilities, whereas the ability to withstand harsh weather is the important consideration for outdoor surfaces.

Determining the best flooring material for use in multipurpose facilities is an area where the materials scientist, architect, and human scientist can usefully cooperate. Sport facility floors should be reasonably resilient, should be nonslip and nonreflective, should give a true bounce and roll, should be of a color appropriate to their use, should be easy to maintain and resistant to all types of footwear used, and should be nonabrasive. Special flooring considerations apply for certain activities: Squash courts should be capable of absorbing sweat falling from players, whereas floors in weight training rooms must withstand possible damage from heavy apparatus. These features are incorporated into modern designs of gymnasiums.

Development of synthetic indoor and outdoor playing surfaces has been directed mainly by their economic and organizational advantages. The early generation of these surfaces ranged in consistency and in the characteristics of top surface and substrate. Artificial surfaces were implicated in injury occurrence (Habberl and Prokop, 1974) and were rarely laid with human characteristics foremost in mind. The indoor running track laid at Harvard University provided the first major exception. The surface covering was polyurethane with a substratum consisting primarily of wood. The compliance of the track was set to match the stiffness of human muscle established after detailed laboratory investigations (McMahon and Greene, 1978). Improved performance and reduced injury rates were thought to result, although the evidence was not conclusive. The majority of tests currently used for artificial turf are more related to materials science than human factors criteria: Tests of resilience, stiffness, and friction should be included as a matter of course for determining effective interfaces. For natural outdoor game surfaces, soil penetrability tests would prove useful in choosing appropriate footwear.

The design of artificial surfaces for soccer has improved considerably since the first generation of synthetic pitches. This improvement has led the world governing body to approve artificial pitches for competition purposes. Ekstrand and colleagues (2006) compared injury risk in elite soccer played on artificial turf with natural grass. They found no evidence of an increased risk of injury when games were played on modern synthetic pitches. The risk of ankle sprains was increased on artificial turf, but the number of ankle sprains was low. The friction characteristics of the surface and the footwear chosen for use on it are also relevant.

Third-generation turf, which consists of a sand- and rubber-infilled structure, has begun to find favor among game legislators in soccer. Several types of infill have been developed, aimed at minimizing injury rates and enhancing performance. The shock-reduction property of third-generation turf has been related to injuries incurred. This applies to running and soccer-specific actions like jumping and landing sequences, heading, and goal-keeping actions. The effect of the impact of a player on a surface may be examined in several ways. In the past, playing surfaces were mostly tested by means of mechanical tests, which have the advantage that they show a high repeatability. For artificial soccer pitches, FIFA (Fédération Internationale de Football Association) uses the "artificial athlete" for shock absorption, simulating the impact of a person during a vertical jump. The device consists of a mass falling onto a spring that rests on the experimental surface: Measurements are made via a load cell and test foot. The force recorded is compared with the maximum force measured on a concrete surface. This ratio is known as the force reduction and is used as a measure for the shock absorption of the surface. An alternative is to conduct biomechanical tests with actual players, which are more readily acceptable because they provide better external validity and are a valuable addition to the current test procedures.

Some sports are played on a wide variety of surfaces. Tennis was traditionally played on lawns, a fact that contributed to its formal name (lawn tennis). The game is now played year-round given the widespread availability of synthetic courts in indoor facilities. Outdoors, the game is played on clay, synthetic courts, and grass. Among the major tennis championships, only the Wimbledon event is now played on grass and only the French championship is played on clay.

Changes in surface characteristics cause players to adopt different strategies, and athletes playing mainly on one type of surface may find it difficult to play on another. O'Donoghue and Liddle (1998) compared performances in women's singles on clay and on grass, finding more points won on serve and at the net on grass than on clay and more baseline rallies on clay-court competition. Surface characteristics are more versatile on grass courts, particularly when they are exposed to the vagaries of weather.

Sports Clothing

Clothing ensembles in occupational settings are subject to material standards. The influence of clothing is affected by various factors that include insulation for protection against cold and heat, vapor permeability or capacity for heat loss, air permeability, vapor resistance, and protection from penetration of pollutants. Liquid protection against chemicals and waterproofing for repellence of water and rain are also important properties, as is fire protection for motor racing drivers. The visibility of the garments and their mechanical properties are also relevant. In outdoor conditions the solar absorptivity of clothing is relevant, although this factor is not included in indices such as WBGT (wet bulb and globe temperature) in measuring environmental heat stress.

The appropriateness of the clothing worn for sport participation has been a neglected aspect of its design, because fashion, facilities, and market forces have outweighed ergonomics criteria. The use of size indicators in clothing accommodates inherent differences in participants, but there are often quite radical differences between sports. Loose-fitting clothing is often used in hot climates to keep the microclimate next to the skin cool. The dynamic air exchange, or pumping effect, keeps the area beneath the clothing cool by means of convection and evaporation. Exposure of the skin surface for evaporative cooling may be important for endurance running. Tightly fitting clothing is preferable for enhancing aerodynamic properties of the body in cycling, sprinting, and downhill skiing, for example.

Design of clothing for sprinters has used information from wind-tunnel tests to reduce drag, with the anticipation of improved performance. A whole-body garment was used by Cathy Freeman when winning the Olympic 400 m gold medal at the Sydney Olympics in 2004, although the added value of the latter in terms of energetics is considered marginal. Similar principles have been incorporated into clothing worn by swimmers and ski jumpers. For this latter group, attention has been given to the appropriateness of the traditional ski-jumping boots when extraordinarily high power output must be generated by the jumper at takeoff (Virmavirta and Komi, 2001).

The design of swim clothing has progressed from traditional trunks for male competitors and single (one-piece) suits for females. Mollendorf and colleagues (2004) examined swimsuits varying in body coverage from shoulder to ankle, waist to ankle, and briefs. They measured passive drag at different towing speeds during starts and push-offs in a swimming pool. They concluded that it is possible for body suits that cover the torso and legs to reduce drag and improve performance of swimmers. In a later study, Chatard and colleagues (2008) demonstrated that swim performance over six distances from 25 to 800 m was improved by 3.2% on average when normal swimwear was replaced by a full-body or waist-to-ankle fastskin suit. The gain was greatest with the full-body suit, attributed to a reduction in passive drag, lower energy cost, and a greater distance per stroke. Individuals without access to the new designs of whole-body suits for training might be at a disadvantage in competition. These types of swimsuit formed the majority of those worn at the 2008 Beijing Olympics even though a sizeable proportion of competitors used the more traditional designs. Nevertheless, the advantage of swimsuit technology to reduce hydrodynamic drag has been emphasized by more than a hundred world records achieved by competitors in swimming in the first 12 months of its introduction. Obvious disadvantages are the costs of the suit and the time taken, about 15 minutes, to don it. Six months after the Beijing Olympics, the international governing body FINA clarified the rules about design of swimsuits, specifying that swimsuits must not cover the neck or extend past the shoulders and ankles. The Federation reaffirmed its intention to continue monitoring the evolution of sport equipment with the main objective of keeping the integrity of the sport.

Special clothing may be needed to combat the specific hazards presented in some sports. Motor racing suits may need to offer cooling as well as fireproofing because of the heat stress and risk of fire involved. Many machine sports also require pit staff and drivers to wear ear protectors because of the high noise levels experienced. Wet suits for aquatic sports enable users to tolerate sustained periods of immersion in cold waters. Development of novel fibers has improved protection against wet and cold conditions outdoors while permitting sweat to flow through the garment (Holmer and Elnas, 1981).

Survival time in ocean temperatures not quite ice-cold is increased by wearing dry suits or wet suits. Dry suits are designed to keep the body dry, whereas wet suits allow a minimal amount of water through the material; the water is then heated by the body and, after equalizing with skin temperature and forming part of the boundary layer adjacent to the skin, prevents further loss of heat from the skin surface. Wet suits are usually made of closed-cell neoprene to a thickness of 3 to 6 mm and a close fit is needed for effectiveness. Suits that cover the arms are most effective because more heat is lost from the arms compared to the legs when each limb is exercised at the same oxygen uptake. The time of useful consciousness in water temperatures of 5°C can be extended threefold compared to wearing normal clothing by the use of a neoprene wet suit 5 mm thick but the time is increased by a further 100% if a dry suit is worn with dry underclothing (Reilly and Waterhouse, 2005).

The study of protective garments in a variety of extremes in sports and industrial contexts, such as on the mountains or deserts or in accidental immersion in water, is still a rich vein of ergonomics research. There is a growing demand for merino wool garments, normally used by mountaineering and skiing groups, as a wicking layer. It promotes evaporation of sweat, enhances thermal comfort, and does not smell afterward—a marketing claim for après ski contexts. Comparatively little attention is given to the added value of gloves and headgear in extreme conditions where choice is largely based on subjective evaluation of prevailing environmental conditions.

Sports brassieres have replaced the conventional fashion bra for females competing in track-and-field athletics, road running, and games such as football, squash, and tennis. The original "jog-bra" was designed to reduce movements of the female breast during locomotion and decrease pain and discomfort. Such problems included "jogger's nipple," an irritation also experienced by male runners attributable to chafing from their clothing. A stretchable absorbent fabric such as Lycra is commonly used in sports brassieres. The products are made either with encapsulation molded cups or compression designs that limit motion by flattening the breasts. Their features are incorporated into the running tops worn by some distance runners and triathletes without an accompanying shirt. A concern addressed by Bowles and colleagues (2005) was that sports bras were too tight and restricted breathing. The investigators observed no effect on respiratory function for subjects who wore a sports brassiere, which was superior to a fashion bra and a no-bra condition. The investigators recommended that active females wear a sports brassiere to reduce breast movement and related breast pain. In view of individual differences in size, a proper fit is important. Encapsulated bras are more suitable for large-breasted joggers, whereas compression bras are preferable for the majority of runners. The superiority of the compression bras was demonstrated by White and colleagues (2009) who reported the least discomfort with the compression design. Both sports bras were more comfortable than an everyday bra, while wearing no bra was the most uncomfortable condition. In their kinetic evaluations, White and colleagues demonstrated the importance of curtailing mediolateral, as well as vertical, displacement of the breasts to provide female runners with sufficient support for their performance and comfort during their runs.

Compression garments have been promoted for use in sport as well as other contexts. Compression stockings are commonly used by airline travelers to reduce the risk of incurring deep-vein thrombosis. In sport, compression clothing has been designed to improve recovery following exercise and training. Although this fashion has gained acceptance among professional athletes, the physiological mechanisms for any positive benefit are not clearly established. A similar concept applies to the tight-fitting shirts worn by a number of the teams in the finals of the 2007 World Cup for Rugby Union, with claims of increasing energy levels through transfer of ions to the body. It is unlikely that such interventions determine team success at this level of competition.

Sport Shoes

The impetus to improved design of sport shoes was provided by the large market for running shoes as road running became a major recreational activity around the world. The improvements were largely attributable to improved vulcanization technology for treating rubber and availability of synthetic material such as ethylvinylacetate (EVA). This polymer is a plastic foam consisting of ethylene (providing moldability), vinyl (providing resilience), and acetate (providing strength and stiffness). These developments enabled shoe manufacturers to design shoes with better shock-absorbing properties for training and lightweight shoes for racing.

The new designs included a midsole material for cushioning the forces incurred on impact with the ground. This effect was first achieved by constructing the midsole of polyurethane that encapsulated layers of other material. Rear foot control was an important factor in preventing injuries, mediated with a heel counter. For individuals

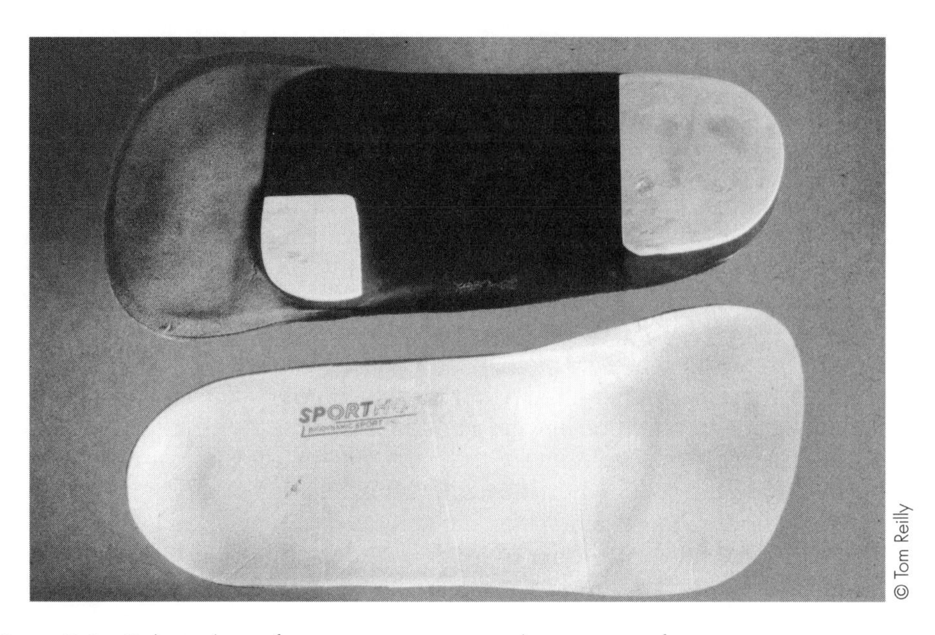

Figure 7.3 Orthotic device for insertion into running shoe to correct for overpronation.

with a tendency to overpronate on landing, antipronation elements were available. The alternative is to use an orthotic device next to the insole (see figure 7.3). The next factor to consider was the fit and comfort of the shoe, a decision made by the purchaser. Wear and functional resistance are important properties for the outside, the part of the shoe that makes contact with the ground. Rubberized materials with good tread patterns have been preferred for this component of the shoe.

The human foot is shaped to provide weight-bearing capacity and to absorb the large forces associated with bipedal locomotion. Differences in foot shape have been found between ethnic groups, and it is assumed that these variations affect shoe choice. Hawes and colleagues (1994) showed that unique foot lasts (models of the foot) for Asian (Japanese and Korean) and Caucasian populations were required. There were significant differences in these populations in the height of the hallux, the shape of the anterior margin of the foot, and the location and angularity of the metatarsal-phalangeal joint axis. Variations within populations are noted in the pattern of toe length, height and arch of the dorsum, ratio between foot length and ball and heel breadths, shape of the heel, and shape and angle of the toes. These variations suggest that one shape of last does not fit the shape of the human foot very closely, with consequences for sport shoe design.

The shoe upper provides the physical link between the foot and the sole of the shoe, giving support and protection to the foot and ankle, contributing to control of foot temperature and humidity levels, and determining the comfort of the fit of the shoe. Comfort of fit is a major factor in the purchase of shoes and is related to the match of shoe shape to the shape of the foot. The shape of the shoe is determined by a last, which is a smooth model of the foot around which the upper material is stretched and stitched or glued to the sole. Typically the upper was made of leather, which is stretchable, and molded with nylon, which is light and "breathable" but does not stretch. Contemporary designs include synthetic leathers, special blends of polyurethane and polyester, or breathable fabrics. Leather was conventionally used in soccer shoes, whereas the lightness of nylon was preferred for running shoes.

Runners and joggers are highly dependent on the quality of their shoes, being especially vulnerable to foot and leg injuries caused by repetitive pounding on hard surfaces. Orthotic inserts to correct for biomechanical malalignments of the lower limbs caused by leg length imbalances, Morton's foot, genu valgus or genu varus (knock-kneed or bow-legged), and overpronation are effective in preventing injury recurrence. Similarly, the use of a cant, inside or underneath the sole of a ski boot, can compensate for anatomical imperfections that would otherwise impair skiing performance.

Discrete laboratory trials have provided the basis for prescribing shoes with elevated cushioned heels and pronation control features to distance runners. Richards and colleagues (2009) questioned this evidence base for prescribing running shoes with elevated cushioning and pronation control features, citing the lack of controlled clinical trials. This absence of preventative studies highlights a gap between the laboratory experiments and field trials quickly marketed by shoe companies and accepted by customers and the prospective epidemiological studies that convince sports medicine specialists.

Soccer shoes are based on leather constructions, generally cut below the ankles, with a hard outsole to which studs are attached. Thin soles provide the shoes with their flexibility, and their hardness allows studs to be attached for firmness. The studs may be detachable to suit different surface characteristics or molded as a part of the shoe. The configuration of the studs varies with manufacturers, but the distribution of pressure should not place undue stress on the metatarsal heads. Most soccer shoes have a foam insock to aid comfort and fit. According to Lees and Lake (2003), the design of the shoe should be related to the demands of the game, protect the foot, and enable the foot to perform the functions demanded of it.

The principle of designing the sport shoe for the demands of the sport has extended to other sports besides running and soccer. Traction is an important criterion for tennis players, and contemporary tennis shoes are a major advance on the tractional "pumps" with canvas uppers. Golf shoes have spikes, a variation on the spikes worn by sprinters, to cope with the large rotational forces when driving the ball. Similarly, shoes for hill walking are designed for purpose, requiring more sturdiness than a normal shoe. Attention has also been given to improving the design of ski boots and their bindings and establishing thresholds for the release from bindings in the case of an accident.

Protective Functions of Sports Equipment

Equipment design can help prevent injuries in three main ways. First, quality control in production minimizes the risks that would be imposed by faulty equipment. Second, risk is avoided if the equipment meets the needs and characteristics of the user: Compatibility may require a range of fabrications within the limits of governing body standards. The physical properties of the equipment, its ease of operation, anthropometry and subjective responses of users, and their age, sex, and level of skill are relevant factors in perfecting the design. Third, effective and comfortable protective equipment can cushion individuals against harmful impact or environmental influences. Guidelines for protective equipment for sport participants are included in

British Standards, BSI PH/3/9, and there are corresponding international standards as well as North American guidelines.

Typical injury patterns are generated by specific sports so that the dominant risks associated with activities can be predicted. Shoulder pads are worn by steeplechase jockeys, for example, because of the high incidence of shoulder injuries in the sport. Various items of **protective equipment** are available for participants in field games, although attitudes must promote their regular use for effectiveness. American football players use helmets, face guards, shoulder pads, and groin and knee protectors, the designs of some of these units depending on positional role. The effectiveness of this equipment depends on the fit to the individual player and the condition of the units. Epidemiological data confirm the reduction of injury incidence since the use of protective clothing became mandatory in American football.

The head is one of the most vulnerable regions of the body in sports, and injuries to it can be critical. Safeguarding participants from head injuries is related to three main factors—skull deformation, intracranial pressure, and rotational motion. Effective headgear reduces the acceleration on the head when it is hit and attenuates compressive forces. Professional boxers wear protective helmets when practicing, yet these are not mandatory in competition despite the frequency and consequences of severe head injury in the sport. Headgear was first worn by U.S. amateur boxers in international contests, following the earlier example of Scandinavian nations—helmets were first used at the Olympics in the Los Angeles competition in 1984 (figure 7.4). None of the bodies ruling professional boxing have advocated the use of protective headgear, although the properties of boxing gloves are controlled with a view to safety.

Helmets are designed and worn for protection in a variety of recreational activities from skateboarding to tobogganing, and they are essential for human–machine

Figure 7.4. Helmet used by Olympic amateur boxers in training.
Lionel Preau/DPPI/Icon Smi.

sports. Special helmets with ear holes were developed for hang gliding so that the pilot can sense air flow for accurate adjustment of aerial speed. Head protection inevitably must include eye defenses, a consideration in sports such as squash where the ball is sufficiently small to enter the orbit. The open-eye type of protector without a lens insert, which is commonly used in squash and racquetball, provides inadequate defense against the ball, even at velocities as low as 22 m/s. Consequently, the establishment of a standard for eye protectors in the United States was welcomed (ASTM Standard F803).

Where ergonomics is applied to the design of protective equipment, other dangers must not be created in its use. If protective helmets are used for butting an opponent in American football or are aggressively handled by an opposition player, the increased leverage may cause neck injuries that might not otherwise have happened. Indeed, not all of the circumstances surrounding the use of protective equipment can be foreseen. The problem of providing head protection in sports such as ice hockey is of such a magnitude that a helmet alone should not be expected to offer complete protection because of the speed and physical contact associated with the sport on an iced floor with surrounding boundaries, and thus modifications of the playing environment may need to be addressed.

Gum shields protect athletes in contact sports by attenuating the force of a blow to the jaw and shielding the teeth. Mouth protectors for use in field games can be individually molded from a model of the wearer's teeth and have been found to reduce cranial, facial, and dental injuries. These are preferable to the conventional type of gum shield used by boxers and purchased off the shelf, the individually molded models now widely used by rugby players, for example.

Rugby Union has a relatively high rate of injuries attributable to physical contact, and players wear protective equipment to mitigate these injuries. Marshall and colleagues (2005) provided supportive evidence for a role of mouth guards and padded headgear in preventing facial and scalp injuries, respectively. Support sleeves had a positive effect in preventing sprains and strains. The use of shin guards was considered ineffective, at least in this sport.

Personal Protective Equipment

Personal protective equipment is designed to protect employees from serious workplace injuries or harm resulting from contact with chemical, radiological, physical, mechanical, electrical, or other workplace hazards. This type of equipment includes face shields, safety glasses, hard hats, and safety shoes as well as devices and garments such as goggles, overalls, gloves, vests, earplugs, and respirators.

Personal protective equipment is mandatory in occupations such as firefighting, defined military scenarios, professional diving, and building and construction work. Protective equipment is also required in motor cycling, motor racing, scuba diving, amateur boxing, and other sports and recreational activities. The scuba diver is reliant on the proper functioning of the breathing system and protection against hypothermia by the wet suit (or dry suit) worn. Protective devices such as shin guards in soccer, scrum caps in rugby football, and helmets in hurling are worn voluntarily. Just as personal protective equipment is sometimes not worn in industrial settings to comply with the traditional workgroup's image of being strong and self-reliant, sport participants often eschew protection to maintain an image of fearlessness.

Overview and Summary

Technological developments for industry have been exploited by designers and modified for sports equipment and sport surfaces. These developments have become possible with the introduction of lightweight and durable material for sports implements, equipment, and machines. In many cases these innovations have induced subtle changes in the execution of skills and the handling of equipment as performers adapt to the new designs.

Aerodynamic considerations have caused design changes in competitive sports apparel and machines. Drag reduction has become a criterion in clothing for sprinters, speedskaters, and cyclists, among others. Drag reduction also has been relevant in the configuration of racing bicycles and racing cars and in the posture of cyclists and ice-skaters. Principles of fluid mechanics have been relevant in the design of racing boats, rowing shells, and surf boards as well as in understanding swimming techniques.

Sport shoes and clothing are subject to fashion as well as design for comfort and performance. Use of ethylvinylacetate in production of sport shoes has improved their protective properties, as have synthetic fabrics in clothing. These developments have occurred without compromising aesthetic factors, given that sport shoes have found their way into general use for casual wear.

Visibility of clothing is an important safety feature, especially among cyclists and runners using the public roads. Color can have performance implications as well as aesthetic consequences. Success over many years in the top soccer league in England has been associated with wearing predominantly red strips, whereas yellow and gray jerseys have had the opposite effect. Goalkeepers facing a penalty taker who is wearing red tend to be more confident about saving the shot than when facing less conspicuous colors. In combat sports such as judo, the alternatives of blue or while seem to offer no advantage to either competitor because the judges tend not to favor either color.

Recreational users have benefited from improved design in various ways. Sailing crafts have been designed for portability, allowing families to transport their boats more easily to different locations. Lightweight cycles can be packed for transport to be used in holidays overseas. Tents can accommodate individuals or whole families overnight during outdoor activities. Such possibilities promote activity that has positive effects on health and well-being.

Ergonomics in Physical Activities

PART III focuses on the applications of ergonomics in physical activities. The constituent chapters consider fitness for work, special populations, and clinical aspects of ergonomics, followed by a holistic and nutritional perspective. Applications to work and sport are covered in a balanced manner according to the evidence available.

Chapter 8 considers occupational stress and different ways of monitoring this stress, such as absence records, fitness deficits, and psychological, psychophysical, and physiological methods. Specific occupations that entail hard physical work are targeted: firefighting, police work, prison work, military services, manual materials handling, motor vehicle driving, and delivery services. Equal opportunities legislation is considered in the context of minimal fitness standards for employment in services and special forces. Attention is directed to the special problems of nocturnal shift workers, especially with organization of their leisure activities. The relation between fitness and performance at work is explored and health promotion in the workplace reviewed in the context of the concept of behavioral change.

The focus of chapter 9 is special populations. The chapter begins by discussing the unique requirements of young people, the ergonomic aspects of talent identification and development being relevant for consideration. Other groups include women, elderly people, and athletes with disabilities. The unique demands, both cognitive and physiological, on match officials are reviewed to conclude the chapter.

Chapter 10 reviews clinical applications of ergonomics. The phenomena of overload and fatigue are analyzed; the physiological basis of fatigue is outlined and means of offsetting fatigue considered. Measurement of musculoskeletal loading is described, specifically the applications of precision stadiometry for assessment of spinal shrinkage. Spinal loading and creep characteristics of spinal responses are described for different sport and training conditions, and gravity inversion systems are discussed in context of restoring normal homeostasis. There is an emphasis on the need to accelerate

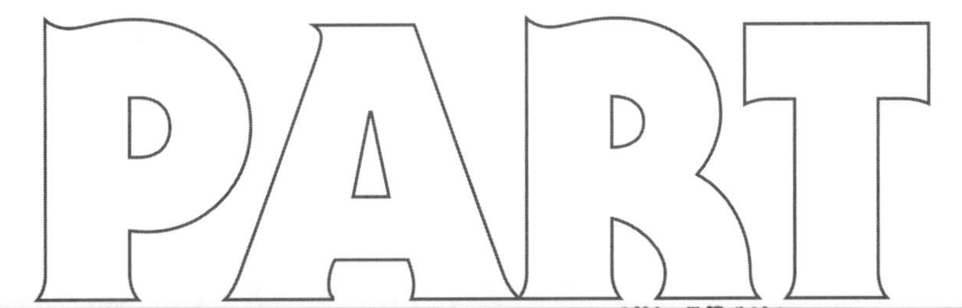

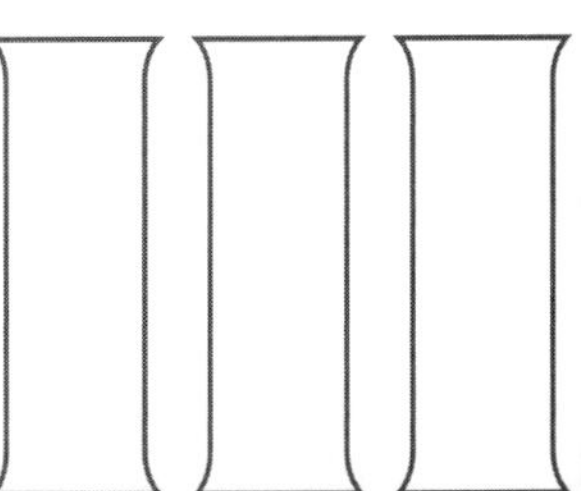

the recovery process following sport competition, and various methods for restoring energy, hydration status, and immune function are described. Alterations in training on recovery days (e.g., deep-water running) are also considered. Attention is directed to principles of rehabilitation and avoiding overtraining. The chapter culminates in coverage of rehabilitation and use of methods to reduce injury risk.

The final chapter in the book, chapter 11, introduces a holistic and nutritional perspective. The emphasis is on considering the whole person in the context of well-being. The text discusses the use of diet in promoting a healthy lifestyle and improving performance and highlights dietary practices known to enhance performance. The nutritional supplement market is discussed, and attention is given to the use and misuse of drugs in both athletic and occupational circumstances as well as for leisure purposes.

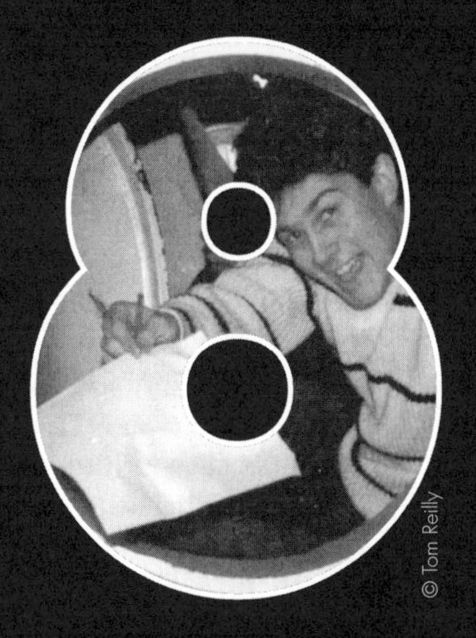

8 Fitness for Work

DEFINITIONS

aerobic deconditioning—Loss of endurance fitness attributable to a lack of training.

body mass index (BMI)—Measure that estimates fitness by considering body mass in relation to height.

nitrogen narcosis—Symptoms associated with accumulation of nitrogen in the tissues during underwater activities.

surface decompression—Procedure for reducing density of air bubbles, usually in a chamber that compresses the gas until it is removed from the body.

test validity—A requirement whereby the property being measured is the one intended.

20 m shuttle run—A field test of endurance performance, designed to estimate maximal oxygen uptake.

A key element of implementing an employment test is to define what the worker needs to do. A task analysis is a crucial aspect of test validation, because this analysis identifies what the worker must be able to do. Another key element is the cut score. This defines the level of fitness a worker must have to be able to meet the demands of the task; this is, it matches the worker to the physical demands of the job.

The nature of the work is a key factor in people's choices about their fields of employment. During people's working lives, about one third of each day is directed by their employer's requirements, whether in the military service, manufacturing, or other industries. Although material rewards may be a high priority for many in determining their occupational choices and professional areas of specialization, other factors of importance include the tasks and challenges to be faced, responsibilities and opportunities for personal and professional development, the social environment, and a host of other job aspects. The physical components may differ with each post as may the psychological stresses to be tolerated. Fitness for work implies an ability to cope with all of the demands imposed by work, encompassing its physical, physiological, psychological, and social elements.

Even when the physical components of different jobs are equal, the environment in which the work takes place and the social milieu may discriminate between jobs. In some workplaces, work processes and their organization are optimized so that individual stresses are reduced. Morale among the workforce, irrespective of the physical demands of the job, is important, particularly with respect to relationships between workers and management. Bullying and harassment from peers or from superiors constitute a source of stress for a number of people; in a sport context, bullying behavior by a coach, manager, or team member is a stress that can hinder cooperation and impair performance. It is essential that such causes of discomfort be identified and order restored if the workforce as a whole is to function effectively. This concept of work is holistic, incorporating the mental, physical, and physiological capabilities to operate at the required competency levels in the context of the work concerned.

The suitability for work is normally determined by selection processes that culminate in an interview. The applicant responds to the job advertisement by matching his or her qualifications, experience, and dispositions to the job specification. Many occupations—firefighters, police officers, prison service officers—require physical assessments including performance tests. Health assessment is a formal procedural requirement before appointment is secured in the United Kingdom, although this is not true of all jobs in the United States.

The health of the workforce may be overseen by an occupational physician or, depending on the size of the organization, a health nurse. Passive health surveillance refers to use of sickness, injury, and absence data, whereas active health surveillance includes staff surveys and questionnaires and invited complaints. Some companies assume a more proactive role by encouraging their employees to engage in health promotion initiatives. These include subsidized membership rates at local gymnasiums, fitness classes on the premises, or simply encouragement of active lifestyles. The assumption is that the employees will have a stronger bond to the organization, absences from work will decline, and productivity will increase as a result.

People who find that they cannot cope with occupational demands might choose to change their employment. The move can occur in association with aging and the natural decline in maximum physical capabilities or because of circumstances at work such as operating on nocturnal shift work systems. People also seek change and opt for alternative employment because of the risk of incurring musculoskeletal disabilities. Those who remain in their jobs cope more easily and form what is known as the healthy worker phenomenon.

In many countries legislation prohibits the adoption of "ageist" criteria in the appointment of personnel. Physical fitness declines with age, a decrease in levels tending to occur in the mid- to late 30s. Fitness programs may arrest the decline in muscle strength and endurance, but there are inevitable physiological changes that ultimately affect physical performance. A consequence is that test batteries used in selection should be neutral to age (as well as sex). Employees in highly demanding posts may solve this problem by moving to less physically demanding jobs, "buying out" their contract as may happen in the armed forces, securing early retirement, or moving to other organizations.

In this chapter the focus is on the physical demands of various occupations. These demands range from the musculoskeletal stresses on health professionals and others engaged in manual materials handling to the heavy daily energy expenditures of traditional forestry work. Special attention is given to the high-intensity requirements of police and prison officers and the thermoregulatory burden on workers in the fire service. The particular requirements of aircrew and nocturnal shift workers are also addressed. Methods of assessing workload in an occupational context are considered, and the value of work-based health promotion schemes is reviewed in the context of maintaining the fitness of personnel.

Military Personnel

The defense of a nation is the responsibility of the military. In extremes, military personnel engage in warfare to secure this end or are deployed on civil initiatives in peacetimes. In the armed forces, personnel have little choice in decision making because they are required to follow instructions. The level of discipline required within the armed forces is unique, being much greater than in civilian contexts.

Soldiers

High levels of physical fitness are required of soldiers for the occupational tasks they are called on to do. These activities include long marches, sometimes in hostile terrain, load carrying, and operation of munitions. Activities vary according to rank, but all ranks must be prepared to engage in armed combat and emergency activities. Because soldiers must be prepared for severe environmental conditions, training maneuvers are often arduous, pushing people to their limits.

Wearing chemical protective clothing can add to the energy cost of military activities. Patton and colleagues (1995) studied 14 active male soldiers (age 20.9 ± 1.0 years; $\dot{V}O_2max$ 55.6 ± 0.9 $ml \cdot kg^{-1} \cdot min^{-1}$) wearing a protective mask, battle dress over garments, over-boots (a protective layer worn over the normal military boots), and battle dress uniform weighing 9.3 kg. The mean $\dot{V}O_2$ was 13% to 18% higher at all exercise intensities when compared with battle dress only. The awkwardness of protective clothing ensembles is partly attributable to additional frictional resistance from clothing layers rubbing over each other and the restriction of joint movements.

The initial training of military recruits may be highly demanding, especially when the training program contains marching exercises and adventure activities not experienced in civilian life. People entering the military with low levels of fitness were found to be more susceptible to injury than their fitter colleagues (Knapik et al., 2001) and were more likely to be disengaged early in their military service (Knapik et al., 2003). It seems that tolerance of high physical loads on a regular basis is an essential requirement for a career in the military.

Activities are set in absolute rather than relative terms, which partly explains the relationship between injury and fitness of military personnel. The reason that injury rates for women are higher than for men is that women must work the same level as men, but their maximum capacity is lower; that is, they must work at a higher intensity, a higher percentage of their maximum.

Female soldiers are usually expected to do the same training as men, although they are not normally expected to fight in the front line. They are, however, required to carry as much weight on their backs as male soldiers. This relatively heavier loading may cause skeletal stress, particularly during long marches. About 10% to 12% of female recruits were found to suffer from stress fractures compared with 1% to 3% of men (Cline et al., 1998). The incidence of stress fractures was higher in amenorrheic women and women who had been less physically active before entering the army compared with the others. Later, Beck and colleagues (2000) concluded that soldiers with stress fractures had weaker lower-limb muscles and smaller thigh muscles and were significantly less fit than their colleagues. Army training may not be the sole, or even the major, reason for stress fractures among recruits, because lifestyle factors are also implicated. Lappe and colleagues (2001) demonstrated that stress fractures were incurred more often in those who drank more alcohol, smoked more cigarettes, were of lower weight, and were less physically active than their colleagues. Female recruits in the United States have different physical fitness standards to live up to compared to recruits in Europe. For the most part they are expected to do the same fighting as men, except for a few key things.

It may be a concern of the government that fitness of its service personnel mirrors that of its civilian population. Knapik and colleagues (2006) reviewed fitness data of U.S. military personnel over a 30-year period. The comparisons were partly confounded by selective availability of data, tests used, and differences in methods. Nevertheless, the authors were able to draw some inferences about fitness trends in this population. Muscle strength—determined as isometric upper-torso strength, isometric lower-body strength, isometric upright pull force, and isoinertial incremental dynamic lift—showed an upward trend over the period 1978-1998. For female staff members, vertical jump performance was 13% higher in 1998 compared with 5 years earlier. Muscle endurance, expressed in performance of push-ups and sit-ups, showed no change in the 20 years prior to 2003. Maximal oxygen uptake was unchanged in male personnel (average 50 $ml \cdot kg^{-1} \cdot min^{-1}$) between 1975 and 1998 while showing a small increase, up from 38 $ml \cdot kg^{-1} \cdot min^{-1}$, among females. There was a corresponding deterioration in aerobic performance as measured by a 1-mile (1.6 km) and 2-mile (3.2 km) timed run. Body composition data showed a slight (16%) increase in body fat (from 15% of total body weight to 17%) from 1978 to 1998 in men and a 7% increase (from 26% of total body weight to 27.5%) in women. These values are close to the 16% and 26% figures, respectively, for the reference male and female in the general population and imply a prevailing secular trend.

Given a worldwide increase in obesity and decline in physical fitness, the secular trend should be evident in new recruits. This concern over the condition of military recruits led to alteration in Finnish law, creating a legal obligation for staff in the defense force to keep themselves in good shape. Those not observing the obligation become ineligible for promotion or participation in international missions.

Physical fitness does seem to benefit military recruits. Rayson and colleagues (2000) reported that the fitter subjects displayed higher levels of performance on both actual and simulated military tasks. Physical fitness tests included static strength (upright pull, arm flexion, hand grip strength, back extension, plantar flexion), maximum lift performance on a dynamic strength test, muscular endurance (consisting of six separate tests), and aerobic fitness as determined by the 20 m shuttle run test (Ramsbottom et al., 1988).

In a prelude to developing physical selection procedures for the British Army, Rayson and colleagues (2000) conducted a job analysis to identify criterion tasks that could be used as a basis for establishing selection standards. The criterion tasks comprised a single lift, a carry, a repetitive lift, and a loaded march. A secondary objective included developing "gender-free" models to provide common physical selection tests and standards for men and women and "gender-unbiased" models to ensure that members of either sex were not disproportionately classified. Although body size and body composition were taken into account in developing predictive models, the accuracy with which criterion task performance could be predicted was variable. Both single lifts were successfully modeled using muscle strength and fat-free mass data. The "carry" model included data for muscle endurance, body size, and composition. The only gender-free model developed was that for the loaded march, and the development of gender-unbiased models proved elusive. Women's scores were poorly predicted and tended to be distributed around the pass standards, causing a greater percentage of women than men becoming misclassified as passing or failing. This error in prediction would amount to unintended discrimination, the opposite to what was envisaged.

Marines

Those operating in armed forces within naval units need to be able to operate on water, underwater, and on land. Candidates for Royal Marines Commando units in the United Kingdom must first demonstrate they can cope with the training course over a 2- to 3-day period. Before they are recruited, they are required to show determination, physical fitness, stamina, mental ability, cool-headedness, and the ability to rise to a challenge. They are advised to acquire good all-around fitness and swim as often as possible before undertaking the mandatory prerecruitment course.

The training course includes a progressive **20 m shuttle run test** to estimate $\dot{V}O_2max$ (Ramsbottom et el., 1988), parade drills, and obstacle and assault courses on the opening day. Gymnasium tests include sit-ups, push-ups, and pull-ups to maximum standard. On the following day candidates are required to complete a 3-mile run as a group in less than 22.5 min in prescribed clothing. Later they complete an activity regimen for 1 hr in the gymnasium. The final day incorporates fitness briefs and task-related activities such as handling weapons and swimming. Those selected for formal training undergo foundation training for 2 weeks, individual skills work for the next 7 weeks, advanced skills training on weeks 10 through 15, and operations of war for weeks 16 through 23. Commando skills including cliff assaults and amphibious operations are experienced in weeks 24 to 26 and full professional training is concluded by week 30.

The U.S. Sea, Air, and Land Special Operation Forces Personnel (SEALs) are used as a stealth platform for military operations. Some missions require their personnel to be deployed aboard submarines for extended periods. Although the performance of these marines on 12 min runs approaches that of some athletes, the prolonged confinement can lead to deconditioning. Fothergill and Sims (2002) found that a 33-day submarine deployment led to a 7% decrease in performance on Cooper's 12 min run test (Cooper, 1968), a decrement they believed could compromise mission success. They recommended the provision of exercise equipment and structured training programs to avoid **aerobic deconditioning** during prolonged submarine deployments.

Prison Officers

Prison officers can be engaged in strenuous physical activity that calls for specific fitness characteristics. Since 2001, all prison officers in the United Kingdom are obliged to pass a battery of fitness tests before joining the service; after joining, they must pass the same test battery each year. Specialist employees including physical education instructors, patrol dog handlers, and those people who undertake training for advanced control and restraint must also take the tests. The main reason for introducing the tests was to ensure that the service was meeting its duty of care by not requiring staff members to carry out tasks that might damage their health or place themselves, their colleagues, and prisoners at risk. The physical fitness tests were developed therefore to ensure that employees were fit enough to undertake the full range of physical requirements of their job safely. Although UK prison officers must pass a test before joining and pass that same test yearly, that is not the case in the United States. Once American officers pass a test, they usually are not tested again. This requirement is not universal across occupations, and a noted exception is a yearly medical test for commercial airline pilots.

The initial test battery in the United Kingdom consists of a series of functional lifts intended to reflect the physical requirement of the prison officer's role. Because the lifts are job-related, they should be neutral to age, ethnicity, and gender. The battery includes grip strength, endurance fitness determined by a progressive 15 m shuttle run, dynamic strength involving pushing and pulling on a dynamometer to mimic control and restraint techniques, an agility test comprising a slalom run in negotiating a series of cones, and a static shield hold (i.e., holding a 6 kg protective shield in front of the body). The physical tests are complemented by pretest screening by means of a questionnaire, body mass index, resting blood pressure, and a lung function test to establish suitability for standard duration breathing apparatus (SDBA).

The prison environment may place staff in the prison service at risk of injury. About 20% of injuries incurred by staff are attributable to falls and trips (Scott and Hallas, 2006). Cleaning regimens, footwear policy, and stair safety were implicated. Behavior such as running to alarm calls often results in falling. Practical preventive measures include removing obstacles that pose hazards and providing suitable lighting so that obstacles can be seen and avoided. These practices reflect basic ergonomics principles for workplace layout.

The prison officer is largely responsible for custodial duties but must be prepared for emergency actions at any time. These might include reacting to alarms and carrying out security checks and search procedures. Officers are also required to maintain close relationships with the prisoners in their charge. The job is therefore more complex than its physical requirements suggest, calling for personal and social skills as well as physical fitness.

Police Officers

Traditional physical standards for recruitment of a police officer included height and weight. These standards became indefensible as job-related criteria, notably when the profession was opened to women. Because job requirements are not based on age or sex, an allowance for these factors is not meaningful. Nevertheless, it makes sense for police officers to maintain good all-around fitness for their duties: An officer's guide to fitness for law enforcement performed in the United States underlines the importance of maintaining physical fitness (Hoffman and Collingwood, 2005). Although it is true that female applicants in the United States were instrumental in getting rid of the height and weight requirement, the reason was **test validity,** not sex. The cases ended up being decided at the U.S. Supreme Court. The basic issue was that height and weight were not valid for the work tasks and discriminated not only against women but also Asians and Hispanics.

Law enforcement agencies apply fitness assessments in one of two ways. One involves a physical fitness test battery of items including 1.5-mile (2.4 km) and 300 m runs, 1RM bench press, push-ups and sit-ups, vertical jump, agility, and flexibility tests or a simulation of job-related scenarios. Percentile values for these batteries of tests are available for individual checks (see Hoffman and Collingwood, 2005). The alternative is an obstacle-type course within which simulated tasks are incorporated. Legislation in the United States prohibits use of different cutoff scores for recruiting employees based on religion, national origin, color, race, or sex.

Firefighters

The tasks associated with firefighting impose high physiological demands. Carrying equipment, operating in protective clothing, working while wearing breathing apparatus, and dealing with tasks at hand entail a large outlay of energy expenditure. The load on the circulatory system is accentuated when operating under high ambient temperatures, causing body temperature to rise and fluid to be lost in sweat. These factors are extended by the stress of operating in emergencies that may threaten those being rescued and the firefighters themselves. Because firefighters are exposed to working conditions that are conducive to incurring heat strain, cardiovascular fitness is required for effectiveness in this job.

Firefighting and emergency service work often takes place at high intensities for relatively brief durations. Excessive core temperatures may be experienced caused by the heat storage attributable to physical activity and the personal protective equipment worn. The energy expenditure is elevated by about 15% above normal when standard duration breathing apparatus (SDBA) systems are worn (Baker et al., 2000). This equipment limits the exposure time, although exposure can be extended by wearing extended duration breathing apparatus (EDBA) clothing, which allows for an increased air supply. Irrespective of the clothing layers worn by the operator, personal protective clothing creates a barrier to heat exchange with the environment, thereby accentuating the risk of heat injury. Thus firefighters must be withdrawn from extreme heat exposure periodically and allowed to cool down before reentering the hot environment.

The stress on firefighters has been studied in responses to real-life events, during training exercises in "fire houses" (a training scenario that mimics a real house on fire) and in laboratory-based and field-based simulations. There is an immediate increase in heart rate and catecholamine concentrations in response to the alarm bell that alerts personnel to an emergency event. The responses may vary in intensity according to the nature of the event, the duration of activity, and the conditions encountered. In addition to heat tolerance, major physical loads are attributable to setting up operations, climbing stairs and positioning ladders to reach fire sites, and carrying casualties to rescue.

Bilzon and colleagues (2001) attempted to identify the minimum cardiovascular fitness levels required to complete simulated firefighting on board ship. The trials consisted of several 4 min tasks including boundary cooling, carrying a drum, carrying an extinguisher, running a hose, and climbing a ladder. The metabolic demand averaged 32.8 $ml \cdot kg^{-1} \cdot min^{-1}$ (both men and women combined) corresponding to about 90% of the maximal heart rate. The results suggest that firefighters would be able to sustain tasks at 80% $\dot{V}O_2max$ for up to 16 min when wearing SCBA.

Reilly and colleagues (2007) studied firefighters' responses to a simulation of firefighting in a fire house. The training exercises consisted of exposure to a heat barrier, a free search of the burning premises, and a hose-reel operation. The mean heart rates were 182, 187, and 194 beats/min for these operations, respectively. The perception of thermal sensation was highest for the hose reel among the three drills.

Lemon and Hermiston (1977a) attempted to quantify the energy cost of firefighting by monitoring $\dot{V}O_2$ and heart rate responses to climbing an aerial ladder, rescuing a victim, dragging a hose, and raising a ladder. These activities were conducted in full fire-kit ensemble but without breathing apparatus. The task corresponded to 70% $\dot{V}O_2max$, the most vigorous component being dragging the hose.

In their simulation of firefighting, Elsner and Kolkhorst (2008) used 10 separate task components representing an actual fire scene. The items included advancing a hose 35 m from a fire engine to a hydrant, carrying an extension ladder 30 m and extending it to a third-floor building, donning SCBA and advancing two sections of a fire hose from an engine to a stairwell, and then using a sledge to pound a large wooden block 5 cm along a concrete floor. The subjects then climbed three flights of stairs, pulled two sections of the fire hose with a rope from ground up to third-floor level, advanced the hose 30 m through a cluttered area, returned to ground level by the stairs while discharging a task en route, and finally conducted a search-and-rescue task to locate a mannequin and drag it 30 m. The protocol took 11.65 ± 2.21 min for the firefighters to complete, being allowed to do so at a self-selected pace. The mean $\dot{V}O_2$ was 29.2 ± 8 $ml \cdot kg^{-1} \cdot mmin^{-1}$ or 62 ± 10% $\dot{V}O_2$ with a mean heart rate of 175 ± 7 beats/min or 95 ± 5% HRmax. The peak $\dot{V}O_2$ achieved during any part of the simulation was 80% $\dot{V}O_2$ and at the end was 31.5 $ml \cdot kg^{-1} \cdot min^{-1}$ when the heart rate was 183 ±-8 beats/min. Those firefighters with the larger values for $\dot{V}O_2max$ (mean 46.2 ± 7.8 $ml \cdot kg^{-1} \cdot min^{-1}$; range 32.2 to 58.4 $ml \cdot kg^{-1} \cdot min^{-1}$) demonstrated the higher values for $\dot{V}O_2$ during the protocol and the shorter times in completing it.

The training instructor may impose additional loads on trainees when supervising live-fire exercises. Personnel must cope with working repeatedly and regularly in very hot environments while wearing both personal protective equipment and self-contained breathing apparatus. Bruce-Low and colleagues (2007) studied this problem in a training facility at temperatures in excess of 120 °C during live-fire training exercises for 35 min and compared physiological responses to conditions in the same premises when the fire was not ignited. This procedure helped to quantify the additional load on the heart attributable to using a breathing apparatus while wearing personal protective equipment and SCBA at moderate workloads. Physiological and subjective responses were elevated considerably when the fire was ignited. The authors concluded that the protective equipment is the significant factor in reducing heat strain during such events. At high exercise intensities, wearing the breathing apparatus may influence physical capacities attributable to a reduction in $\dot{V}O_2max$, as reported by Dreger and colleagues (2006).

It is not surprising that the maximal aerobic power ($\dot{V}O_2max$) has been considered relevant to this occupation. The Home Office (Scott, 1988) recommended a value of 45 $ml \cdot kg^{-1} \cdot min^{-1}$ as the standard, and most studies of firefighters tend to show figures close to this level (see table 8.1). In many instances the aerobic fitness measures were deemed inadequate to provide the firefighter with reserve capacities for operational duties, and a program of in-house physical training was recommended. Such measures would also arrest age-related deteriorations in aerobic fitness that might occur. Indeed, Puterbaugh and Lawyer (1983) reported a 20% increase in aerobic power with 12 weeks of in-house training, irrespective of age. Such measures might be important in view of the observations that aerobic fitness and general strength of UK firefighters declined after 18 months of service. This finding suggests that operational duties alone are not sufficient to maintain aerobic fitness levels.

In the UK, the Home Office recommends a $\dot{V}O_2max$ of 45 $ml \cdot kg^{-1} \cdot min^{-1}$, a value that would not work in the United States. Although it is true that fighting fires is an aerobic task, many firefighters have an aerobic power below this figure and still work effectively as firefighters. This is an issue of setting a cut score congruent with the physical demands of the job. Besides, many workers with a $\dot{V}O_2max$ greater than 45 $ml \cdot kg^{-1} \cdot min^{-1}$ would not be able to meet firefighter tasks, because they do not have sufficient strength. The ability to cope with the variety of tasks that firefighters experience is the crucial element.

People who possess superior $\dot{V}O_2max$ values are able not only to work harder but also to work for longer durations compared with those with a lower aerobic power (Davis at al., 1982; Sothmann et al., 1990). Similarly, Lemon and Hermiston (1977b) noted that people with a $\dot{V}O_2$ value in excess of 40 $ml \cdot kg^{-1} \cdot min^{-1}$ appeared to perform all the tasks in a shorter time than the others. These observations highlight the importance of maintaining aerobic fitness for firefighters.

Aerobic power determined in a progressive exercise test to volitional exhaustion may overestimate the functional margin available to firefighters when they wear their full assembly of protective apparatus and clothing. Dreger and colleagues (2006) showed that $\dot{V}O_2max$ was decreased by 17% when firefighters wore a personal protective ensemble and self-contained breathing apparatus. These investigators suggested that a logical alternative to measuring aerobic fitness in normal exercise clothing is to conduct the assessment with subjects wearing both breathing apparatus and protective clothing. This approach should provide a more functional assessment of aerobic work capacity in firefighters. Nevertheless, the traditional approach provides a useful reference for longitudinal comparisons or cross-reference to normative data.

Table 8.1 Values for $\dot{V}O_2max$ and Percent Body Fat Reported in Various Groups of Firefighters

Group	*n*	Age, years	$\dot{V}O_2max$, $ml \cdot kg^{-1} \cdot min^{-1}$	Percent body fat (mean SD)	Reference	Note
68 males; 4 females	72	31	46.4 ±	17.0 (10.0-27.0)	Love et al. (1996)	Measured by submaximal treadmill test
Full-time firefighters (male)	12	31.6 ± 1.3	50.3 ± 1.2	—	Baker et al. (2000)	Measured on treadmill
Merseyside staff (male)	10	40.1 ± 1.7	43.6 ± 5.2	19.2 ± 3.1	Reilly et al. (2007)	Measured on treadmill
Ontario staff	45	35.0 ± 2.5	40.6 ± 5.3	20.4 ± 5.0	Lemon and Hermiston (1977a)	Predicted from submaximal heart rate on cycle
U.S. staff (Los Angeles)	17	32.3 ± 6.7	48.5 ± 9.1	15.3 ± 3.0	O'Connell et al. (1986)	Measured on cycle ergometer
U.S. staff	38	35	43.0 ± 1.4	—	Ben-Ezra and Verstraette (1988)	Measured on treadmill
UK firefighters	291	32 ± 6.4	43.7	19.0 ± 4.5	Scott (1988)	Predicted from submaximal $\dot{V}O_2$ on cycle
Royal Navy personnel	34	26 ± 6.9	52.6 ± 5.2	16.7 ± 3.5	Bilzon et al. (2001)	Measured on treadmill
U.S. staff	100	33.1 ± 7.6	39.6 ± 6.4	21.1 ± 6.7	Davis et al. (1982)	Measured on treadmill
UK staff (4 females)	10	29 (21-37)	47 (36-61)	19.0	Carter et al. (2007)	—
UK staff	17	31 (21-38)	48 (36-65)	19.0	Carter et al. (2007)	—

Note: Values for age, $\dot{V}O_2max$, and percent fat given as mean ± standard deviation or mean (range).

Bus Workers and Postal Workers

Many occupations require ambulatory activity that is not highly strenuous but maintains cardiovascular health. The classical comparison of bus drivers and bus conductors by Morris and colleagues (1953) provided compelling evidence of how occupational activity can promote health. The drivers were largely sedentary and were more prone to cardiovascular disease than the conductors, whose occupational roles required low-level activity, including climbing the stairs to the top deck of the bus. With automation of ticket purchases, the bus conductor has disappeared from the working population in most countries.

Postal delivery is another occupation for which locomotion is an intense part of the job. Although a large volume of routine communication is now transmitted by e-mail, there is still a considerable amount of mail transported by conventional means. Traditionally the rural postman cycled around the delivery route, whereas the urban counterpart operated on foot. Durnin and Passmore (1967) cited a mean value of 9.4 kcal/min (39.3 kJ/min) for a simulated postal delivery, which constitutes heavy work. Even though motorized vehicles are used for mail delivery, delivery still requires periodic locomotion to the householder's mailbox. Although the work is not intense at any given time, it is sufficient to raise metabolic rate and indirectly provide health-promoting activity. Carrying a delivery bag contributes to this stimulus.

The majority of mail delivery staff carry mail in a bag, irrespective of the mode of travel. A typical delivery walk takes about 2 hr, with the bag loaded to a maximum of 16 kg. Parsons and colleagues (1994) described the redesign of a delivery bag to make it more effective and more comfortable to carry (see figure 8.1). The redesign process incorporated a task analysis of delivery activities, a questionnaire survey, and suggestions from personnel involved. These were followed up in both laboratory and field studies of three alternative designs, using spinal shrinkage, biomechanical analysis, and subjective responses as criteria. A modified simple pouch design was the most popular during the field trials; the double-pouches caused asymmetries in the load as the delivery progressed because the remaining mail was not distributed equally between the two pouches.

Drivers

Train drivers spend prolonged periods of their working hours in a sitting posture that is constrained by the driving task. These drivers are exposed to whole-body vibration for extended periods, and high-mileage driving on the road has also been associated with a high precedence of musculoskeletal pain (Porter and Gye, 2002). Truck drivers are periodically required to complete strenuous physical work at the start and end of their journeys and at stops in between. These tasks include loading heavy goods, jumping up and down from cabins and trailers, decoupling the trailers, and conducting routine repairs. In their study of 192 train drivers, Robb and Mansfield (2007) reported that 81% experienced musculoskeletal pain and 60% had low back pain in the previous 12 months. Manual handling and seat discomfort were associated with musculoskeletal patterns. The mean **body mass index (BMI)** of 28.6 kg/m^2 was above the mean (25-26 kg/m^2) for adult males in the United Kingdom and is indicative of overweight. Given this mean BMI, coupled with an unhealthy diet and lifestyle, with 41% of respondents being smokers and a further 11% former smokers, attention to the fitness of this occupational group was deemed urgent.

The driving of motor vehicles on the public roads is subject to government regulations that in most cases require passing a driving test. Such tests require knowledge of the rules of the road and competence in a formal driving test. Apart from these requirements, visual capacity is important. The vehicle can be adapted for disabled drivers. Temporary incapacitation through sleep deprivation, alcohol use, or fatigue as a result of a long journey can cause accidents. Consequently, a legal limit for blood

Figure 8.1 Alternative backpacks for postal deliveries (from Parsons et al., 1994). The designs illustrated offer options according to the prevailing work cycle.

Adapted, by permission, from C. Parsons, G. Atkinson, L. Doggart, A. Lees, and T. Reilly, 1994, Evaluation of new mail delivery bag design. In *Contemporary ergonomics* 1994, edited by S.A. Robertson (London: Taylor and Francis), 236-240.

alcohol concentration applies, although the exact level that breaches legislation varies from country to country. Commercial truck drivers have their hours at the steering wheel logged automatically to avoid infringing on permitted working hours. Ingestion of medications may also alter fitness to drive, antianxiety drugs and melatonin causing drowsiness (Reilly, 2005). In nonregulated jobs, the driver is responsible for assessing his or her fitness before undertaking a journey.

Ambulance Workers

Ambulance work can place a high degree of physical strain on personnel. Carrying patients can entail heavy dynamic activity for leg muscles and loading on the shoulders, arms, hands, and trunk. Tasks such as carrying a loaded stretcher downstairs require isometric strength and cardiorespiratory fitness. Like a number of health care professionals, ambulance drivers must be capable of manual handling of loads.

In an attempt to reduce fatigue during task-related activities, Aasa (2008) implemented a 1-year physical training program among ambulance personnel. The exercise sessions were conducted three times each week and consisted of basic resistance training combined with cardiorespiratory work. Training was effective in decreasing perceived exertion and blood lactate concentration in response to a simulated work task and in improving performance in some strength tests. Maximal oxygen uptake values remained at a modest level, reaching 42.8 ± 7.3 ml · kg^{-1} · min^{-1} on average.

Training three times each week was more effective than twice a week in a majority of the tasks used. The decreases in blood lactate and rated exertion were taken as evidence that fatigue in a work-related context was reduced. Follow-up of the volunteers indicated poor compliance with continuing the physical training program. It seems that the motivation for exercise, particularly among those with low physical capacity, remains a challenge for those charged with health promotion in an occupational setting.

Forestry Workers

Forestry work traditionally required physical fitness, the tasks of using an axe and saw calling for strength and skill. Cutting down and dismantling trees and clearing forest areas entailed huge energy expenditures. The mean daily rate of energy expended by forestry workers was estimated to be 3,670 kcal (15.36 MJ) by Durnin and Passmore (1967), although values as high as 5,700 kcal have been reported for workers in the Swedish forests (Lundgren, 1946). These figures compare with 2,520 kcal (10.55 MJ) for office workers and 3,000 kcal (12.56 MJ) for building workers (Durnin and Passmore, 1967).

The heavy rates of energy expenditure incurred in traditional forestry work were reflected in the high energy intake of workers. Surveys indicated that workers' intakes typically exceeded 4,000 kcal/day (16.74 MJ/day). Intakes as high as 7,000 kcal/day (29.3 MJ/day) have been recorded during lumbering competitions (Karvonen and Turpeinen, 1954). This form of competition is maintained in forested areas as a strength contest.

Forestry work has changed as mechanization has reduced the heavy physical burden that was involved in felling and harvesting trees. Many different tasks are associated with contemporary forestry work, such as managing the wooded environment and growing and harvesting timber. Tasks such as planting saplings, digging in the forest environment, and hewing branches constitute moderately heavy work. Very little of the working day in this environment is spent in sedentary activity, so workers must be physically fit to perform forestry work.

Beach Lifeguards

Beach lifeguards are responsible for the safety of a large number of swimmers at public beaches. Although similar standards operate internationally in the selection of lifeguards, these standards are not based on task analysis. A beach lifeguard should be able to reach a casualty within 3.5 min to reduce the likelihood of the victim's drowning.

Reilly and colleagues (2006a) identified towing a casualty, board paddling with the victim, and casualty handling as the three most demanding activities. After conducting a series of tasks that included running on the beach, swimming freestyle and underwater, and board paddling with the victim and casualty handling, the investigators concluded that if only a rescue board is available, the area out to sea patrolled by a lifeguard should be reduced from the standard 400 m to a maximum of 300 m.

In a follow-up study, the same group (Reilly et al., 2006b) investigated the validity of a fitness standard based on the physical demands identified earlier. The distance paddled to sea in 3.5 min was significantly correlated with 400 m time in front crawl, $\dot{V}O_2max$ determined during towing, and deltoid circumference. They concluded in their recommendation for fitness testing that a swim time of 7.5 min or less over 400 m in a pool should enable the lifeguard to paddle 310 m in the sea in less than 3.5 min.

Professional Divers

Professional divers may spend prolonged periods under increased ambient pressure underwater. Operating at greater depths than atmospheric pressure with normal gas mixes can cause a host of problems that include oxygen toxicity, CO_2 toxicity, and **nitrogen narcosis** and at much greater depths the symptoms described as high pressure neurological syndrome. To avoid these difficulties, divers replace part of the O_2 in the air supply with an inert gas like helium or a combination known as trimex. The physiological mechanisms and risks are described in detail by Reilly and Waterhouse (2005).

Nitrogen narcosis refers to the feeling of drunkenness that divers experience at depth as nitrogen accumulates in the nervous tissues. The feelings are described by divers as "the raptures of the deep" and are accompanied by a decrease in cognitive function once a depth of 40 m is reached. Thomas and Reilly (1974) monitored mental performance and mood states in eight amateur divers in a compression chamber at a simulated depth of 46 m for 65 min. They noted a large decrease in concentration, an increase in elation, and an improvent in the rating of fatigue. These changes in mood states were accompanied by increased errors in mental addition. The authors concluded it is likely that more complex mental operations are adversely affected at a similar depth.

A major health risk is associated with the return of divers to the surface. Gases dissolved within the body come out of solution, expand with the reduction in pressure, and may cause embolisms in major arteries, including in the lungs, heart, or brain. Divers need to undergo gradual decompression according to generally accepted schedules (figure 8.2).

A great deal of time must be spent in decompression in cases of deep and lengthy dives. Because divers eventually attain a new equilibrium of saturation with gas under pressure, divers can remain at depth for many weeks. This allows professionals to operate from pressurized cabins for sustained spells, at the end of which a single, albeit prolonged, decompression routine suffices. It has been suggested that this form of so-called saturation-diving may have application if pressurized underwater

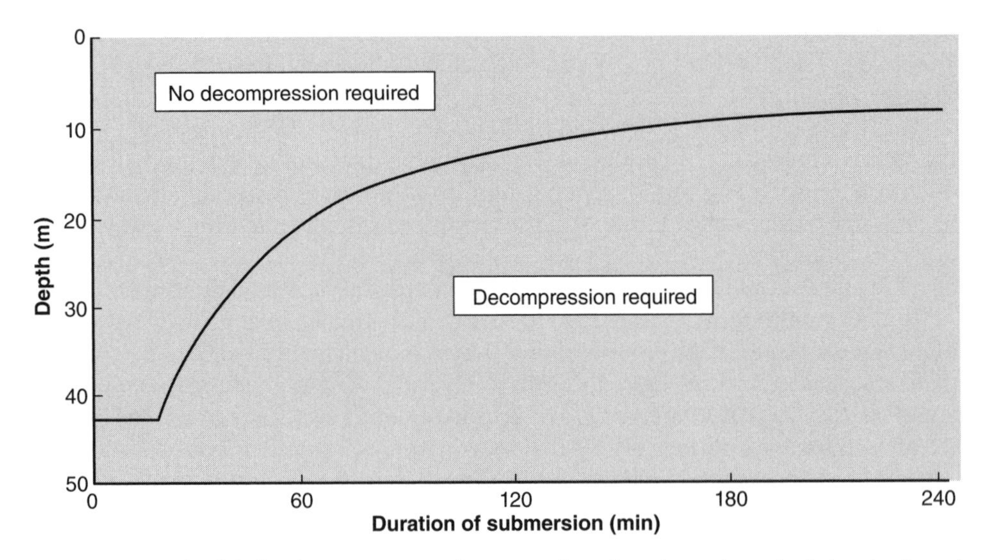

Figure 8.2 Schedule for decompression: The time to be taken depends on both the depth and the duration of the dive.

Reproduced from *Sport, Exercise and Environmental Physiology*, T. Reilly and J. Waterhouse, pg. 84, Copyright 2005, with permission from Elsevier.

holiday camps in undersea areas like the Australian Barrier Reef become a tourist attraction.

Several injuries and illnesses are associated with faulty decompression. The first risk is pulmonary barotrauma, or burst lung, which can occur when the ascent is too rapid. The air confined in the chest at depth is compressed. As the surface is approached the pressure of this air decreases and volume increases according to Boyle's law. As the diver relaxes the chest muscles, the sudden expansion and escape of air may tear lung tissue. This allows air to pass into the surrounding tissue, producing emphysema or air emboli that enter the circulation and can block vital arteries to the heart or brain. The treatment is instant repressurization in a decompression chamber or return to depth in water if this facility is unavailable.

Formation of air bubbles on ascent can lead to various symptoms; the most common is pain in the joints and limbs, which is usually referred to as "the bends." The bends entail severe pain; minor pain around the joints is known as "niggles." "Staggers" refers to involvement of the spinal cord or brain with varying levels of muscular or sensory paralysis. Permanent disability may result from bubbles being released within the brain or spinal cord, and careful therapeutic decompression is called for. "Chokes" refers to respiratory distress associated with bubble formation within the pulmonary alveolar circulation, although in this case recompression is not essential. Finally, bone necrosis at the end of long bones, which may cause severe arthritis, can occur some time after exposure, although this happens to the professional rather than the recreational diver.

Surface decompression procedures have been used among professional divers. The diver is brought to the surface and almost immediately transferred to a decompression chamber (see figure 8.3). Although there are few immediate observable symptoms when divers use this procedure, long-term problems associated with miniature bubble formation are likely. In particular, damage to nerve and bone cells can occur when

Figure 8.3 Novice diver entering a compression chamber for experience of mood changes at depth.

Reproduced from *Sport, Exercise and Environmental Physiology*, T. Reilly and J. Waterhouse, pg. 76, Copyright 2005, with permission from Elsevier.

bubbles are formed within these tissues. Such adverse effects may remain latent for months or even years but are immediately evident with contemporary medical imaging techniques.

Fitness to dive is determined under normal medical screening for cardiovascular risk. Upper respiratory tract infections are a contraindication to diving because the ability to equalize pressure through the eustachian tubes or between the sinuses and the respiratory system is compromised. Where physical activity is undertaken under high atmospheric pressure, the load on the oxygen transport system is increased. A normal cardiovascular and ventilatory response to submaximal exercise is typically considered in assessments of divers rather than absolute standards of maximal aerobic power. Freedom from neurological consequences of previous faulty decompression experiences is also important but more difficult and costly to assess.

Workplace Fitness Programs

Many companies place intrinsic value on the fitness of their employees. This commitment is reflected in the installation of on-site training and sport facilities. Alternatively, employers can subscribe to membership in commercial gymnasiums or local sport centers on behalf of their employees. The employer is likely to benefit in a material way from the improved physical fitness of personnel.

The value of worksite fitness programs was reviewed by Shephard (1988). The evidence supported the contribution of these programs to human well-being. Benefits to employers were apparent in increased worker satisfaction, improved corporate image, and facilitation of employee recruitment; there was some evidence for increased production, decreased absenteeism, reduction in turnover of staff, decreased health care costs, and decreases in numbers of occupational injuries.

Overview and Summary

Occupational fitness means that the employee can cope with the demands of the job, especially with its physical components. In many occupations these demands can be identified by highlighting the most difficult tasks likely to be experienced. The ability to cope is then calculated with a battery of performance tests; when appropriate, adjustments are made for age and sex.

Many occupations have specific stresses and hazards, yet fitness can be determined largely by a general health screening. Professional divers, for example, undergo general fitness assessments but their health status may be affected by chronic neurological consequences of serial decompression schedules underwater and on the surface. In other occupations, people may opt out of employment if the demands are unduly heavy for them. Jobs such as furniture removal and building construction constitute heavy work, and those employees who remain in such jobs demonstrate the "healthy worker" syndrome. The assessment of fitness for work in these instances may be made by a physician.

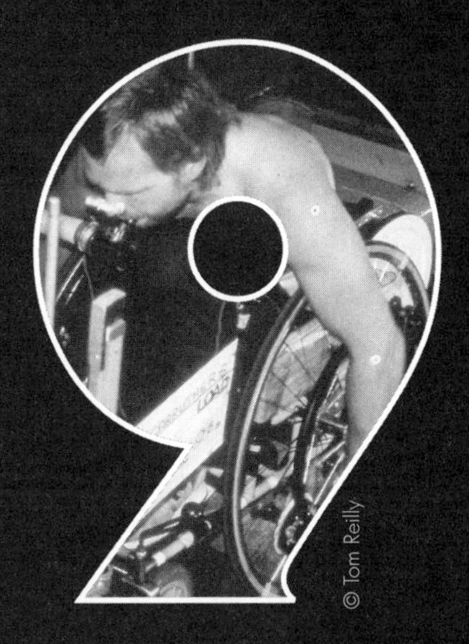

© Tom Reilly

9 Special Populations

DEFINITIONS

aging—The entire life cycle process from growth and development to functional decline, particularly the period following peak performance potential.

athletic amenorrhea—Absence of the normal menstrual cycle.

disability sport—Activities organized on a regulatory basis for participants with physical or mental disabilities.

menstrual cycle—The changes in physiological systems occurring roughly each 28 days to support the woman's childbearing role.

referee—The main match official in field games, often working in conjunction with designated assistants and sometimes supported by decisions based on video technology.

veterans—Athletes 35 or 40 years old; a competitive category for sport.

nine

HUMAN structure and function are characterized by variability. This diversity must be considered in the design of clothing, equipment, and facilities for special populations who may differ from norms of the general adult population. Children are not adults with scaled-down requirements but should be considered in their own right. Equal opportunities legislation notwithstanding, design criteria for females differ in important respects from those of males of similar age. Age needs to be taken into account as individuals become frail and gradually lose functional capacity. Any disabilities need to be taken into account when people's needs and fitness requirements are being assessed.

This chapter begins by discussing young people as a special population. Issues related to talent identification and development for sport are addressed with attention to criteria for selection. Impact of the reproductive cycle on functional performance of females is considered along with the influence of training loads on the normal **menstrual cycle.** The cessation of the menstrual cycle at menopause is an important milestone in the life cycle of women, characterized by a disproportionate decrease in muscle strength and bone mass. Exercise is important in offsetting the declining capacities that accompany aging. Athletes who have disabilities or use wheelchairs constitute an athletic population in their own right, the Paralympic Games representing the pinnacle of their competitive calendar. Their training needs and assessment modes can be dictated by the nature of their disabilities and matched to their capabilities.

The attention of sport spectators is mostly directed toward participants, unless decision making by match officials becomes contentious. Similarly, training manuals and fitness needs appear focused on participants with complete disregard for the tasks of the officials. Match regulations are implemented by a designated number of officials, according to the specifications of the ruling body. Typically the **referee** is the main arbiter and thus is in the public light and exposed to psychological stress. Match officials are therefore included as a special category in this chapter.

Young People

Our future sport stars are the adolescents or youths of today. In many sports individuals with the potential for success must be identified at a young age if they are to benefit from the best available training programs. Young people's physical capacities are inferior to those of adults, and many of their physical features alter with growth and development.

A special area of application is pediatric ergonomics. There is a recognition that children are not miniature adults but rather are growing, developing, and evolving in individual ways. Although ergonomics aspects of the classroom, school equipment, and furniture have been concerns in the profession, pediatric issues in this section are focused on exercise and sport contexts. To follow these issues, the chapter offers background information on growth and development and factors that promote individual differences. Special attention is given to the relative age effect, a potential dissociation between biological and chronological age.

Growth and Development

Key stages in the development and maturation of boys and girls affect their capability for physical performance. The development of the reproductive system is described in five stages according to Tanner and Whitehouse (1976), whose scheme is useful in characterizing a child's pubertal status. It requires visual inspection by an expert on self-rating using photographs of each stage. In view of the personal embarrassment that the self-assessment can cause to some young people, maturation offset provides an alternative. This requires several comparisons of sitting height and total height (stature) and isolating the differential rates of growth in the legs and in the trunk.

People differ in their rates of maturation so their chronological age may not truly reflect biological age. Biological age can be determined by looking at closure of growth plates in the wrist. Skeletal age can then be established using the Fels, Tanner-Whitehouse, or Greulich-Pyle technique (Malina et al., 2003). Any one of these three methods provides an estimate of the degree of early or late maturation with an error of ±6 months.

Growth curves are used to illustrate the rate of change in physical features with chronological age. The attainment of peak height velocity is a prelude to sexual maturation, characterized by menarche in girls and increased testosterone levels in boys. After this phase there is a relative increase in muscle strength, accompanied by improved performance in a range of physical activities and a relative increase in lean body mass.

During adolescence there is not a parallel growth in metabolic systems. Anaerobic capacities tend to lag behind aerobic power and capacity, and lactate production is considerably below the values observed in adults. These differences need to be taken into account when planning curricular activities for physical education classes and designing physical training programs according to age.

Talent Identification and Development

Many countries have attempted to set up systems for talent "detection" and selection into specialized development programs with the aim of nurturing the elite athletes of the future. Talent detection implies spotting people with outstanding abilities and drawing them into sport, whereas identification refers to recognizing talent characteristics in those already engaged in sport. The development of elite athletes implies harnessing their endowed characteristics with optimal training programs that will allow them to reach their full potential. The process, as represented schematically in figure 9.1, can be repeated several times before culminating at maturity in a drive toward

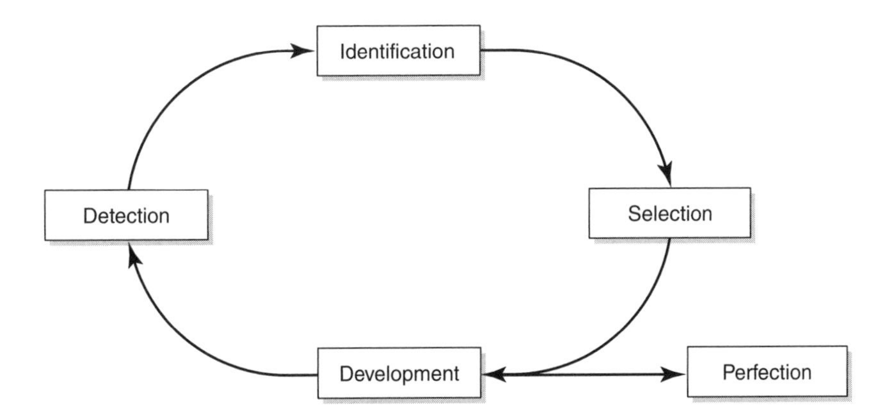

Figure 9.1 Diagnostic representation of talent identification and development. Perfection is the ideal to which all goals are oriented but represents an aspiration rather than a realistic target to be met.

perfection. Perfection is the ideal to which all goals are oriented but represents an ultimate aspiration rather than a realistic target to be met. Few top performers reach this idealized state and the majority of talented youngsters encounter obstacles in their development, experience opportunities seized or missed, or benefit from (or are hampered by) their personal circumstances.

Many investigators have attempted to develop a systematic approach to talent identification and development. These have varied in sophistication, depending on the sport concerned, the aspirations involved, and the criteria used in assessment. Matsudo and colleagues (1987) provided a hierarchical model of selection that could be applied at different levels of participation and to different sports. In contrast, Pienaar and colleagues (1998) focused on young rugby players to isolate the anthropometric, physiological, and performance characteristics of success in the game. A limitation of this approach is that characteristics of success on youth sport are not necessarily those that distinguish elite adult champions, and many physiological functions do not track completely across the years of growth and development.

Reilly and colleagues (2000a) considered the degree to which anthropometric and physiological factors were influential in determining success in soccer. They based their assessments on a range of measures related to work rate in the game, describing the degree to which each variable is influenced by hereditary or environmental factors or the interaction between them. The influence of genetic endowment was expressed as a heritability coefficient, values ranging from highly genetically determined (such as height) to a moderate degree of determination (skills), as shown in the highlight box below.

Analyzing data from young English soccer players selected for specialist training at the national Football Centre of Excellence, Reilly and colleagues (2000b) failed to show a distinction between those who later played professionally at the top clubs

Heritability Coefficient for Various Characteristics and Aptitudes Related to Performance in Sport*

Anthropometry

- Height 0.85 ± 0.07
- Leg strength 0.80 ± 0.10
- Height^3/weight 0.53 ± 0.19
- Skinfold 0.55 ± 0.26
- Ectomorphy 0.35 – 0.50
- Mesomorphy 0.42
- Endomorphy 0.50

Physiology

- Maximal oxygen uptake 0.30 – 0.93
- Slow-twitch muscle fibers 0.55 – 0.92
- Anaerobic power 0.44 – 0.97
- Muscle endurance 0.22 – 0.80

Field and Performance Tests

- Sprinting 0.45 – 0.91
- Jumping 0.33 – 0.86
- Flexibility 0.33 – 0.91
- Balance 0.24 – 0.86
- Static strength 0.30 – 0.97

*mean ± SD or range

and those who played at a lower level. It seems that when athletes are homogenous at a high level of skill, anthropometric, physiological, and psychological measures cannot distinguish between the extremely successful and the professionally competent. Participants may need to satisfy a threshold level of fitness in a variety of components to be able to play at elite level; although a multivariate test battery could discriminate elite young players from their subelite counterparts on a range of test items, discrimination among the elite performers was not significant.

Anthropometric, physiological, and psychological factors are influential in individual sports and to different degrees. In games, practice history and exposure to optimal coaching methods and learning environments seem important. In the multidimensional assessment of young soccer players, Reilly and colleagues (2000b) highlighted the importance of game-related skills. These skills were assessed in ecologically valid anticipation and decision-making tasks that comprised game intelligence. This factor complemented the physical measures of agility, speed, endurance, and slalom dribbling and the ego orientation and task orientation to address and tolerate arduous training programs. This schematic model for predicting talent in soccer included physical, physiological, psychological, and sociological predictions, as shown in figure 9.2.

Broadly similar results were reported for the relationship between multidimensional performance characteristics and level of performance in talented youthful

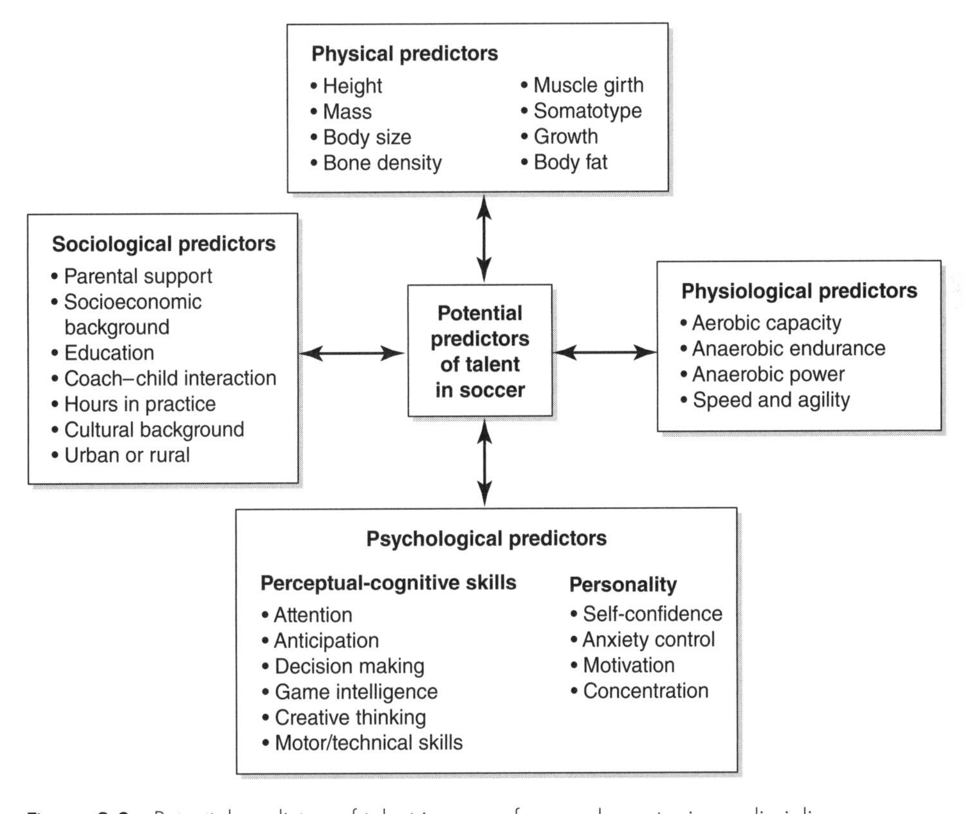

Figure 9.2 Potential predictors of talent in soccer from each sport science discipline.

Reprinted, by permission, from A.M. Williams and T. Reilly, 2000, "Talent identification and development in soccer," *Journal of Sports Sciences* 18(9): 657-667, Taylor & Francis Ltd, www.tandf.co.uk/journals.

hockey players. Working with 13- to 14-year-old Dutch hockey players, Elferink-Gemser and colleagues (2004) distinguished elite from subelite levels on the basis of technical (dribbling a ball at speed), tactical (general tactics), and psychological variables. The most discriminating variables were motivation, tactics for possession of the ball, and performance in a slalom dribbling test. The authors advised that more attention be paid to tactical qualities, motivation, and specific technical skills for guidance of talented hockey players at this age.

Relative Age Effect

Children are separated into 1-year or 2-year age bands for competitive sports, depending on the sport concerned. Those who mature early have an advantage over their late-maturing counterparts in view of the greater maturity-related development in body size, strength, speed, and endurance. In a number of circumstances the early maturer who has greater physical development for his or her chronological age will be prioritized for selection purposes. By the time the late maturers have caught up in development, they will have missed out on the learning and coaching opportunities provided for the early maturers.

In sports that have 1-year chronological age bands, participants born just after the cutoff date for eligibility will have almost a whole year's advantage in age over those born near the end of the selection period. Although the difference of 12 months is unimportant in adults, it can amount to pronounced anthropometric and physiological variances in children. This phenomenon is known as the relative age effect and can be accentuated by maturation status. An early maturer with a 1-year relative age advantage over a relatively small 10-year-old child can be approximately 0.2 m taller and 2.7 kg heavier (Tanner and Whitehouse, 1976).

The relative age effect has been observed in a number of performance domains. It is not clear whether the advantage is attributable merely to the extra experience associated with the early birth date, the higher motivation to perform as a result of dominance over peers, the mediation of self-image and self-esteem attributable to physical development, or to other factors. The phenomenon has been observed in early scientific creativity, academic achievements in primary education, and access to university (Bell and Daniels, 1990). Its greatest impact has been observed in sports.

Helsen and colleagues (2005) provided compelling evidence for a relative age effect in youth soccer across a range of European countries. The investigators found an overrepresentation of players born in the first quarter of the selection year (from January to March) for all the national youth teams of the under-15, under-16, under-17, and under-18 categories. The effect was evident also at Union of European Football Associations (UEFA) under-16 tournaments and other international competitions. Players with a greater relative age are more likely to be identified as talented because of the accompanying physical advantages they have over their peers of younger relative age. Altering the cutoff date for eligibility is not a solution: When this change was made in Belgium, there was an associated change in selection that corresponded to the relative age effect.

Ice hockey is another sport where physical size can influence performance, particularly in underage competitions. Various authors have shown that successful ice hockey players are more likely to have birthdays early in the selection year than at later times (e.g., Musch and Grondin, 2001). In a study of 619 Canadian male ice hockey players aged 14 to 15 years, Sherar and colleagues (2007) showed that selectors for the Saskatchewan provincial team preferentially chose early-maturing boys who had birth dates toward the beginning of the selection year. The players selected for the final squad were significantly taller, heavier, and more mature than both the unselected players and the 93 age-matured control boys. Age at peak height velocity, an indication of biological maturity, predicted those selected for the first and second selection stages, accentuating the relative age effect.

The relative age effect may not be as pronounced in females as in males, at least in soccer, where gender differences were formally investigated by Vincent and Glamser (2006). They studied 1,344 female and male players considered by the U.S. Olympic Development Program in 2001 to be the most talented soccer players born in 1984. The investigators found only a marginal relative age effect for regional and national teams and no effect for state teams among the female players. In contrast, a strong relative age effect was found for the 17-year-old boys at all three levels of representation. Vincent and Glamser concluded that a complex interaction of biological and maturational variables with socialization influences contributed to the gender differences.

It would seem important that selection for specialized training be based on skills rather than body size and physical development. Coaches must find ways of accommodating late developers. There should also be opportunities for youth who missed out on early selection to be given a second chance as they catch up in growth. The relative age effect can clearly influence decision making with respect to developing athletes, when, for example, strict criteria about selection and specialization are applied in young people. The consequences may later track into adult working life.

Women

Characteristic anatomical and physiological differences distinguish males and females. Consequently, standards derived for males do not necessarily apply to females. Nevertheless, many occupations including emergency, military, and other services must be open to women. In this circumstance the key concern for fitness assessment is the cut score, or lowest value on a list that demonstrates the applicant can cope with the task demands.

Differences between males and females are recognized in competitive sport by the separate contests for each event. A variety of physiological measures explain the differences between male and female athletes, the magnitude being reflected in comparison of record times or distances for running, swimming, and throwing or jumping events. The main reason for the characteristic structural and functional features of women is their ability to bear children. The consequences of the menstrual cycle on exercise responses and the effects of strenuous training on the normal menstrual cycle are explained next.

Menstrual Cycle

The normal menstrual cycle has an average length of 28 days but varies between individuals and between cycles from 23 to 33 days. Menstruation refers to shedding of the surface portion of the endometrial wall and the bleeding that accompanies it. Menstruation (menses) lasts 4 to 5 days during which about 40 ml of blood is discharged with about two thirds of the endometrial lining. Blood losses usually vary from 25 to 65 ml but can exceed 200 ml, after which the woman may be anemic as a result of the heavy loss of blood. The endometrial wall is renewed under the influence of estrogens (mainly estradiol), and follicle-stimulating hormone (FSH) promotes the maturation of an ovum into a graafian follicle. This follicle ovulates at about midcycle (day 14), triggered by a surge of luteinizing hormone (LH). The ovum has to be fertilized within 24 hr for conception to occur. The wall of the ruptured follicle, from which the ovum has burst, then collapses, and the follicle now forms the corpus luteum, which produces increased amounts of progesterone and characterizes the luteal phase. The corpus luteum regresses if implantation has not occurred, usually by day 21, and progesterone levels decrease premenses. The endometrium is shed in menses, after which the next cycle commences.

The menstrual cycle is regulated by a complex system incorporating the hypothalamus (gonadotropin-releasing factor or GnRH), anterior pituitary (FSH and LH), ovaries, follicles, and corpus luteum (estrogens, progesterone, and inhibin) with feedback loops to the pituitary and to the hypothalamus. Contraceptive pills, composed of estrogen and progesterone combinations, prevent ovulation by inhibiting LH release. The control system as a whole is referred to as the hypothalamic–pituitary–ovarian axis (see figure 9.3).

Hormonal changes during the course of the normal menstrual cycle influence other aspects of human physiology that are important in exercise performance. An increase

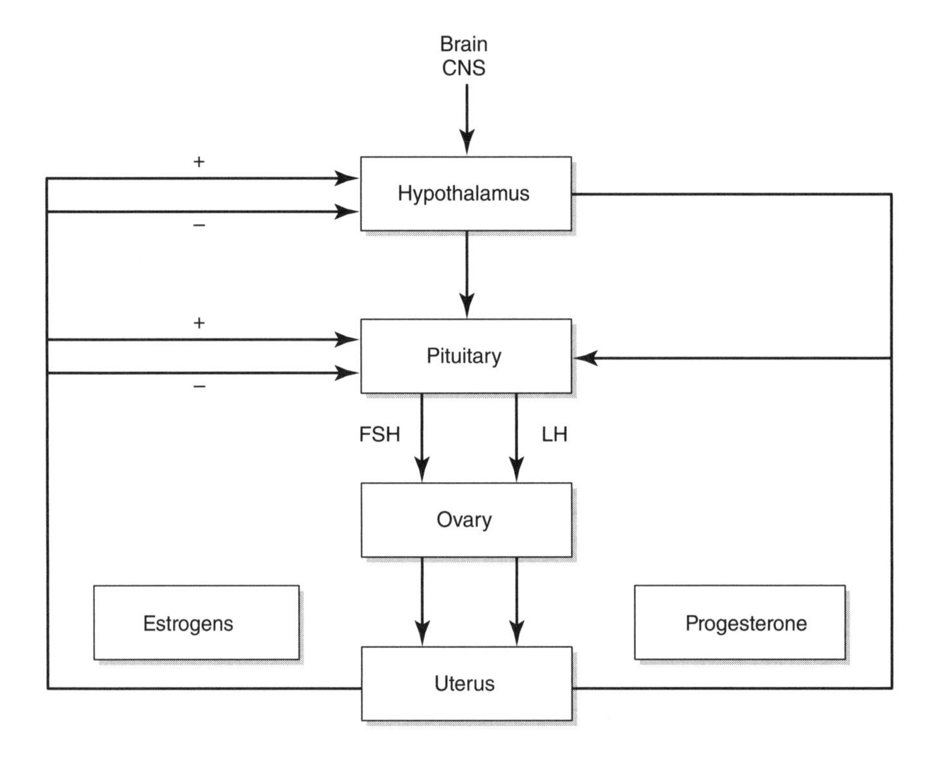

Figure 9.3 The hypothalamic–pituitary–ovarian axis with its controls over female reproductive hormones.

Reprinted, by permission, from T. Reilly, G. Atkinson, and J. Waterhouse, 1997a, *Biological rhythms and exercise* (Oxford, United Kingdom: Oxford University Press), 115.

in core temperature of about 0.5 °C coincides with ovulation. Body weight increases premenses attributable to the storage of water and altered potassium-to-sodium ratios. Loss of this premenstrual weight starts with menses. Some women experience abdominal cramps that are caused by increased prostaglandin production premenses. Administration of prostaglandin inhibitors helps to reduce this problem—the lowered incidence of menstrual-related cramp noted in athletes is probably attributable to lowered levels of prostaglandins. Others suffer painful menstruation known as dysmenorrhea. The many hormonal changes that occur during the menstrual cycle can alter metabolism and thus affect responses to exercise. A preferential use of fat as a fuel for exercise during the luteal phase would have a glycogen-sparing effect and enhance endurance performance. It is possible also that the elevations in steroid hormones (estrogen and progesterone) affect muscular strength in an analogous manner to the steroids used illegally by some athletes.

Exercise and the Menstrual Cycle

In many cultures taboos have been attributed to the menstrual cycle, particularly menses. In the context of sport, menstruating women were traditionally discouraged from swimming for hygienic reasons, and for a long time it was thought that they should not take part at all in strenuous exercise during menses. It was also thought that participation in endurance races, and in certain jumping and throwing events, impaired reproductive functions. The Olympic track-and-field program was extended to 800 m only at the 1964 games in Tokyo. The first Olympic marathon race for women took place at the Los Angeles Games in 1984, and the first 10K gold medal was contested by female runners at Seoul in 1988. Nowadays, female participation in sport is socially acceptable in most countries and menses is no bar to training or competing.

Olympic gold medals have been won, and world records set, at all stages of the menstrual cycle, and so exercise performance is not necessarily impaired during the menstrual cycle. Responses to submaximal exercise may be subject to changes; for example, an increase in ventilation at a set exercise intensity has been reported during the luteal phase (O'Reilly and Reilly, 1990). This increase was associated with the surge in progesterone noted at the same time. This elevation in ventilation would increase CO_2 output but does not seem to affect the maximal oxygen consumption ($\dot{V}O_2max$), as determined by an incremental exercise test to exhaustion.

The fuel used for oxidative metabolism can have a significant influence on performance in prolonged sustained exercise. In endurance athletic events lasting 90 min or more, the level of performance may be determined by prestart stores of glycogen in liver and muscle. Mechanisms that increase these depots or spare existing stores by increasing fat oxidation can improve overall performance. The elevated levels of progesterone and estrogen during the luteal phase of the menstrual cycle might benefit submaximal exercise of long duration by diminishing the use of glycogen. This view is corroborated by the finding of increased free fatty acids during exercise in the luteal phase (De Mendoza et al., 1979) and lowered levels of blood lactate (Jurowski-Hall et al., 1981). Eumenorrheic athletes have a lower respiratory exchange ratio (RER) in midluteal compared with midfollicular phases of the menstrual cycle during exercise at 35% and 60% of $\dot{V}O_2max$ (Hackney et al., 1994). The mechanism for altering fuel use is thought to be a hormone-sensitive lipase that stimulates lipolysis and is activated by the hormonal changes in the luteal phase. The time span over which performance might be enhanced is likely to be only a matter of 3 or 4 days, after which progesterone decreases premenses. This possible enhancement may be counteracted in a competitive context at other phases of the menstrual cycle when catecholamine secretion, which leads to similar effects, is increased in the course of competitive stress.

Low estrogen levels can adversely affect human strength. The effect has been demonstrated in ovariectomized mice whose force production was impaired after surgery. The ergogenic effect of estrogen has also been demonstrated in postmenopausal women when the adductor pollicis muscle (which draws the thumb in over the palm of the hand) was isolated for measurement of isometric force under experimental conditions (Phillips et al., 1993). The active stretch force—the tension within the muscle in response to its being stretched—is not impaired and the weakness can be offset by hormone replacement therapy. This loss of strength in the muscles of aging women may accentuate the loss in bone strength attributable to demineralization. A cyclical variation of muscle strength with changes in estrogen levels during the normal menstrual cycle is difficult to show. This is because performance in gross muscular function is influenced by a variety of factors other than circulating reproductive hormones.

The quality of sport performance depends on psychological factors such as attitude, motivation, and willingness to work hard. The most dramatic effects of the menstrual cycle may be observed in mood factors: Positive moods tend to be more evident during the follicular and ovulatory phases; negative moods are prominent preceding and during menses (O'Reilly and Reilly, 1990). These variations should be taken into account by sport coaches when structuring the training programs of female athletes.

Variations in mood are most pronounced in those who suffer from premenstrual tension (PMT). In its extreme form, PMT is characterized by irritability, aggression, abnormal behavior, and confusion. Bouts of irritation may be affected by the time of day, linked with fluctuations in blood glucose, with irritability peaking in the late morning if breakfast is missed (Dalton, 1978). In less extreme forms, sufferers may feel anxious and tired but unable to relax. Although the incidence of PMT is probably less common in athletes than in nonparticipants in sport, women in general seem to be more susceptible than normal to injury during premenstrual days. Swedish female soccer players were found most likely to sustain injury immediately prior to menses (Möller-Nielsen and Hammar, 1989). The mechanism for this increased susceptibility has not been fully resolved.

Athletic Amenorrhea

Female athletes on strenuous training programs may experience disruption of the normal menstrual cycle **(athletic amenorrhea).** One irregularity is a shortened luteal phase (Bonen et al., 1981). Secondary amenorrhea, or absence of menses for a prolonged period, has also been reported. So-called athletic amenorrhea is linked with low levels of body fat, low body weight, and high training loads; psychological stress is also implicated because of the influence of catecholamines on the hypothalamic–pituitary–ovarian axis.

Although amenorrhea is associated with low values of body fat, the mechanism responsible has not been clearly established. Endurance training lowers body fat, which in turn leads to a reduced peripheral production of estrogens through aromatization of androgens, catalyzed by aromatase in fat cells. The peripheral production of estrogens is thought to be important in stimulating the hypothalamic–pituitary–ovarian axis. Hard exercise or extreme weight reduction will lower pituitary FSH secretion, prevent follicular development and ovarian estrogen secretion, and decrease progesterone secretion.

Exercise-induced amenorrhea occurs in 20% of female athletes, compared with a prevalence of 5% in the general population. In runners, the prevalence increases linearly with training mileage to nearly 50% in all athletes covering 130 km (80 miles) per week. This linear increase is not found in swimmers and cyclists (Drinkwater 1986), because these athletes do not have to support body weight during exercise and their bones are not subject to the same repetitive loads as in the runners.

Psychological stress has been implicated in the occurrence of amenorrhea (Reilly and Rothwell, 1988). A sample of British international, club, and recreational distance runners was divided into those with amenorrheic, oligomenorrheic (irregular), or regular (normal) menstrual cycles. The amenorrheic athletes were younger and lighter, had less body fat, experienced more life stress, had a higher training mileage, and trained at a faster pace than the other groups. A high frequency of competition was the most powerful discriminator of the amenorrheic athletes from the other groups. This finding supports the possibility that increased outputs of catecholamine, cortisol, and endorphins interfere with the normal menstrual cycle by affecting the hypothalamic–pituitary–ovarian axis. Training-induced amenorrhea does not necessarily mean that the athlete is infertile, because ovulation can occur spontaneously and fertility can be restored after a long absence of menses. When high training loads are reduced, exercise-related menstrual disturbances are quickly reversed and there is a marginal increase in body fat.

The problems associated with severe athletic training include a negative energy balance. This problem may arise in sports with an aesthetic component where loss of body mass is indirectly encouraged. A hormonal imbalance that follows may lead to loss of bone mineral content. Athletes with low body mass appear to be more vulnerable than heavier runners. Although a moderate level of exercise stimulates bone growth and reverses bone loss in older women, overtraining in younger women leads to decreased bone density and risk of stress fractures. Reducing the training load and decreasing the frequency of competitive racing help restore the normal menstrual cycle. However, the interactions between training parameters and risk of osteoporosis have yet to be fully explored.

Oral Contraceptive Use

The reproductive process can be prevented by oral contraceptives. These work by blocking the normal hormonal feedback mechanisms and inhibiting ovulation. Oral contraception is also used to treat menstrual discomfort and to stabilize the menstrual cycle. Athletes may use oral contraceptives to ensure that important competitions do not coincide with menses.

The administration of estrogen or progesterone in appropriate amounts in the follicular phase prevents the preovulatory surge of LH secretion that triggers ovulation. Contraceptive pills contain combinations of estrogen in small amounts and progestins (substances that mimic the actions of progesterone), because excess of either type of hormone can cause excessive bleeding. Medication is started early in the cycle and is continued beyond the time that ovulation would normally have taken place. Administration can then be stopped to allow menstrual flow to occur as usual and a new cycle to commence. The use of contraceptive pills demonstrates that the menstrual cycle is produced by a series of feedback loops rather than originating from some internal master body clock.

Whether the use of oral contraceptives affects the performance of exercise is inconclusive. The reported adverse effects of body-water retention may have been attributable to the particular combination of hormones used in early contraceptive pills. If use of the contraceptive pill prevents adverse menstrual effects, the outcome is clearly beneficial. Their main influence is more likely to be a more consistent performance than normal, because they stabilize the fluctuations in peptide and steroid hormone concentrations linked with the menstrual cycle.

Antiprogestins offer control over menstruation without the risks of long-term exposure to estrogen and progestin. These synthetic agents block secretion of progesterone, the hormone responsible for the buildup of blood vessels in the uterus as the menstrual cycle progresses. This effect prevents the growth of the uterine lining so there is nothing to shed at the end of the cycle. Antiprogestins also inhibit ovulation, making them suitable for purposes of contraception. Given that few antiprogestin substances have been tested adequately in humans, their role in reproductive medicine is yet to be clarified.

Menopause

The end of the female reproductive cycle is manifest at menopause, when endogenous secretion of estrogen and progesterone ceases. The perimenopause period is characterized by hot flashes, attributable to periodic pulses of luteinizing hormone causing dilatation of skin arterioles and a feeling of intense warmth. Menopause occurs around age 50 as ovarian function gradually ceases.

In the absence of reproductive hormones, muscle strength is lost but can be maintained through hormone replacement therapy. Greeves and colleagues (1999) reported decreases of 10.3% and 9.3% in isometric and dynamic leg strength of women experiencing menopause compared with a group on hormone replacement therapy who maintained their levels of performance (figure 9.4). The findings support the view of a role for these hormones in determination of muscle strength.

Another consequence of menopause is a decrease in bone mineral density and a resulting risk of osteoporosis. Vertebral bone mass may decline at a rate of 6% per year, compared with an annual loss of 1% in males as an aging effect. The accentuated loss is attributable to the lack of estrogen, which acts on receptors in bone cells and stimulates the thyroid gland to secrete calcitonin. This hormone maintains the integrity of bone by moving calcium from blood to the bone and also blocks the action of parathyroid hormone. Both substances regulate bone homeostasis by balancing bone formation and bone resorption. After about 40 years of age, bone resorption tends to exceed bone formation so there is a gradual decline in bone mineral content with age that is accentuated at menopause in females.

The accentuated bone loss at menopause increases the risk of fracture upon falls and impacts. Winner and colleagues (1989) reported a higher incidence of falling in females at menopause. Although hormone replacement therapy can correct bone loss, its use is controversial. This debate arises from an increased risk of breast and endometrial cancer thought to be associated with this treatment, but incorporation of progestins is believed to reduce this risk.

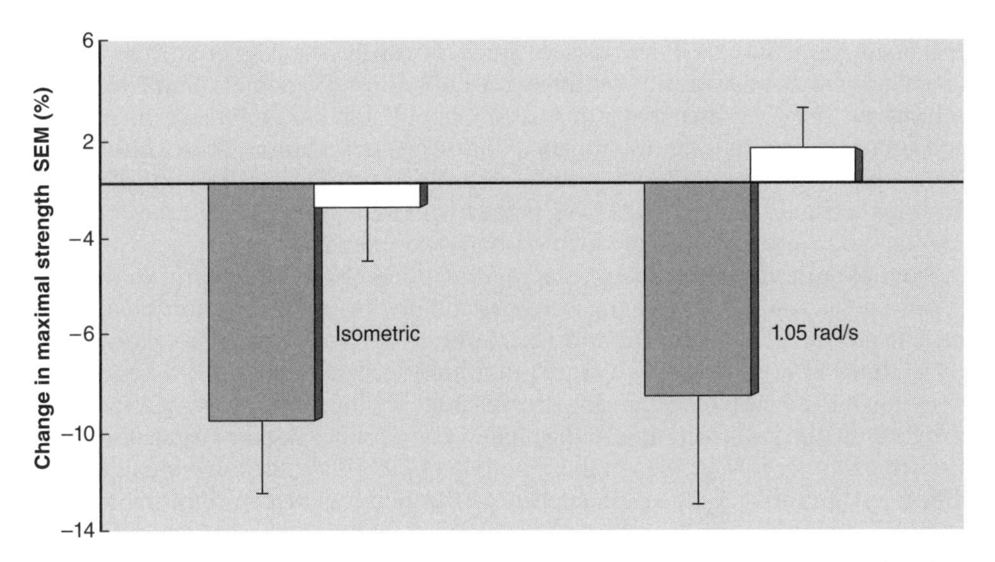

Figure 9.4 Changes in maximal isometric and dynamic strength (at 1.05 rad/s) between baseline and 39 weeks in postmenopausal women with (empty bars) and without (shaded bars) hormone replacement therapy.

Reprinted, by permission, from J.P. Greeves, N.T. Cable, T. Reilly, and C. Kingsland, 1999, "Changes in muscle strength in women following the menopause: A longitudinal assessment of hormone replacement therapy," *Clinical Science* 97: 79-84.

Exercise During Pregnancy

The physiological changes that occur during pregnancy impinge on exercise performance or the continuance of training. These include altered hormonal concentrations, particularly progesterone and human chorionic gonadotropin (hCG); increased plasma volume and red blood cells; increased ventilatory response; and associated changes in PCO_2, the partial pressure of carbon dioxide in the blood. Body weight increases during pregnancy, altering the woman's center of gravity. The average weight gain during pregnancy is about 10 kg; the gain is slowest in the first trimester.

Pregnancy itself does not prohibit the performance of exercise. Top athletes have competed in marathon races while pregnant, and pregnant athletes won medals at the 1956 Olympics in Melbourne. It is likely that female athletes unwittingly train and perform in the first trimester before the pregnancy is detected.

Women are advised not to start exercise programs during pregnancy, apart from dedicated regimens that aid in delivery at birth. The same caution does not apply to pregnant women who participated in physical exercise before pregnancy. Many female athletes wish to continue their physical training during pregnancy, albeit at a reduced level. There appears to be no compelling reason why they should not do so. Indeed, the balance of evidence is in favor of exercise in uncomplicated pregnancies, with benefits for mother and fetus. The evidence is provided in studies where physiological measures have been recorded at rest and during exercise throughout the course of a normal pregnancy.

Women may be obliged during pregnancy to maintain manual handling and lifting activities while in domestic or occupational roles. Changes occurring during pregnancy include weight gain, alterations in the center of mass and body shape, and adaptations in gait. Physiological and anatomical changes occurring during pregnancy are not necessarily detrimental to physical performance. Sinnerton and colleagues (1994) reported that lifting performance (isometric endurance lift; vertical and asymmetrical lifts) was not impaired during pregnancy up to approaching term. Reilly and Cartwright (1998) reported that isometric endurance in particular was improved postpartum and that lifting performance was not seriously compromised throughout pregnancy when the load was self-determined.

Physiological responses to exercise were examined by Williams and colleagues (1988) at low, moderate, and high exercise intensities in 10 pregnant women at each trimester. The heart rate and metabolic responses (oxygen consumption, ventilation) were compared with those in control subjects. The exercise consisted of stepping onto a bench for 5 min at each exercise level; the intensity was increased by speeding up the rate of stepping. At the highest exercise intensity, the pregnant subjects were able to sustain heart rates in the region of 155 to 160 beats/min without undue discomfort, except for local muscular fatigue in the third trimester. At comparable heart rates, the control subjects were able to do more work, but no deterioration was apparent in the fitness of the experimental subjects as pregnancy progressed. The higher metabolic responses compared with the control subjects were largely accounted for by the increased body weight and the physiological adjustments to pregnancy found at rest.

A woman's mental state may also vary during pregnancy and can be monitored using a standard mood-adjective checklist. There is a general trend toward increased fatigue and friendliness but decreased vigor and activity as pregnancy progresses. When reviewed alongside the physiological responses, these observations suggest that there is no compelling physiological reason for inactivity during pregnancy, and psychological factors may be more responsible for the decline in habitual activity usually associated with pregnancy.

Exercise should be promoted in healthy women with uncomplicated pregnancies. Clearly, some activities are unsuitable and exercise in hot conditions should be avoided because high temperatures can damage the fetus. Serious runners will need to curtail their training, if only because of the discomfort attributable to the alterations in gait and the extra body mass to be lifted against gravity. Warning signs include pain, bleeding, rupture of membranes, and absence of fetal movements.

Although it is hard to make individual prescriptions, many women restore their training to moderate levels soon after delivery. It is difficult to return to serious athletic training soon after delivery given disrupted sleep patterns and extra hydration needs for lactation. Nevertheless, many women have returned to serious competition within 6 months of giving birth and some athletes produce career-best performances once their family has been started. Both Liz McColgan (1991) and Jana Rawlinson (2007) won world championships at 10,000 m and 400 m hurdles, respectively, 8 months after delivery of their first child. Indeed, there may be a residual effect of hCG, a hormone secreted profusely during pregnancy, that benefits exercise performance after the baby's birth.

Elderly People

Aging is an inevitable part of the life cycle. Physical performance increases to a peak at a given age, depending on the capacity concerned, beyond which deterioration sets in.

Impairment is curtailed by physical training until some processes become irreversible as body cells die. This process is usually under way by the mid-50s but the individual may partly compensate by using experience. Retirement age is usually 65 years but in many countries retirement is not mandatory. Ageist criteria may not be applied per se in selection of employees, but the impact of aging on performance capability is relevant. This section provides the background for the aging process to demonstrate the complexity of the decline in elderly people. As functional capacities decrease with age, the relative demands of set tasks increase, a change that has consequences in sport and ergonomics contexts. Categories of sport competition are based on age groups in many sports and accommodate all categories from **veterans** (usually 35 or 40 years) to the oldest participants. This focus on competition between age-related peers acknowledges the decrease in performance capability with aging, a correction that has no strict parallel in occupational settings. Any evaluation of physical capabilities for work must be tempered by a consideration of effects of age as well as sex. Muscle strength and endurance tend to reach a peak value in adults before declining in middle age and decreasing progressively with further aging. The start of the decline may be attributable in part to disuse, so that the loss in performance capability may be arrested for a time or offset by physical training. The loss of strength is associated with a reduction of about 30% in total muscle mass between the ages of 30 and 75 years. There is a roughly similar decrease in aerobic power so that at the typical age of retirement, a 65-year-old man would have V.O2max value 70% of what he possessed at 25 years of age.

The aging adult loses muscle contractile cells that cannot be compensated by cell renewal systems or reactive myocyte hypertrophy. A negative balance between protein synthesis and breakdown causes skeletal muscle to atrophy. Chemical signals from hormones and growth factors and mechanical factors such as stretch or different patterns of activity are associated with exercise and can influence protein turnover irrespective of age. As the functional reserve of heart and skeletal muscle is decreased with age, this decline can be attenuated by physical activity. For exercise to be effective in this context, the intervention should occur before there is irreversible loss of motor units or cardiomyocytes. Speed deteriorates before muscle endurance, because loss of fast-twitch fibers precedes irreversible loss of slow-twitch fibers. There are also fewer small motor units in senescent muscles, leading to a loss of muscle steadiness and fine control of movements. These factors are worsened by a slower ability to recover from loss of footing, increasing the likelihood of falling after a stumble (Goldspink, 2005).

The aging effect is evident in all of the tissues associated with the oxygen transport system. The heart muscle, in particular the sinoatrial node, loses many of its myocytes. Calcification and accumulation of fat deposits occur in both the nerve conduction system of the heart and its ventricular walls. The maximal heart rate declines with age, the formula 220 minus age providing an estimate of maximal heart rate at any age. Compliance of the smooth muscle in blood vessels decreases, leading to an increased stiffness in arteries and veins (Goldspink, 2005). There is a gradual reduction in both systolic and diastolic blood pressure and a decrease in peripheral blood flow. Total-body hemoglobin is decreased, causing a reduction in oxygen supply to active muscles. The overall effect is a reduction in the functional reserve capacity. This decline becomes problematic when its margin above resting values leaves little leeway for physical activity.

Athletes With Disabilities

Physical or mental disabilities do not necessitate an absence from sport or physical training. The growth of sport for people with disabilities is underlined by the importance now attached by many nations to success in this domain. The categories of competition for **disability sport** are based on an assessment of the capacities that are impaired, the degree of impairment, and modification of rules of play. As examples, wheelchair sports and amputee soccer are given special attention. The use of prosthetics, modified for the sport, can provide a competitive edge and, subject to appropriate biomechanical assessment, allow individual athletes (such as South African 400 m runner Oscar Pistorius in 2008) to compete in races for nondisabled people.

Wheelchair Athletes

The Paralympic Games are the pinnacle of sports competition for participants with physical disabilities. The Games are held every 4 years at the same venues as the Olympic Games and in the weeks immediately following them. Besides the Paralympic Games, sports competitions for athletes with disabilities are organized at club, regional, national, and international levels in a variety of sports. This recognition of sports for people with disabilities mirrors the recognition of the needs of the participants at large within the community for access to buildings, facilities, and general amenities.

A wide range of disability is recognized in the competitive categories of Paralympic events. The standards for each sport are functional and reflect an attempt to have fair competition between contestants. The level of disability is a function of the spinal cord lesion, injury to C6, for instance, causing tetraplegia (i.e., affecting all four limbs). Wheelchair basketball, for example, can accommodate players with different degrees of disability, graded as 1 for the more serious disabilities to 4.5 for the least serious. With 5 players on court, the total count should not exceed an aggregate grade of 14.

Body composition patterning differs between wheelchair athletes and nondisabled counterparts. Increased lean tissue mass and increased bone mineral content are evident in the upper limbs of wheelchair athletes, in contrast to atrophy and demineralization in the lower limbs (Sutton et al., 2009). The forces associated with wheelchair propulsion seem to provide a training stimulus to bone as well as to the skeletal muscles engaged in the actions.

Success in wheelchair sports depends not only on the athlete's performance capacity but also on the design of the wheelchair. It is not surprising, therefore, that considerable engineering efforts have been made to improve vehicle mechanics and the wheelchair–user interface (Van der Woude et al., 1995). Such improvements have included the incorporation of lightweight material for vehicle construction, better weight distribution, proper alignment, and reductions in rolling resistance and air drag. Improvements in propulsion have included enhancements in hand rim propulsion mechanisms, variations in gear ratios and rim diameters, and improvements in hub cranks that allow a continuous motion of the hand around the wheel hub of a track or racing wheelchair.

Conventional ergometry must be modified to suit the needs of wheelchair athletes when their fitness is being assessed (see figure 9.5). Goosey and colleagues (1995)

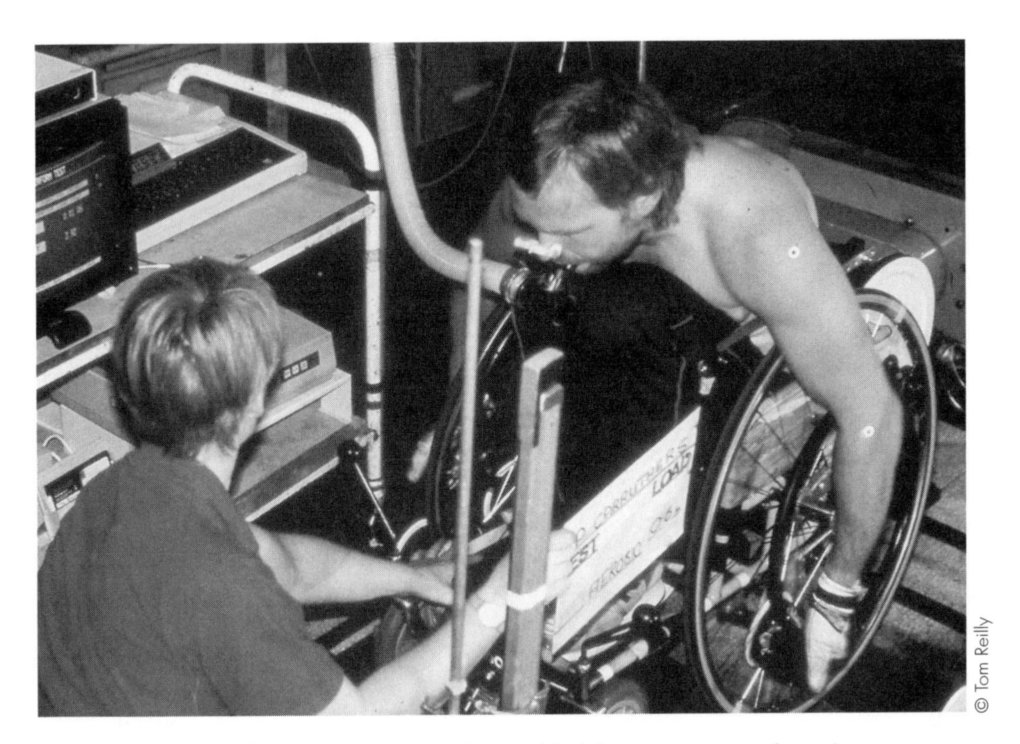

© Tom Reilly

Figure 9.5 Wheelchair ergometers can be modified for assessments of aerobic power or anaerobic power.

developed a treadmill test to examine the physiological responses of wheelchair athletes to submaximal exercise. The investigators suggested that a treadmill gradient of 0.7% was sufficient to stimulate an increased physiological demand without affecting the movement pattern of wheelchair propulsion. Athletes could be assessed using this protocol and their own racing wheelchairs at realistic race speeds. Similarly, Lees and Arthur (1988) described the validation of an anaerobic capacity test for a wheelchair ergometer. The test was modeled on the Wingate anaerobic test of Dotan and Bar-Or (1983), and both mean and peak power output were closely correlated with sprint performance times.

Amputee Soccer

Amputee soccer involves players who have classes A2/A4 and A6/A8 amputation. An A2 amputation is above the knee and A4 is below the knee of one leg. Conversely, A6 denotes that one arm is amputated above or through the elbow joint, and an A8 indicates one arm amputated below the elbow, but through or above the wrist joint. The A2 and A4 classes make up the outfield players and A6/A8 players can only play in goal. During match play participants are not allowed to wear prosthetic devices, and all outfield players must use crutches. Furthermore, the game includes Les Autres players, players with congenital limb impairments to either a single leg or arm. To ensure equality between various types of disability, outfield players and goalkeepers are not permitted to control or touch the ball with the residual limbs. In addition, crutches may not be used to advance the ball. Blocking, trapping, or touching the ball with a crutch, and any intentional contact between nonplaying limb and ball, are considered the same as a "hand ball" in nondisabled soccer.

The game consists of two halves each of 25 min with a 10 min halftime interval. Game surfaces include synthetic turf, grass, and indoor (sports hall) flooring. FIFA rules apply with some minor alterations to accommodate the disability of the players. Officiating is carried out using two referees, adjudicating half a pitch each. The offside rule does not apply, and kick-ins are used instead of throw-ins. Goalkeepers are not permitted outside the goal area, and slide tackling is prohibited. Substitutions are unlimited and any one player can be used on several occasions during a game. Finally, both teams can use one time-out in each half lasting 1 min (similar to basketball).

Wilson and colleagues (2005) described the physiological responses of the England national team playing seven consecutive games in the World Cup competition. The mean heart rate was 182 beats/min for the first three games before declining to a mean of 169 beats/min for the last game. It was evident that amputee soccer is an intense activity and that fatigue can develop over the course of a tournament.

Bioengineering Advances

Competitive runners with lower-limb amputation have benefited from developments in design of prosthetics. Athletes with bilateral amputations run on artificial legs that return to the runner the energy that is stored as the prosthetic foot hits the ground, in a similar manner to how tendons function in a natural ankle. The devices lack the ability of skeletal muscles to generate their own power and therefore provide less energy overall than natural legs. Implanting battery-powered motors into prosthetic legs helps to return more energy with each stride, an area of current work in bioengineering. Technological innovations also include the use of brain signals to manipulate prosthetic arms by capturing and transmitting the signals by means of implanted electrodes. These prosthetic limbs can be altered to provide an ergonomic advantage, such as helping an athlete run faster or a worker perform his or her job better.

The design of prosthetics used by participants at the Paralympics has benefited from developments in materials science and design technology. Participants also attract scientific support for analysis of their technique. Nolan and Lees (2007) described the adjustments to posture, kinematics, and temporal characteristics of performance made by athletes with lower-limb amputations during the last few strides prior to long-jump takeoff. Athletes with transtibial amputations appeared to adopt a technique closely resembling that of nondisabled long jumpers, although the former were less able to control their downward velocity at takeoff. Athletes with transfemoral amputations displayed a large downward velocity at takeoff attributable to an inability to flex the prosthetic knee significantly. This adjustment, combined with a relatively slow approach velocity, was considered a restriction to their performance.

Match Officials

A consideration of match officials is based on matching the workers (officials) with the physical tasks of the job. The job requirements have been documented with a task analysis—the documentation of distances covered, exercise rate, amount of high-intensity efforts, and time available to recover from them. In these ways, the strain on referees and other match officials can be viewed as an alternative application of a classical ergonomics approach. People in these posts tend to be neglected because the focus of spectator attention is generally on the athletes participating in sport competitions. Responsibility for implementing the rules of the sport and guaranteeing that competitors comply with the regulations is allocated to officials or referees. These officials may act individually, operate within teams, or work in concert with assistant referees. In the majority of sports the officials are active, needing to change their positions constantly to keep up with play. In other sports, such as tennis, boxing, and gymnastics, the officials are sedentary but are required to maintain vigilance and attention in the administration of their duties.

Sports officials are typically drawn from the population of former participants, bringing with them their experience of competitive engagement. Consequently, officials tend to be older and less well trained than current competitors. Nevertheless, the physical and physiological demands on match officials, notably in field games where large distances may be covered, may equal those imposed on participants. Referees must undergo training programs to meet these demands, and in many cases the ruling bodies set fitness standards as targets to be achieved. The advent of professional referees in the major games at an elite level of competition has promoted a more scientific approach to preparing match referees for their roles.

Soccer referees cover in excess of 10 km in an average professional game, with mean heart rates around 160 to 165 beats/min and $\dot{V}O_2$ close to 80% $\dot{V}O_2$max. Assistant referees cover about 7.5 km in the same time with mean heart rates of 140 beats/min, the corresponding oxygen uptake being about 65% $\dot{V}O_2$max. The relative strain on the referee is similar to that incurred by players who typically are about 15 years younger (Reilly and Gregson, 2006). For major tournaments and for operating at the top level, professional referees are obliged to pass threshold levels of fitness each year. Specific training programs enable them to meet these fitness targets. The main emphasis is on aerobic training, but anaerobic elements, agility, and backward and sideways movement should also be included.

Match officials become the focus of spectator attention when their decisions are contested. Apart from these sources of psychological stress, officials must quickly make decisions—visible and invisible to the audience—to retain command over the participants. Officials may feel pressure from the noise of a partisan home crowd or by prior knowledge of a team's reputation for aggressive behavior (Jones et al., 2002). Only in certain sports is the official's job eased by use of video technology. Decisions about line calls are assisted by optical sensors in tennis, and the Rugby Union referee can call for video replays when uncertain about whether a try was scored. At lower levels of sport, officials have recourse to the personal development programs of their sport, combined with their individual sport-related educational program.

Overview and Summary

Differences between groups are reflected in the rules and organization of competitive sports. Underage competitions are conducted in different age groups and do not necessarily lead to equality among participants. Similarly, males and females have separate competitive categories even though they may participate together in recreational activities. Veterans also compete in discrete age categories, an acknowledgment of the effects of aging on physical capabilities. People with disabilities are attracted to sports at a level where they can pit their abilities and skills against their peers. Match officials are not competitive in their roles but experience considerable physiological and psychological stress. Professional referees therefore need targeted training programs to help cope with these demands.

Coaches and trainers who prepare individuals for participation in sports must account for their needs and their characteristics. This consideration applies to their fitness training programs and the need for appropriate ergometry and test protocols. Ergonomics principles apply to the choice of shoes and clothing, training apparatus, and sport-specific equipment. Such purposive design should support individual aspirations and enhance satisfaction with participation.

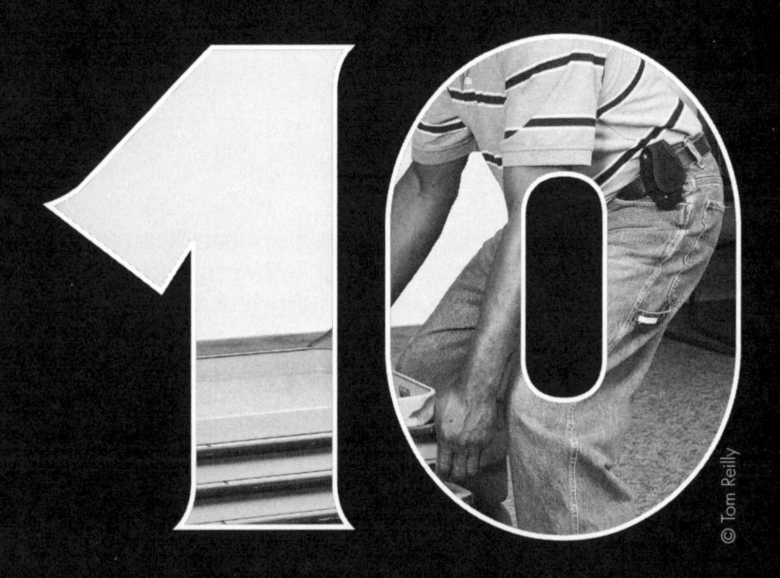

© Tom Reilly

10 Clinical Aspects

DEFINITIONS

contrast bathing—Water therapy that entails varying warm and cold treatments in a systematic manner.

cryotherapy—Any cold treatment, such as contrast bathing or immersion in water.

delayed-onset muscle soreness—Subjective symptoms and biological markers of microtrauma to muscle that peak 2 or 3 days following exercise that contains eccentric or stretching components.

Morton's foot—Condition characterized by an unusually long second toe and associated with overpronation.

predispositions—Imperfections in demographic and physical characteristics that if unresolved increase the risk of incurring trauma.

prophylactics—Preventive approaches to reduce the injury risk.

proprioception—The sensory system concerned with maintaining balance in static postures and in dynamic movements.

repetitive strain injury—Injury caused by fine movements at a high frequency, usually associated with the tendons of the small muscles in the hand.

cool-down—Light activity undertaken postexercise before terminating the session.

ten

ERGONOMICS has many applications in a clinical context and as a preventive tool. Safety and prevention measures are more cost effective than treatment. Such prevention measures emphasize the correction of weaknesses by implementing tailored training programs or using orthopedic devices, designing appropriate equipment to fit the individual, or improving tolerance to stressful contributors. As athletes strive for greater and greater improvements to performance in their quest for excellence, overload is in some cases inevitable. A failure to adjust to overload may induce maladaptive syndromes associated with overtraining, overreaching, and acute or chronic fatigue.

For the competitive athlete it is important to recover between contests or strenuous training sessions. Athletes use various methods to recover, regenerate, and restore hydration status. Physical contact sports may lead to contusions and bruises in participants and microtrauma to the body's tissues. The **delayed-onset muscle soreness** caused by eccentric contractions or stretch-shortening cycles of muscle actions can last for days following exercise. Methods of alleviating this form of soreness have included cool-down, various forms of **cryotherapy,** massage, and physiotherapeutic modalities.

A major role for the physiotherapist and physical trainer is the restoration of normal function in athletes recovering from injury. The stages in the rehabilitation process will depend on the nature and severity of the injury. Monitoring the restoration of function helps the trainer assess when the participant is ready to return to training and subsequently to competition. Without such guidance from the physiotherapist, there is a risk of reinjury.

A systematic program of physical training improves the outcome of surgery. Exercise as soon as feasible after surgery reduces atrophy processes and detraining effects attributable to inactivity. Exercise is advocated to ameliorate the effects of disease processes and reduce the risk of cardiovascular events recurring.

Predispositions to injury can be identified in a comprehensive assessment of the athlete. These factors include anthropometric, fitness, or psychological entities as well as a history of previous injuries. The isolation of individual weaknesses for improvement during training can reduce injury rates, especially among team participants.

In this chapter, individual predispositions to injury are first described. Attention is then given to musculoskeletal loading. Overtraining and its consequences are then considered before recovery strategies postexercise are addressed. The final section reviews nutrition interventions that help delay fatigue, enhance performance, and aid recovery.

Predispositions to Injury

Athletes can have anthropometric, physiological, or psychological profiles that predispose them to injury. These **predispositions** may be innate or acquired. In many cases, the injury may not manifest until the volume and intensity of training are too high for the participant to tolerate. If the predisposition is unresolved through remedial training or physical therapy, the participant who carries any existing weaknesses into competition is at an increased risk of incurring trauma.

Anthropometry

It is mostly assumed that individuals are symmetrical in limb lengths and in gait kinematics. This duplication between left and right sides is not always observed, because people tend to have a preferred body side that they favor. Because the lower limbs absorb force on ground contact while the upper limbs are used to manipulate objects, anatomical anomalies in the legs have received attention of researchers.

Leg-length discrepancies have been identified as a cause of injuries in distance runners. The difference is accentuated in an athlete running for a long period on an uneven road surface. The longer leg tends to be the one most pronated, with excess load on the medial aspect of the knee joint. Building up the shorter leg with a shoe orthosis can prevent injury. Even when no leg-length discrepancy is found in runners who overpronate, using orthotic inserts in the shoes can reduce injury risk.

Athletes on high training loads (particularly runners) are vulnerable to metatarsal injuries if they have **Morton's foot.** This anatomical characteristic is represented by an unusually long second toe. About 30% of the population have this feature, which often only comes to light when injury is incurred. The risk is reduced if training is conducted wearing shoes with good shock-absorbing properties and predominantly on compliant surfaces such as natural grass.

Muscle Strength

A difference between limbs in muscle strength can be a factor in injury. Typically the weaker limb is most vulnerable, especially in the case of hamstring injuries (Reilly, 2007). Regular assessment of muscle strength by means of isokinetic dynamometry highlights such weakness (see figure 10.1). Remedial action can be taken by targeting the weaker muscle groups for specific strength training.

Asymmetry can be manifest as an imbalance between joint flexors and extensors. At the knee joint, for example, the quadriceps muscle group is often emphasized in strength and weight training to the relative neglect of the hamstrings. The hamstrings need to be sufficiently strong to withstand a fast stretch, and so a high hamstrings to quadriceps strength ratio is important.

Traditionally it was thought that a hamstrings-to-quadriceps ratio of less than 0.6 rendered the hamstrings liable to injury. This proportion referred to the respective isometric strength of the knee flexors and extensors. With the use of isokinetic dynamometry, the index was expressed as the dynamic control ratio. This measure refers to the peak torque of the hamstrings in eccentric mode divided by peak torque determined concentrically for the quadriceps. A dynamic control ratio of 1:0 is deemed acceptable, although at high angular velocities a high ratio may be desirable (Fowler and Reilly, 1993).

Core stability is a function of the isometric strength of the trunk stabilizers, which act statically while the limbs move the body or change its orientation in space. Strong core muscles help to maintain balance and counter whole-body rotation in many sport actions. Core stability also protects participants in contact sports where they may be otherwise brushed aside or felled on contact.

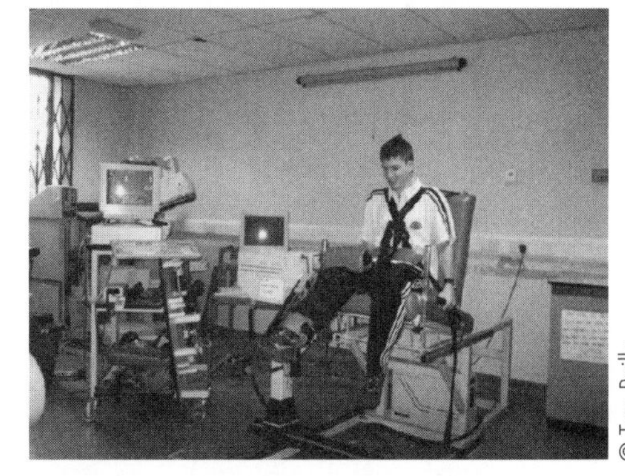

© Tom Reilly

Figure 10.1 Isokinetic dynamometer used to measure knee extension strength.

Aerobic Fitness

Aerobic fitness is typically determined by measuring $\dot{V}O_2max$ (aerobic power), the maximum ability to consume oxygen. Many sports are performed at submaximal intensities and so the person capable of exercising at a high percentage of $\dot{V}O_2max$ has an advantage. There are various methods of determining this exercise level, which is normally associated with accumulation of lactate in the blood. Various indices of this threshold level for maintaining exercise at the highest pace possible have been developed. Such measures are commonly referred to as anaerobic threshold, lactate threshold, or onset of blood lactate accumulation (Jones and Doust, 2003).

The greater the aerobic fitness, the better the athlete is able to withstand or delay fatigue. Fatigue refers to a decrease in performance despite attempts to maintain it. Fatigue has both peripheral and central components that have been implicated in injury occurrence. Injuries in football show an increased incidence toward the end of the game (Hawkins et al., 2001). The disrupted locomotion pattern in runners is also thought to be associated with the soft-tissue trauma experienced by endurance athletes when fatigued.

Repeated measurements of aerobic fitness help trainers monitor the recovery of athletes following injury. By comparing present values against baseline data for the participant, the trainer can effectively time the athlete's return to competition and avoid reinjury.

Musculoskeletal Loading

Musculoskeletal disorders have been implicated in a range of occupational injures, causing absences from work and health problems among employees. Lifting and manual handling, poor working postures, and faulty operational techniques have been implicated (figure 10.2). Among the more common disorders are low back pain, **repetitive strain injury**, and other upper-limb disorders. These problems have been a persistent source of research projects in the United States, Europe, and throughout the world. Musculoskeletal disorders occur among sport participants as well.

The cause of low back pain has been studied extensively in occupational ergonomics. Actions implicated in the causal nexus include the method of lifting and lowering heavy loads, bending or twisting during load carriage, and inadequate recovery periods between bouts of exertion. Educational and promotional programs have been launched to encourage correct lifting techniques and to lighten the loads being handled. The most effective interventions have been the replacement of human effort with technological aids. Another solution is to find a means of lightening the load. The use of hoists in handling bed-ridden patients, for example, has eased the musculoskeletal burden on heath care personnel (Beynon and Reilly, 2002).

Low back pain has been reported among sport participants. Runners may be at increased risk when training on hard surfaces or when they supplement their normal program with weight training exercises. Most injuries to the back during weight training occur because of faults in technique or lifting loads that are too heavy. Occasionally, inattention to floor characteristics and footwear is also implicated. Back pain among golfers and tennis players is associated with the high forces during driving and serving that are accompanied by spinal rotation. Cricket bowling seems to be especially demanding of the low back region when the chosen technique causes high torsion (Elliot, 2006).

Repetitive strain injury is associated with fine manipulative movements at a high frequency using the muscles of the hand. The syndrome includes severe wrist pain attributable to damage to the retinaculum. Scheduling of work-to-rest ratios can help prevent this condition. In occupational contexts, repetitive strain injury is linked with activities such as typing, certain factory tasks, and plumbing activities that involve frequent and rapid manipulations of lightweight tools.

Each sport has its own characteristic distribution of injuries: Ankle, knee, and lower-limb injuries tend to predominate in soccer, whereas upper-limb injuries are more common in rugby. Injuries in contact sports tend to occur more often during competition than training, the opposite applying in sports such as running. Training-related injuries are attributable to multiple factors, and modifications to the training program can reduce injuries.

The causes of injury ascribed by the athlete may not agree with the assessment of an informed observer. In a study of Swedish League soccer players, just under half of the injuries were attributed to factors related to the player (Ekstrand, 1982). In a study of a mixed group of games players (soccer, field hockey, and Rugby Union), close to half of the subjects ascribed their injuries to chance (Reilly and Stirling, 1993). In both cases, rule infringement was deemed causal about equally (see table 10.1).

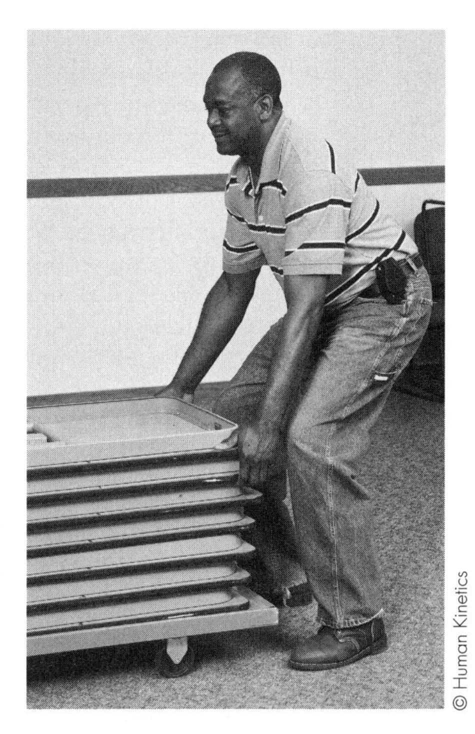

Figure 10.2 Heavy manual and handling activity is common among many jobs.
Photo by Neil Bernstein.

Overuse injuries in different sports have been named according to the sport involved. Tennis elbow refers to lateral epicondylitis (Hennig, 2007), whereas golfer's elbow affects the origin of the elbow flexor muscles on the medial side of the elbow. Injuries to golfers also include posterolateral elbow impingement (Kim et al., 2006). Swimmer's shoulder injuries include capsular laxity and tendinitis (Weldon and Richardson, 2001). Jumper's knee is attributed to patellar tendinitis, and footballer's groin has been reported in a variety of soccer codes (Slavotinek et al., 2005). Equivalents in occupational settings include housemaid's knee (inflammation of the prepatellar bursa), tailor's ankle (pain over the lateral malleolus from sitting cross-legged), and writer's cramp (related to carpal

Table 10.1 Mechanisms of Injuries and Their Attribution in Games Players (% of Total)

Swedish Football League (Ekstrand, 1994)				Games players (Reilly and Stirling, 1993)	
Player factors		Non–player factors		Ascribed causes	
Incomplete rehabilitation	17	Surfaces	24	Chance	47
Joint instability	12	Equipment	17	Poor warm-up	19
Muscle tightness	11	Rules	12	Poor rehabilitation	17
Lack of training	2	Other factors	5	Foul play	17

tunnel syndrome). Advice from a clinical team is often needed to devise appropriate treatments and to modify training practices.

Various instruments have been designed to assess incidence of musculoskeletal disorders among the workforce. The Nordic questionnaire has gained widespread acceptance for use within ergonomic studies and has been validated against clinical examinations. Descatha and colleagues (2007) concluded that Nordic-style questionnaires exploring prior symptoms can be useful tools for monitoring work-related musculoskeletal disorders to the upper limbs, especially if the questionnaires include numerical ratings of symptom severity. These questionnaires are used to uncover sources of injury so preventive actions can be taken. Methods of risk assessment and postural analysis (see chapter 3) provide another means of identifying potential mechanisms of injury and triggering preventive actions. In many instances kinematic analysis, electromyography, and behavior analysis are needed to provide insights into particular injury causes.

Overreaching

In an attempt to reach peak fitness, participants sometimes undertake training loads that are beyond their capabilities of adaptation. In such circumstances the performance level declines rather than improves. This condition is variously referred to as underperformance, overreaching, or overtraining syndrome. Overtraining reflects the athlete's attempt to compensate for a decrease in performance. This decline can be associated with incomplete recovery from illness or failure to recover from previous strenuous training. It sometimes results from inadequate nutrition during periods of strenuous training when energy expenditure is high. The increased effort in training becomes counterproductive and the athlete enters an underperformance spiral.

There is no universal formula that prescribes the perfect path to peak fitness. The theory underpinning the design of training programs to engender peak fitness levels is based more on empirical observations than experimental evidence, attributable largely to the many factors involved that are too difficult to model fully in laboratory investigations.

With respect to overuse injuries, Dvorak and colleagues (2000) considered various possibilities for reducing the rate of injuries in soccer, referring to perspectives of trainers, medical staff, and players. The trainers' perspectives included structured training sessions, appropriate warm-up, appropriate ratio of training to games, and reduction of physical overload. The medical perspective included adequate rehabilitation, sufficient recovery time, sufficient regard for complaints, and routine taping of weak ankle joints. The players' perspective embraced performance and lifestyle factors. The former included flexibility, skills, endurance, and improvement of reactions. The latter included personal habits such as smoking and alcohol use, nutrition, and fair play. Other issues were related to implementing the rules, observing regulations for play, and improving them where necessary.

Warming Up and Cooling Down

Warming up is important both for reducing injury risk and preparing for impending competition. The warm-up should be structured so that it provides specificity for the sport. Flexibility exercises form a relevant component of any warm-up practice. Preexercise activity raises muscle and whole-body temperature in preparation for the more strenuous exercise to follow. Injury prevention is most effective when the warm-up routine includes exercises specific to the sport in question (Reilly and Stirling, 1993).

There is a physiological basis for the desirability of cooling down, yet it is only rarely practiced. Its main benefit may lie in the acceleration of recovery processes. The main resistance to its universal adoption is likely to be the psychological state of those involved in the game, who may be emotionally affected by the game's events for some hours afterward.

The **cool-down** allows the players a few moments of reflection after competition or training. It may be a group activity after training or an individual activity after a game. In the latter case, the athlete can reflect on things done well or focus on aspects that were less successful. Emotions can be held in check while composure is regained should there have been any critical incidents in the game. By the time the athlete has changed clothing in the dressing room, the arousal level is restored to normality and the activities of the day can be evaluated in an objective manner. In this way the cool-down draws the professional aspects of the day's work to a close.

In a study of soccer players, Reilly and Rigby (2002) demonstrated the benefits of an active cool-down. The benefits were evident in a lower decrease in muscle performance over the 2 days following a competitive match and a reduction in the muscle soreness perceived. Muscle homeostasis was restored more quickly when the active cool-down was conducted.

Cool-down represents good hygiene and partly offsets a temporary depression of the immune system. The J-shaped relationship between exercise and the immune response suggests that strenuous exercise has a depressant effect on immune system function, whereas light exercise has a beneficial influence. The open-window theory proposes that the body is more vulnerable than normal to infection, particularly upper respiratory tract infections, for 4 to 6 hr after exercise. Finishing off the session with a light cool-down may offset the detrimental effects of the preceding intensive exercise on the immune system. It could prevent a sudden change in thermal state that appears to be implicated in picking up a common cold.

Absence from infection is necessary for elite athletes to maintain their status and reach new peaks. Malm (2006) modified the J-shaped curve to suggest an S-shaped relationship between exercise load and risk of infections. When the entire range of sports participation is considered, the relationship may resemble the shape of W (see figure 10.3). An elite athlete must have an immune system able to withstand infections even when coping with severe physiological and psychological stress.

Recovery Processes

Impairment in performance capabilities is assumed to occur temporarily after strenuous training. The recovery of physiological and mental resources varies according to the function impaired and can range from 24 hr to 3 weeks. Athletes are interested in accelerating recovery so that training can proceed according to plans. There is no single answer to this problem, and ergonomics interventions can be applied as a combination of measures. These include restoring energy and fluid levels, decreasing muscle soreness, modifying training, and having sports massages.

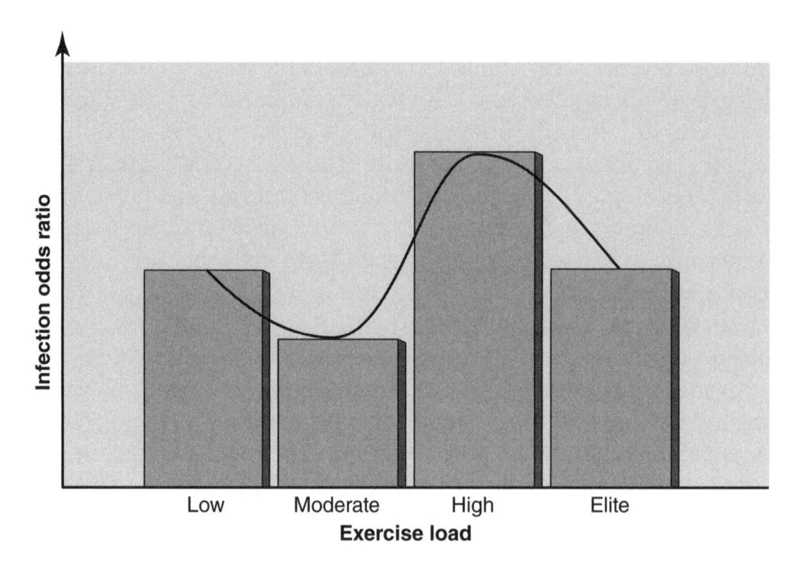

Figure 10.3 The S-shaped curve proposed to link training load and infection rate.

Adapted from C. Malm, 2006, "Susceptibility to infections in elite athletes: The S-curve," *Scandinavian Journal of Medicine and Science in Sports* 16: 4-6.

Energy Level Restoration

Competitive sport can reduce energy levels, in particular depleting carbohydrate stores. The resynthesis of glycogen in muscle and liver is therefore a priority soon after competition or strenuous training ends.

The optimal time for beginning the replacement of energy is in the first 2 hr after exercise ceases because the enzymes associated with glycogen synthesis are most active during this period. Glucose sensitivity and muscle GLUT4 expression are increased in the early period postexercise (Dohm, 2002). Nutritional guidelines suggest a carbohydrate intake of 1.5 g/kg body mass over the first 30 min of recovery. This figure would amount to 120 g of carbohydrate for an individual weighing 80 kg. The glycogen resynthesis rate is itself limited (Coyle, 1991), suggesting that an intake exceeding 50 g of carbohydrate beverage can initiate the recovery of energy and compensate for the transient suppression of appetite attributable to strenuous exercise. Solid foods with a high glycemic index can be provided in the dressing room for athletes to eat after showering. Foods with a high glycemic index are listed in table 10.2. The inclusion of essential amino acids along with carbohydrate provides a suitable means of enhancing protein synthesis, especially if ingested 1 to 3 hr postexercise (Rasmussen et al., 2000). Protein degradation is often increased after exercise, so imposing a net protein gain is important (Tipton and Wolfe, 2001). The restoration of energy must be continued the next day to be effective. The protocol may entail ingesting 10 g/kg or more of carbohydrate over this day, representing a proportionate carbohydrate intake of 60% of the daily energy intake.

Euhydration

The fluid deficit incurred during training or competition should be reversed as soon as possible afterward. Drinking pure water in the period after finishing hard exercise lowers plasma osmolality and plasma sodium levels; this reduces thirst and increases

Table 10.2 Classification of Some Foods Based on Their Glycemic Index

High index	Moderate index	Low index
Bread	Pasta	Apples
Potato	Noodles	Beans
Rice	Crisps or chips	Lentils
Sweet corn	Grapes	Milk
Raisins	Oranges	Ice cream
Banana	Porridge	Yogurt
Cereals	Sponge cake	Soup
Glucose		Fructose

urine production, both of which tend to delay effective rehydration (Maughan, 1991). Drinks that include electrolytes, notably sodium, facilitate absorption of water through the intestinal wall. The electrolyte content of sweat varies between people; some may be in particular need of electrolyte replacement, notwithstanding that sweat tends to be hypotonic. If sodium and chloride are not included in the drink, some of the fluid ingested is lost again in the urine. As body water content begins to decrease, the secretion of vasopressin (a posterior pituitary hormone) stimulates renal retention of fluid while the adrenal glands secrete aldosterone in an attempt to preserve sodium. Nevertheless, the body's total electrolyte reserve can tolerate some short-term losses without any evident effect on physical performance. Most meals that cover the daily energy expenditure also include enough electrolytes to compensate fully for losses that occur during training. Because there is a marked variation between people in the sodium content of sweat and in the amount of sweat lost during exercise, some additional salt (added to food or in drinks ingested) may be needed for those athletes incurring high salt losses.

The loss of water from the body during exercise (dehydration) can impair performance in training and competition. Because thirst is satisfied before body water is fully restored, athletes need to drink more than they believe they need. The deficit may carry over into the following day and affect performance in training, especially when matches or training is held in hot conditions. Monitoring urine osmolality, or its specific gravity or conductance, provides a good indication of hydration status (Pollock et al., 1997). There is no gold standard measure of hydration status, although a urine osmolality greater than about 900 mOsmol/kg has been recommended as a reasonable indication of a hypohydrated state (Shirreffs, 2000). The urinary measures are deemed more appropriate than blood markers, and urine osmolality is the preferred method. Simple measurements include monitoring of body mass (in the morning) or assessment of urine using a color chart (Armstrong et al., 1998).

Regeneration of normal metabolic reserves following exercise is a priority, and practices to promote recovery should commence once the competition (or formal training) is finished. The benefits of physical therapy are unclear, except where minor soft-tissue trauma has occurred. The training program should be modified to take into account the transient reduction in physical capacities and include incorporate recovery sessions in the weekly regimen. Unless complete recovery is achieved, the athlete will enter the next contest at a disadvantage. To avoid this possibility, recovery strategies should incorporate specific hydration, nutritional, psychological, training, and lifestyle factors. There are numerous interactions among the factors influencing recovery, and so knowledge of the principles involved is important.

Massage

Massage has a rich history of use in sports, particularly in cycling and the football games. Massage is believed to be helpful in preparing muscles for strenuous exercise and accelerating the recovery from microtrauma. It is thought also that the relaxation effects of massage alleviate psychological stress. There are different schools of massage but this form of physical therapy can be applied in generic terms.

Surface massage has a twofold effect. It stimulates blood flow within the underlying muscle and promotes reabsorption of any hematoma that is present. Relief from aches attributable to minor trauma or microdamage can result from the soothing effects of massage. It has been found to increase muscle temperature more effectively than ultrasound, although both of these therapeutic modes have only limited effects on deep muscle temperature (Drust et al., 2003).

Hilbert and colleagues (2003) reported that massage offered subjective relief of symptoms associated with delayed-onset muscle soreness but had no effect on reducing muscle damage and the inflammatory response. Although massage may have a role in physiotherapy, there is no compelling evidence that it facilitates the recovery of physiological processes following intense sport competition.

Cryotherapy

Exercise in water has long been used for both training and rehabilitation. The resistance to motion provided by water and the buoyancy to reduce impact loading are the underlying reasons for training in water. Hydrotherapy pools are common in sports medicine clinics and spas. Deep-water running has been found to relieve muscle discomfort following stretch-shortening exercise without affecting the creatine kinase response, which is indicative of muscle damage (Reilly et al., 2002). Water-running drills tend to be conducted in swimming pools where the water temperature is compatible with thermal comfort. Cryotherapy, contrast bathing, and ice baths use water at much lower temperatures, usually around 8 °C.

Eston and Peters (1999) reported only limited benefits for cold-water immersion and no effect on the perception of tenderness or strength loss following damage-inducing eccentric exercise of the elbow flexors. Their cryotherapy technique entailed submersing the exercised arm in a plastic tub of ice water for 15 min. Howartson and Van Someren (2003) found that although ice massage reduced creatine kinase, it had no other effects on signs and symptoms of exercise-induced muscle damage. Bailey and colleagues (2007) reported that immersion for 10 min in a water temperature of 10 °C immediately after prolonged intermittent exercise reduced some (but not all) indices of muscle damage.

Immersion in cold water induces a host of physiological responses that include hyperventilation, bradycardia, and alterations in blood pressure and blood flow

(Reilly and Waterhouse, 2005). Heat is lost much more quickly in water than in air, so immersion in cold water must be short term, a matter of minutes. The use of ice baths to promote recovery from strenuous muscular exercise has gained credence in team sports, notably Rugby Union and soccer teams. Others have preferred **contrast bathing,** again a conventional practice within physiotherapy, whereby periods of immersion in cold water are interspersed with exposure to water at a temperature close to mean skin temperature.

Contrast bathing, alternating immersion up to the trunk in barrels of cold water with warmer water or air, is practiced by Australian Rules football players but without convincing evidence of its benefits. Dawson (2005) used a protocol of alternating between standing in a hot shower (~45 °C) for 2 min and standing waist-deep in icy water (~12 °C) for 1 min, repeated until five hot and four cold exposures had been completed. The England Rugby Union team first used the strategy of immersion in ice water following its matches in the 2003 International Rugby Union World Cup tournament. Typically, a water temperature of 8 to 11 °C is used in ice-water baths for games players, but the duration and frequency of immersions seem to be more determined by local and individual preferences or by trial and error than by scientific evidence. Where contrast baths are used in preference to ice baths, a protocol of 1 to 2 min in each medium for four or five successive immersions is common.

Whole-body cryotherapy is the practice of exposing individuals to very low temperatures for short periods of time in a special environmental chamber. In a typical session the subject is exposed to −60 °C for about 30 s and later −110 °C for 2 min. Despite this extreme cold stress, subjects report improvements in general well-being and reductions in inflammation and pain. The technique has been used for various conditions, especially rheumatoid disease. It has also been used by professional Rugby Union and soccer players, apparently with positive subjective responses, but controlled studies to clarify the effects are still awaited. In their study of 10 players from Italy's national Rugby Union team, Banffi and colleagues (2008) concluded that whole-body cryotherapy cannot be considered as an unethical means of blood boosting because they found it did not alter any hematological measures.

Compression Garments

Compression garments have been promoted as a means of boosting recovery following strenuous exercise. They are usually provided in the form of tight-fitting stockings over the full length of the legs and are worn postexercise and overnight. They are worn by some travelers to safeguard against developing deep vein thrombosis attributable to spending a prolonged period on board flight in a cramped posture.

Montgomery and colleagues (2008) studied the effectiveness of compression garments in hastening recovery by monitoring players during a basketball mini-tournament in which three games were played on three consecutive days. The compression stockings provided no reduction in the cumulative fatigue occurring during the tournament matches. In contrast, the decrease in performance was attenuated by a regimen of water immersion consisting of five 1 min exposures up to midsternal level in 11 °C water with a 2 min break between immersions. The authors concluded that cold-water immersion restored physical performance more effectively than did carbohydrate plus stretching routines or compression garments.

Prophylactic Measures

It is a truism to state that prevention is better than cure. The elimination of weaknesses identified in advance provides the individual with a strengthened base for coping with training and competitive demands. Many of the **prophylactic** measures used in contemporary sport have their origins in clinical contexts. Among these are the training or retraining of balance and the use of high-technology platforms such as manifest in the CAREN (computer-assisted rehabilitation environment) system. The following sections also address the use of exercise referral schemes in public health, the potential of artificial neural networks across a spectrum of applications, and the use of selected supplements as preventive measures.

Balance Training

Balance is needed in some form in all sports; in activities such as gymnastics, skiing, track-and-field athletics, and court and field games, being able to correct the orientation of the body's segments in space or on the sport's apparatus is crucial to success. This facility is a complex function of neuromuscular coordination involving the body's **proprioceptive,** vestibular, and visual systems.

Balance training is normally achieved by means of sport-specific drills. A controlled form of retraining balance is needed in rehabilitation from injury. Traditionally, this restoration has been achieved by using a wobble board, where the athlete tries to recover as a perturbation causes the board to move. Recently, computer-aided systems have been found to be effective for balance training, and these systems allow training to be tailored to the individual athlete.

The CAREN system is a computer-controlled system designed for clinical applications. It consists of a platform that can be perturbed in each of six degrees of freedom. Originally a device for postural and balance research, it has many and varied applications (see figure 10.4). These include retraining of movements in patients with neuromuscular disorders, rehabilitating patients after stroke, studying the effects of exercise on relearning movements, and training of gait following injuries. Its potential for applications to sport skills has yet to be exploited (Lees et al., 2007).

The kinematic response characteristics of the CAREN system to sine and ramp input functions were described by Lees and colleagues (2007) for its six transitional and rotational axes. The investigators concluded that this moving platform device was appropriate for postural and balance research and had some unique features that can be used in research.

The coupling of computer-controlled moving platforms to visual displays enables virtual reality scenarios to be presented to a subject, patient, or learner. Activation of specific muscle groups while balancing on the platform can alter certain features of the display. This principle has been applied in neurological investigations by targeting specific muscles for training. It should be applicable in the future to a variety of sport actions as well as to clinical conditions.

A potential downside of immersion in a virtual environment is the production of symptoms similar to those of motion sickness. The three-dimensional impression is obtained by the use of spherical lenses in a headset when monitoring the display on screen. With repetition, the symptoms are generally reduced in prevalence and severity, and the onset of nausea is delayed. Howarth and Hodder (2008) found evi-

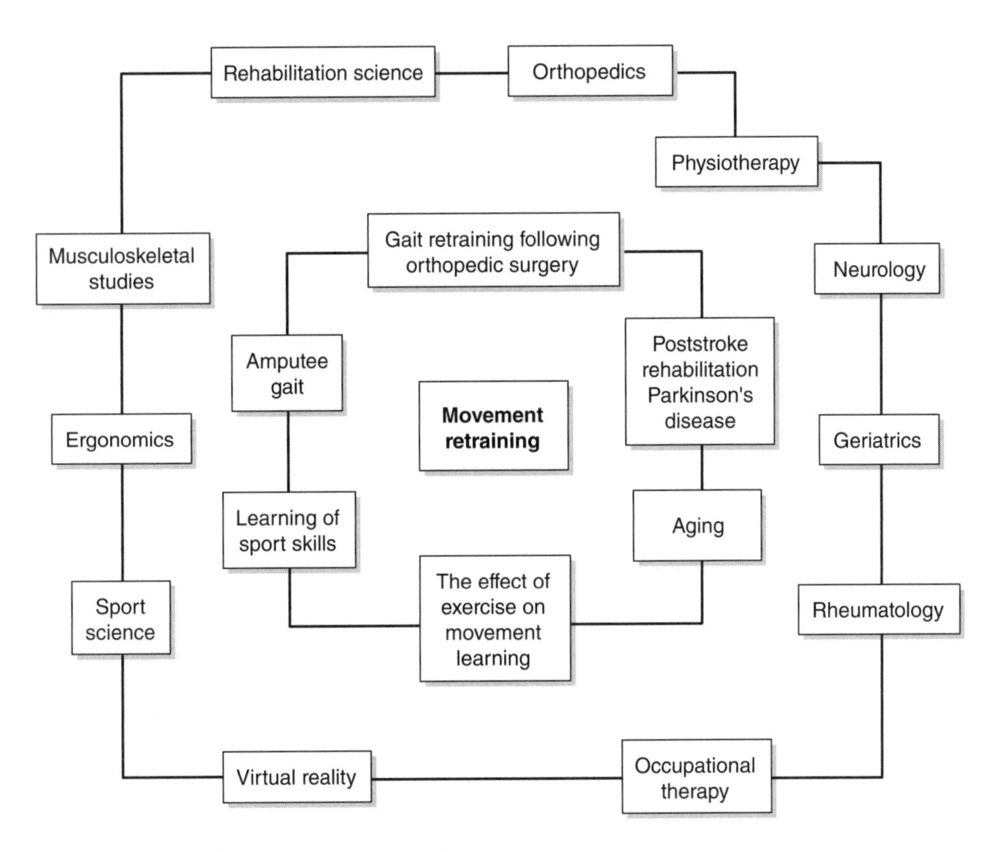

Figure 10.4 Applications of moving-platform devices.

dence of habituation in participating in a computer-based racing game, the number of exposures being more important than the interval between exposures. There were small improvements in visual acuity and some modifications of behavior, although head movements alone did not seem to elicit a universal effect on visually induced motion sickness. The time scales of habituation to different symptoms (general discomfort, fatigue, boredom, dizziness, vertigo, visual flashbacks, faintness, confusion, vomiting) indicated differences from simulator sickness and true motion sickness (Howarth and Hodder, 2008).

Exercise for Prevention

Preventive approaches are preferable to curative treatments wherever possible. Training can be geared toward eliminating factors predisposing to injury, targeting weaknesses for improvement, and avoiding training errors. A systematic program of flexibility can reduce muscle tightness and reduce the incidence of muscle tears (Ekstrand, 1982). Special attention is needed when the training load is increased or stretch-shortening cycles are introduced into the conditioning program. A sound conditioning base is necessary before an athlete undertakes plyometric training, and transient muscle damage is reduced when stretch-shortening cycles of exercise are repeated. This enhancement of training tolerance is known as the repeated bouts effect of exercise (Cleak and Eston, 1992).

Alleviating Delayed-Onset Muscle Soreness (DOMS)

- Stretching
- Cold application
- Ultrasound
- Transcutaneous electrical nerve stimulation (TENS)
- Anti-inflammatory creams
- Pharmacological agents
- Exercise
- Massage
- Training

Note: Although some success was found by a few authors using stretching, transcutaneous nerve stimulation TENS, and topical anti-inflammatory creams to alleviate DOMS, the majority view was that there was no single effective way of reducing the soreness once it has occurred. Prevention is best secured by training according to the repeated bouts effect.

Reviewed by Cleak and Eston (1992).

Strength and conditioning work provides the best means to reduce the impact of delayed-onset muscle soreness (see the box above). This type of soreness occurs if the participant incurs contusions and bruises via physical contact. Pharmacological means (such as nonsteroidal anti-inflammatory drugs) of treating muscle soreness attributable to eccentric exercise have proved largely ineffective (Gleeson et al., 1997). Ultrasound may be effective therapy for other muscle complaints but has not proved beneficial for DOMS.

Exercise Referral Schemes

Exercise programs for preventing cardiovascular diseases have been promoted with guidelines for intensity, frequency, and duration of training. The recommendations of the American College of Sports Medicine (1998) have been adopted worldwide and used as a standard for determining training quality. Attention has also been directed toward using exercise during rehabilitation from coronary heart disease. Instructors overseeing and supervising these training sessions are specifically qualified to deal with this population.

The role of exercise in preventing other diseases and disorders is increasingly acknowledged. These include diabetes, metabolic syndrome, a number of cancers, fibromyalgia, and back pain. Personal trainers are advised by their professional body with respect to unique needs of clients with spinal cord injury, multiple sclerosis, epilepsy, and cerebral palsy (La Fontaine, 2004). Pregnant women and people who have specific orthopedic concerns should get approval from their physician (with the addition of a consultant's advice where appropriate) before participation.

The exercise referral scheme in the United Kingdom was implemented to promote physical activity among susceptible people to decrease ill health among the population. The scheme reflected its global strategy of the World Health Organization to address health problems associated with physical inactivity. The effectiveness of the initiative relied partly on the source of referral, participants referred from cardiac and practice nurses having greater adherence than those referred from their general physician (practitioner). There are clear benefits when participants comply with program requirements: It seems that the nationwide exercise referral scheme has value for certain segments of the population but not necessarily all (Dugdill et al., 2005).

Supplements in Clinical Contexts

Among the complex psychological responses to physical activity, strenuous exercise is associated with the production of free radical oxygen species. Antioxidant defenses located within body water pools and in lipid stores are used as scavengers to prevent damage to cells caused by free radical species. It may be that antioxidant resources are enhanced by use of nutritional supplements. However, Malm and colleagues (1997) found that supplementation with a lipid-soluble antioxidant (co-enzyme Q10) had an unanticipated negative effect on young soccer players during a period of increased training load. It seems that supplementation with this particular antioxidant is undesirable.

It has been thought that other antioxidants may benefit recovery processes after sports competition. Thompson and colleagues (2003) investigated whether postexercise supplementation with 200 mg of vitamin C influenced recovery from 90 min of shuttle running designed to correspond to the average exercise intensity of a soccer match. No differences were found between the group receiving supplementation and a placebo group in the rate of recovery for the 3 days following exercise. Serum creatine kinase activities, myoglobin concentrations, muscle soreness, and recovery of muscle function in leg extensors and leg flexors were similar between the two groups. Plasma concentrations of malondialdehyde (reflecting oxidative stress) and interleukin-6 increased postexercise equally in the placebo and supplemented groups. It was concluded that either free radicals are not involved in delaying recovery from such exercise protocols or the consumption of vitamin C immediately after exercise is unable to deliver an antioxidant effect at the appropriate sites with sufficient expediency to improve recovery. It seems that the benefits of antioxidant supplementation may be long term rather than short term.

Carbohydrate supplementation during periods of heavy training is a potentially effective countermeasure to illness susceptibility attributed to the open-window period. Carbohydrate supplementation has been shown to attenuate increases in blood neutrophil counts, stress hormones, and inflammatory cytokines. Nevertheless, carbohydrate does not convey immunity because it is largely ineffective against other immune components, including natural killer cell and T-cell function (Nieman and Bishop, 2006).

There has also been an interest in using centrally acting muscle relaxants to relieve painful muscle spasm. Bajaj and colleagues (2003) investigated the role of tolperisone hydrochloride, hypothesizing that its use as a prophylactic agent relieves muscle soreness, based on the spasm theory of exercise pain. The investigators found no difference between the treatment and a placebo in perceived soreness for 48 hr after eccentric exercise, but the decrease in isometric muscle force was accentuated. The prophylactic use of muscle relaxants could not therefore be supported in athletes.

Artificial Neural Networks

With developments in bioinformation technology, computer-based techniques are increasingly used in clinical contexts. Artificial neural networks represent an attempt to simulate how the neurons of the central nervous system work together to process information. The activities of an input layer determined by the raw information fed into it, a hidden interconnecting muscle layer, and the output units in the network correspond to the dendrites, cell body, and axons of neuronal organization. A neural network tends to be configured for a specific application such as recognizing patterns or classifying data. The network learns to recognize patterns among its interconnecting elements, much as adjustments to synoptic connections occur with human learning. This approach is an extension of computer science and artificial intelligence.

Neural networks are used to model parts of living organisms such as the cardiovascular system or the mechanisms of the brain. These networks have therefore found applications in physiology, psychology, neuroscience, and movement science. They are also suitable for identifying trends or patterns in data and so have found use in risk management. Their potential for application to sport has not been realized, but the most likely applications are in sports medicine and performance analysis.

Neural networks have been used as tools for decision making in a variety of clinical contexts. They have been used for diagnosis, prognosis, and survival analysis in the medical domains of oncology, critical care, and cardiovascular medicine (Lisboa, 2002). Neural networks learn from examples of the variations of a disease and in this way are said to be trained. Diagnosis can be achieved by building a model of the relevant biological system for an individual and comparing it with real measurements from the patient. In this way potentially harmful conditions can be detected early.

Barton and colleagues (2006) used self-organizing neural networks to reduce the complexity of joint kinematic and kinetic data in conducting an assessment of gait. Three-dimensional data for joint angles, moments, and power values were projected onto a topological neural map, and patients were positioned on the map in relation to each other. This means of comparing gait patterns was considered a step toward an objective analysis protocol for clinical decision making.

Perez and Nussbaum (2008) provided an example of how a neural network model could be used to predict postures during manual materials handling tasks. They developed a model to predict full-body posture at the start and finish of a lift in two and three dimensions. Their models provided what they considered reasonably accurate predictions, outperforming previously available computational approaches. A challenge for future developments is to apply the information generated to lifting strategies that suit individual anthropometric and strength characteristics.

Because neural networks can extract meaning from imprecise data, they have the potential to clarify patterns not readily apparent to the human eye. Although these networks have found applications in biomechanics and sports medicine, their potential for analyzing sport performance has not been thoroughly investigated. Passos and colleagues (2006) used artificial neural networks to reconstruct three-dimensional performance space by focusing on one-on-one encounters in the game of rugby. Their conclusion was that neural networks may be instrumental in identifying pattern formation in team sports generally. Neural networks might also have value in formulating patterns in fitness and anthropometric data that would suggest predisposition to injury.

Overview and Summary

Exercise participation has numerous benefits in eliciting physiological adaptations and an accompanying sense of well-being. It also has attendant risks of acute and chronic disorders associated with overload or extrinsic factors. A preventive approach is effective in reducing injuries but is often insufficient on its own.

Physical and clinical assessments can identify individual weaknesses and provide remedial guidelines. Assessment is especially relevant during rehabilitation with a view to avoiding reinjury. In the short term various methods are available to accelerate the recovery process and help the athlete return quickly to full training. These measures require a systematic approach in engaging exercise therapies that are more effective than pharmacological means. Exercise programs have value in ameliorating metabolic and cardiovascular disorders and can ease the symptoms of neurological conditions. In all these instances special attention must be given to individual cases, both in designing exercise regimens and monitoring their effects.

A Holistic and Nutritional Ergonomics Perspective

DEFINITIONS

alkalinizers—Substances used to buffer the body's accumulating lac tate levels during high-intensity exercise and thereby delay the onse of fatigue.

amphetamines—Drugs that increase arousal by stimulating the centr nervous system.

blood doping—Use of artificial means of boosting the body's oxyge transport system by altering the red blood cell count.

human enhancement technologies—Scientific approaches to increa ing human capabilities.

neuroergonomics—The study of brain and behavior at work.

participatory ergonomics—System whereby the individual takes an active part in decisions about ergonomic aspects of task design, workstation layout, and other factors.

population stereotype—The expectation that movement of a control device causes a compatible motion or change of direction denotec by the display.

supplements—Pills or foods used with the intention of raising the body's energy stores or recovery to normal homeostasis.

World Anti-Doping Agency (WADA)—The agency set up with formal authority to police drug abuse across the international sports

A holistic perspective must take into consideration the well-being of the individual in both occupational and leisure contexts. These separate domains interact in that enjoyable leisure-time activities can enhance a person's motivation for and satisfaction with work. Ideally, the outcome of greater work satisfaction is an increase in productivity. Indeed any ergonomics intervention in the workplace should enhance output. The success of such an intervention depends on the involvement of workers or their representatives.

Sport participants train to raise their current limits and transcend their assessed capabilities. Practice based on scientific principles can improve both physical and cognitive aspects of performance. Innovations in the training environment influence these improvements. Such changes include redesign of training tasks or use of simulators to enrich the experience of training.

Human endeavor has always been directed toward overcoming challenges. This spirit led to the conquest of Mount Everest in 1954, the manned flight to the moon in 1969, and the achievement of human-powered flight in 1978. Whether on land, at sea, in air, or underwater, achievements have surpassed what was hitherto deemed improbable. Notwithstanding these outstanding achievements, the frailties of humans must also be placed in the balance. Individuals are subject to viral and bacterial infections that alter health and behavior. They need adequate sleep and rest to function effectively in work and sport. They also need the right nutrition and balance of macronutrients for their specific sports. Dietary practices for healthy living include energy intakes commensurate with habitual energy demands of activity.

Training programs are most effective when combined with sound nutritional principles complemented by a suitable lifestyle. Performance can be enhanced by nutritional means and pharmacological agents, both in occupations such as military posts and in sports; many of the drugs designed for clinical purposes have been used to enhance physique, offset fatigue, increase nervous system stimulation, and aid recovery. Such uses are considered to infringe on principles of fairness and are banned in the major sports, at Olympic Games, and at championship events. Nevertheless, some dietary practices and nutritional supplements have a scientific basis for their use in preparing athletes and promoting healthy lifestyles. This chapter focuses on participatory ergonomics and human enhancement. The use of drugs and nutritional supplements is considered alongside other enhancement technologies.

Participatory Ergonomics

A negative perception of ergonomics is that it is a management tool for changing working conditions. This impression can be counteracted by involving workers in making decisions about improving working conditions or accepting innovative working practices. Engaging the workforce in decisions that directly affect them on a daily basis is a central feature of participatory ergonomics. It is relevant also in the context of empowering members of sports teams to implement game strategies.

Participatory ergonomics approaches include interventions at macro (organizational, systems) levels as well as at micro (individual) levels where workers are given the opportunity and power to use their knowledge to address ergonomics problems related to their own working activities (Hignett et al., 2005). The understanding is that the end users (the beneficiaries of ergonomics) are integrally involved in developing

Level and Form of Participation

1. Information from management to workers about plans for action
2. Gathering of information and experience from workers
3. Consultation where workers can make suggestions and present points of view
4. Negotiations in formalized workshops at meetings
5. Joint decision making in agreement between parties involved

A typology of participation modified from Jensen (1997).

and implementing any new technology (Jensen, 1997). Cost-effectiveness is usually demonstrated in reduced absences from work and fewer compensation claims. Other outcome measures include increased productivity, improved communication between staff and management, and development of new processes and new designs for work environments and activities. A typology for participatory ergonomics is shown in the highlight box above.

For this approach to be effective, participants' input must be obtained in a directed way. In a focus group, members are interviewed together for a targeted discussion of issues. The information obtained from the interaction and discourses is essentially qualitative. Topics are provided by a facilitator, who ensures that critical questions are addressed. Members convened to form the focus group are invited to take part and encouraged to express their views by commenting on their individual experiences. The focus group has a role to play when an organization adopts participatory ergonomics, and in many instances this technique is used alongside quantitative methods.

Vink and colleagues (2006) provided evidence of higher productivity and greater comfort after enabling end users to participate with management in discussions about plans for improvement. Key factors in success were a good inventory of initial problems, a structured approach, a steering group responsible for guidance, and involvement of end users in testing ideas and prototypes. The chances of success increase with empowerment of users, such as giving them some responsibilities for deciding the next step in the process.

Participatory ergonomics has been used to deal with work-related musculoskeletal disorders. Rivilis and colleagues (2006) used a quasi-longitudinal research design in a large Canadian courier depot. Changes in work organizational factors improved health outcomes. Improved communication was correlated with reduced musculoskeletal pain and improved work rate. The investigators concluded that a participatory ergonomics approach can reduce risk factors for work-related musculoskeletal disorders and that meaningful participation of workers in the process is important for the success of such interventions.

Although participatory ergonomics does not guarantee solutions to problems, the benefits of the approach have been noted in a range of settings. Kogi (2006) reviewed participatory ergonomics projects in small enterprises, home workers, construction workers, and Asian farmers. A good-practice approach produced positive results and low-cost improvements when multiple technical areas were addressed together. Typical areas include materials handling, workstation design, physical environment, and work organization.

The philosophy of participatory ergonomics was first developed in Volvo's car assembly factory in Sweden but since then has been adopted in many different occupations. The principles apply equally to sport, particularly with respect to how coaches and sport science support staff communicate with the playing members of their professional teams. Coaching styles vary between individuals. The team captain can be used as a communication filter between players and management and can assume a leadership role on the playing field. Nevertheless, individual team players must play some role in consultation about choice of tactics because their comprehension of the tactics adopted is critical for success.

Participation of coaches and players in an intervention program to reduce injuries was shown to have a beneficial effect in a study by Junge and colleagues (2002). Seven youth soccer teams took part in a prevention program focused on education and supervision of coaches and players; further seven teams training and playing as normal acted as controls. The program included improvement of warm-up and cool-down, taping of unstable ankles, adequate rehabilitation, and promotion of the spirit of fair play. A set of 10 exercises designed to improve the stability of ankle and knee joints and the flexibility and strength of trunk, hip, and leg muscles was also incorporated. Over a 1-year period the intervention led to a 21% decrease in injury incidence. The greatest effects were observed for mild injuries, overuse injuries, and injuries incurred during training. The prevention program was more effective in low-skill than in high-skill teams. The conclusion was that coaches and players need to be better educated regarding injury prevention strategies and should include such intervention as part of their regular training.

Linked to participatory ergonomics is the notion of individual and organizational change. The stages of change model offers a way of improving interventions. The model assumes that a change in behavior entails moving systematically through various steps: (1) precontemplation, when the individual or the organization may resist acknowledging the need to change; (2) contemplation, or thinking about changing but not prepared to act; (3) preparation, where the individual or organization intends to modify behavior and plans to do so over the next 30 days; (4) action, reflecting that behavior has been altered in the last 6 months; and (5) maintenance, where behavior was altered more than months ago and the individual or organization is working on avoiding a relapse into the problem behavior and toward consolidating the gains made.

The individual or organization's current stage determines the receptiveness to and the effectiveness of the proposed intervention. The model has been applied to behaviors such as drinking, smoking, and exercise and to work-related musculoskeletal disorders (Whysall et al., 2007). In his review of how ergonomics interventions in the workplace can be best targeted, Haslam (2002) considered that ergonomists can usefully draw on a stages of change framework as applied in health promotion. Areas recommended as appropriate included manual handling, upper-limb disorders, tripping and falling, and plant safety. The model could also be adopted by physical trainers when attempting to increase compliance to exercise programs.

Human Enhancement Technologies

Human enhancement technologies refer to a gamut of ways for improving human performance. They range from optimal methods of training according to scientific principles to transient elevations of capabilities. Scientific approaches include use of simulators, virtual reality, and modified equipment. Cognitive enhancement is a means of improving mental performance. Some of these technologies constitute legal ergogenic means of securing a competitive edge over opponents. Others are pharmacological agents that are banned in sport but have gained use in situations where armed forces are being prepared for combat and in a range of occupational contexts by individual workers to overcome fatigue. A select list of ergogenic aids is considered here to illustrate how human limits may be extended beyond those assumed in classical ergonomics.

A fundamental concept in ergonomics is that there is a limit to human capacities. If humans are stretched beyond their limits, failure will result, being manifested as fatigue, error, or injury. These limits are thought to apply to physiological systems as well as cognitive functions, and various ergonomics models have been designed to establish the point of breakdown in performance. An example is the secondary task used to assess mental workload while engaged on a primary task (Young and Stanton, 2007), a model applicable to tasks such as driving and psychomotor skills. For ergogenic aids to be considered effective, performance of the primary task must improve.

Human capabilities are determined both by heredity and by environmental influences. Among the latter is training, which can alter the upper limits of physiological determinants of performance. Practice and training are essential for skills acquisition and the realization of potential. The degree of enhancement that is possible represents the interaction between heredity and environment and constitutes trainability. It may also be dependent on external factors such as the quality of coaching and mentorship available. Consequently, practice is the most effective means not only of acquiring sport skills but also of enhancing performance. Human performance can be improved by measures other than practice, but a solid foundation of training and conditioning married to expertise in the sport is imperative before nutritional strategies can work.

Performance and Cognitive Enhancement

Mental as well as physical components of performance are amenable to improvement. This scope for enhancement is applied at various stages of mental operations from awareness of stimuli to making decisions. Analogous to physiological dimensions, such mental capacities have limits. With appropriate skills training, strategies, and technology, these limits may be extended.

Performance

Many research programs on improving human capabilities have been concerned with harmonizing sports equipment and the participant's characteristics and capabilities within an ergonomics framework. There is also a considerable effort made to optimize training practices and procedures with a view to producing peak performance. The details of interventions are fed into counseling services provided for practitioners and in their continuing professional development programs. The most effective means of individual improvement are observed with performance analysis methods, when outcomes are presented as feedback to the performer.

Neuroergonomics has been described as the study of brain and behavior at work (Parasuraman and Rizzo, 2006). It combines the disciplines of neuroscience and human factors in examining how to match technology with the capabilities and limitations of people to promote safety and effectiveness at work. This emerging area is focused on the factors that limit mental performance at work: Cognitive enhancement technologies are concerned with raising these limits, by training practices, behavioral changes, or pharmacological means.

Cognitive Enhancement

The ability to make correct decisions under pressure is fundamental to elite sport performance. There is growing awareness that perceptual–cognitive skills such as anticipation, decision making, and situational awareness are prerequisites for skilled performance. Elite performers develop sophisticated, task-specific knowledge structures as a result of extensive practice that enable them to deal with situations in a more effective and efficient manner than less elite counterparts. Elite participants search the visual scene in a selective manner, focusing their attention on relevant rather than irrelevant sources of information. Moreover, experts are more aware than novices of the likely events that may unfold in any given situation and are better at picking up key contextual cues (e.g., postural cues from an opponent), which facilitates situational awareness and anticipation of events. These skills are coupled with an extensive knowledge of available strategies and tactics and how these may be implemented quickly in any given situation. It is also evident that successful performance in high-level sport depends on an athlete's ability to deal effectively with stress. Although researchers have shown that stress can lead to deterioration in perceptual, cognitive, and motor performance, the evidence suggests that elite performers are less inclined to suffer the negative consequences associated with stress by developing effective emotional control strategies.

The prototypical experimental approach has been to capture the performance environment using either film or virtual reality simulations. Performers are typically asked to imagine themselves in the real-world situation and to make the correct decision promptly and accurately. Performance measures are recorded during performance, including response time and accuracy, heart rate, galvanic skin response, eye movements, and concurrent verbal reports. These measures indicate how skilled athletes differ from less skilled counterparts and provide a principled basis for designing systematic training programs to enhance such skills. Film and virtual-reality-based training simulations, instruction, and feedback allow the athlete to experience the demands typically faced in the competitive situation under controlled laboratory conditions. The level of instruction can vary (e.g., explicit instruction, guided-discovery learning, discovery learning) to create optimal conditions for learning and transfer of skills.

Not all techniques advocated for human enhancement have complete scientific approval. Neurolinguistic programming provided a useful model in the field of counseling psychology that gained some favor in the 1990s, notably in the United States. The approach has been applied intermittently in the context of motor skills acquisition, normally for improving motivation, task orientation, and pattern recognition. With only limited scientific support (Gallese et al., 1996), this technique has not gained acceptance as a productive approach to cognitive enhancement.

Virtual reality environments are used to improve both physical and cognitive capabilities. Training programs incorporating computer games can be adapted from clinical contexts for use by athletes via the addition of sophisticated simulations.

Such computer-based systems are used alongside film-based training simulations to enhance decision-making capabilities in sports.

Just as drugs have been used to augment physical aspects of performance, cognitive enhancement technologies have included pharmacological elements. Such drugs include amphetamines, modafinil, and pemoline. Amphetamines were implicated in the death of a British cyclist competing in the Tour de France in the 1960s. Modafinil, a drug administered to treat narcolepsy, was used by a female sprint finalist in the World Championships for track and field in 2003. Pemoline has been used in a military context to stave off fatigue but would be unacceptable as a means of supporting sport performance. Sport participants and gymnasium users are bombarded with promotions for over-the-counter, Internet, or "underworld" sources of supplements but must exert caution in their purchases because they are ultimately responsible for what they ingest.

Human performance can be improved by other measures. Some nutritional substances have ergogenic properties, particularly with respect to improving power and endurance. Pharmacological manipulation can have profound effects on immediate performance and on the adaptations caused by training. Recognition of the advantage that different classes of drugs provided for users led the sporting bodies to ban certain categories of drugs. The major initiative was provided first by the International Olympic Committee, but the widespread abuse of drugs for ergogenic purposes led to the establishment of the **World Anti-Doping Agency (WADA),** which now polices drug testing throughout sport. Yet the abuse of drugs is not a modern phenomenon, having a long history both in sport and in society at large.

Historical Perspective on Drug Use

Athletes have embraced all means possible for improving their chances of success in sport. These encompass a variety of forms of training, massage, nutrition, and drugs. Competitors in the ancient Olympic Games took their training seriously, especially in light of the adulation the victors received. More substantive evidence is available on the use of drugs and supplements since the Olympic Games were revised in 1896, and the rules for the major games were formalized just before that.

The kola nut from Africa and coca plant from South America received attention at the end of the 18th century for their sustaining and strength-conserving powers, kola being produced soon in commercial drinks (Dimeo, 2007). Strychnine was used in the St. Louis Olympic Games marathon in 1904 by the winner Thomas Hicks from the United States, whose handlers gave him the drug with brandy during the race. Rivers (1908) considered that strychnine, along with caffeine, coca, and cocaine, boosted performance by stimulating the central nervous system but that author doubted the beneficial effects of alcohol. Many of these drugs were used in society at large although narcotics such as opium were considered the source of social problems.

Drugs used to sustain alertness of fighter pilots during the Second World War had properties that were soon identified among athletes as ergogenic. Amphetamines were adopted by cyclists, football teams, and athletes as a means of raising their performance level for single events. Benzedrine and methedene, found to be used by German war pilots, were identified as ergogenic substances (Cuthbertson and Knox, 1947). As stimulants found their way into recreational use, concern emerged about their long-term effects. In more recent times, modafinil—a drug designed to treat narcolepsy—was used by sprinters at world track-and-field championships.

Whereas **amphetamines** are ergogenic for some activities short term, anabolic steroids have properties that aid sustained preparation for sprint and power events. The first laboratory test for amphetamines was conducted at the 1966 World Cup Finals for soccer, but testing for use of anabolic androgenic steroids was not available until the 1976 Olympic Games. Only later was it possible to develop tests for testosterone and growth hormone. The use of new doping methods was not restricted to sprint and muscle power events. Blood doping in various forms was used since the 1980s, initially as autologous transfusion and later by use of the hormone erythropoietin. Only at the 2000 Olympic Games in Sydney was it possible to test for blood doping.

With the systemization of sport and the unrelenting emphasis on success, many of the principles of amateur sport became compromised, including the principle of fair play. The use of drugs to enhance performance was central in this debate, which

Prohibited Classes of Substances and Prohibited Methods

I. Substances and Methods Prohibited at All Times

- S1 Anaerobic agents
  1. Androgenic anabolic steroids
  2. Other anabolic agents
- S2 Hormones and related substances
  1. Erythropoietin
  2. Growth hormone, insulin-like growth factor, mechano growth factor
  3. Gonadotropins
  4. Insulin
  5. Corticotropins
- S3 Beta-agonists
- S4 Agents with anti-estrogen activity
- S5 Diuretics and other masking agents
- M1 Enhancement of oxygen transfer
- M2 Chemical and physical manipulation
- M3 Gene doping

II. Substances and Methods Prohibited in Competition

- S6 Stimulants
- S7 Narcotics
- S8 Cannabinoids
- S9 Glucocorticoids

III. Substances Prohibited in Particular Sports

- P1 Alcohol
- P2 Beta-blockers

IV. Specified Substances

eventually led to the formalization of substances deemed to be illegal. Those classes of substances currently banned are shown in the highlight box on page 226. Methods of policing entail out-of-competition testing, compliance with which is mandatory. The system has governmental support through WADA. After its establishment in 1999, this anti-doping body immediately set up a scheme of educational and scientific projects.

Concerns remain about the use of synthetic substances and new pharmaceutical products that are unknown at the laboratories accredited for drug testing. This problem was underlined with the use of THG, a synthetic anabolic steroid, which came to light after 2004 only when a syringe with the substance was provided anonymously by an international athletics coach to the director of an accredited laboratory. The fear of "gene doping" and cognitive enhancement technologies prompted a House of Commons Science and Technology Committee (2007) to review the ethics and state of the art in these areas. Among its concerns was the separation of what should be deemed illegal and what is permitted.

Clearly drugs are used in society, in some occupational settings, and in sport to enhance the user's sense of well-being and combat fatigue. Drug addiction has become a huge and costly social problem. In work settings caffeine has clear positive effects in sustaining attention, whereas alcohol (notably at lunchtime) has depressant effects. Athletes ingest a wealth of substances, purportedly to improve their performances. Some of the more common ones that are not on the list of banned substances are now considered.

From Clinic to Gymnasium

Pharmacological advances for alleviating diseases have been used in sports for some decades. The anabolic effects of androgenic anabolic steroids were exploited in preparation for those sports where muscle cross-sectional area and the generation of high power output are relevant. These include bodybuilding, weightlifting, weight throwing, and sprints. Substances such as dianabol, clenbuterol, stanazolol, and testosterone became widely used and showed up in positive drug tests on a regular basis. Anabolic steroids were used by other athletes to speed recovery processes following hard training, Later, growth hormone and insulin-like growth factor were adopted for ergogenic purposes while tests for their detection were unavailable.

Endurance athletes benefited similarly from procedures originally designed to improve the well-being of patients with kidney disease. Erythropoietin, a hormone produced by the kidneys, stimulates the production of red blood cells and boosts oxygen transport capacity in a variety of conditions including cancer and renal dialysis. Its synthetic version was used by endurance athletes prior to a test for its detection introduced at the Sydney Olympic Games in 2000. An alternative illegal strategy was **blood doping,** either reinfusion of autologous blood or infusion of compatible blood from a donor. Such procedures were found to improve performance in running events such as 1,500 m and 10,000 m (Brien and Simon, 1987). Over the years many athletes in professional cycling, distance running, and cross-country skiing have admitted use of or tested positive for these manipulations of oxygen transport capacity by artificial means.

The scientific advances in genetics and molecular biology have led to hopes for the early identification of major diseases and remedies for them. It has been thought that the genetic technologies available to health specialties would be used for top-level sport. The fear of gene doping prompted a House of Commons Science and Technology Committee (2007) to explore the risks of genetic engineering for sport, members being aware of the unethical nature of this possible use.

Athletes are continually in tune with the potential of ergogenic aids to help them achieve their goals. The adverse health effects of anabolic substances would render such drugs unethical, even if they were not illegal. A similar problem arises with central nervous system stimulants that have addictive properties; competitors may be prepared to compromise long-term health for short-term glory in sport. Athletes are also swayed in use of over-the-counter drugs by whether ingredients are included in the doping list. Such labile attitudes were evident after caffeine was removed from the list of prohibited substances in the early part of the current century. Use of creatine loading is considered ethical by practitioners, the rationale being that creatine is found naturally within the body and is available in the normal meat-eaters' diet.

Nutritional Supplements and Over-the-Counter Drugs

The **supplement** market is highly profitable for industrial suppliers, in particular in North America, Europe, and Asia. The marketing information is seldom backed up by scientific evidence but this lack does not inhibit athletes, weight trainers, and bodybuilders from using supplements. In this section, some attention is given to supplements that do work: Athletes are deemed responsible for the contents they ingest, which may be contaminated with banned substances. A selection of substances in common use is identified for a more detailed consideration; they are illustrative rather than comprehensive for the area.

Supplement Context

The search for nutritional supplements to complement training seems to be ingrained in the mindset of practitioners. It is inevitable that athletes seek substances that have ergogenic properties, especially when faced with aggressive marketing that makes strong claims for the products concerned. One problem is that some of the supplements advocated for use by athletes are contaminated by substances that are on the banned list of the IOC. This problem has been manifest in a number of the positive tests experienced by athletes who subsequently were banned from sport for using prohibited drugs. In some instances, individuals may unwittingly ingest banned substances. Contamination of supplements was implicated in the spate of positive tests for nandrolone among runners and football players at the beginning of the century.

The use of over-the-counter (OTC) stimulants by athletes was prohibited in sport competitions until recently. In 2004, many OTC stimulants were removed from the World Anti-Doping Agency Prohibited List. Common OTC stimulants such as pseudoephedrine, phenylephrine, phenylpropanolamine, and caffeine were placed on a monitoring program, but their use by athletes was not restricted. The aim of the WADA monitoring program is to observe the use of such substances by athletes via in-competition drug tests carried out by WADA-accredited laboratories. Data presented to the House of Commons Science and Technology Committee (2007) suggest that 36.2% of athletes believed OTC stimulants to have performance-enhancing properties and that pseudoephedrine was the most popular OTC stimulant used by athletes (23.3% of athletes having used it in the previous 12 months). Although the ergogenic properties of OTC stimulants are equivocal, the removal of many OTC stimulants from the prohibited list led to a sharp increase in their use by athletes, seemingly for performance enhancement.

Although athletes and the general public may seek benefits from vitamins, minerals, and other supplements that are not realized, dietary manipulations can enhance performance, notably in endurance events. Emphasis on a carbohydrate-rich diet for 2 to 3 days prior to competition boosts performance in cycling and distance running events. The benefits of carbohydrate loading extend to work rates that are sustainable in sports involving intermittent exercise for sustained periods, such as soccer. The level of muscle glycogen stores prior to competition also influences work rate during such games (see figure 11.1). The combination of a high-carbohydrate meal approximately 3 hr before competing and a sports drink ingested during exercise increases exercise capacity more than does a carbohydrate meal alone (Williams and Seratosa, 2006).

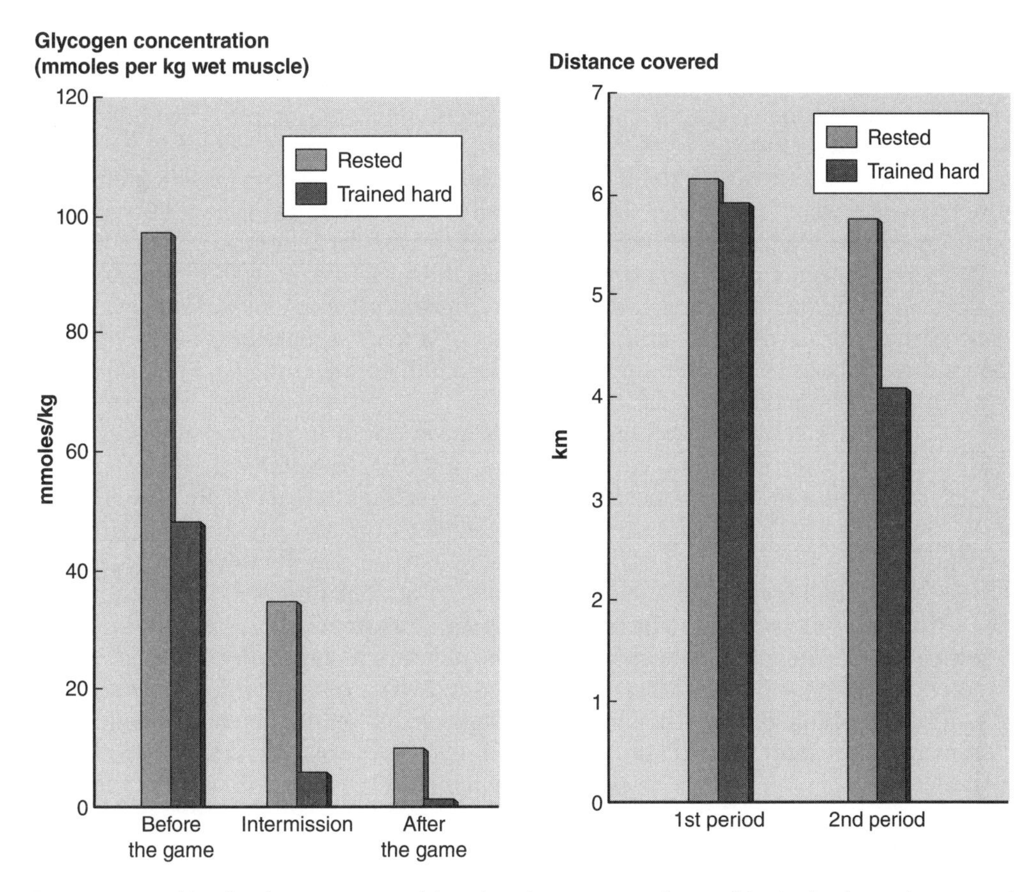

Figure 11.1 Muscle glycogen stores *(a)* and performance in players *(b)* who had rested or trained hard in the days prior to playing a match.
Data from Saltin 1973.

Caffeine

Caffeine is particularly open to misuse in sport because of its apparent ergogenic properties and following its removal from the WADA prohibited list. Indeed, the increase in the energy drink market in recent years was accompanied by a concomitant increase in the association of such products with sport. In a specific group of athletes ($n = 83$), 63% used caffeine for performance enhancement reasons and 17.5% had increased consumption after 2004. Coffee, energy drinks, and caffeine tablets were the most prevalent of the caffeine products used (House of Commons Science and Technology Committee, 2007).

Caffeine and other methylxanthines (theobromine and theophylline) have long been known for their ergogenic properties. Even though caffeine is widely used by the general population, the evidence suggests a significant increase in its use in sport across all levels following its removal from the list of banned substances in 2004. Evidence of adverse health effects of caffeine is limited. Given the deregulation of other stimulants, it is extremely likely that athletes will combine OTC stimulants to enhance performance.

Because the original studies of caffeine showed that endurance running was improved (Costill et al., 1978), the mechanism was thought to be promotion of lipolysis and sparing of muscle glycogen. More recently, research has focused on the effects of caffeine in blocking receptors in the brain (Doherty and Smith, 2005). This mechanism would explain the persistent ergogenic effect of caffeine in the absence of metabolic alterations.

The dose of caffeine originally found to be effective was 5 mg per kilogram of body mass. Recent studies have shown that half of this dose and as little as 90 mg can improve performance in well-trained cyclists exercising for 2 hr (Cox et al., 2002). Pure caffeine seems to be more effective than in the form of coffee and is available as a gel. Its stimulant effects are reduced in desensitized coffee drinkers who may need larger than normal doses for an ergogenic effect. Caffeine intake assists performance in industrial tasks by maintaining levels of arousal that might otherwise fall too low to support continued performance.

Games players have decision-making tasks superimposed on physiological loading. Hespel and colleagues (2006) considered evidence that the caffeine dose needed to yield an optimal effect on visual information processing is substantially lower than that inducing optimal endurance performance. Both functions respond according to an inverted-U curve (figure 11.2). The optimal doses differ between individuals because habitual uses of caffeine need higher than normal amounts.

Theobromine was used at the beginning of the 20th century as a stimulant for offsetting fatigue. Theophylline is another methylxanthine with properties that suggest its potential ergogenic benefit. Observations on six subjects cycling to voluntary exhaustion indicated that theophylline increases arousal of the central nervous system and induces an elevation in heart rate. Its metabolic effects are not as great as caffeine and its alteration of physiological responses to exercise is relatively minor (Reilly, 1988).

Creatine

Creatine phosphate is an immediate source of energy during exercise at very high intensity. The substrate that is broken down during brief intense activity is resynthesized in the following recovery period, whether this involves rest or low-intensity

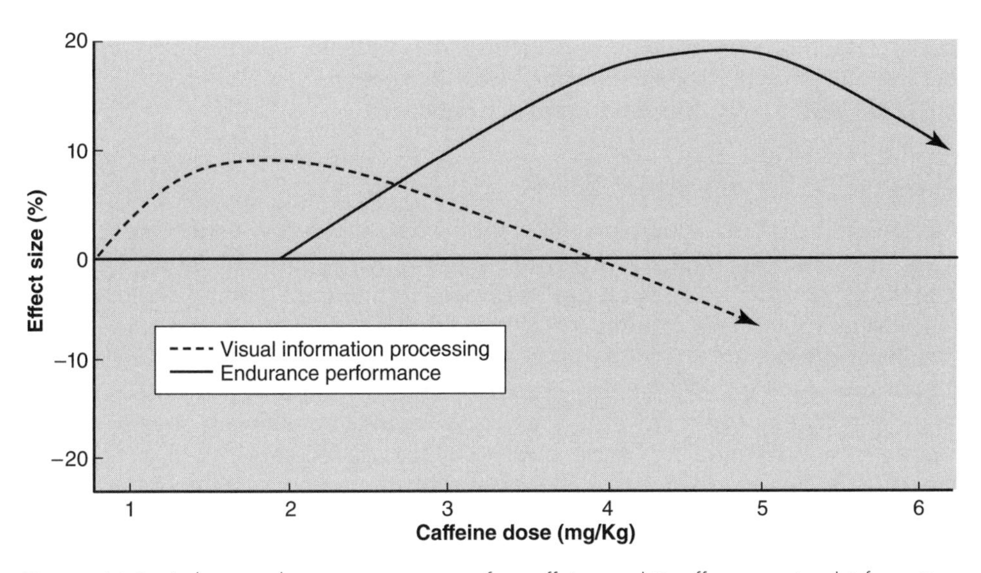

Figure 11.2 Indicative dose-response curves for caffeine and its effects on visual information processing and endurance performance.

activity. When four intense bouts of activity are repeated with only very short intermission between them, creatine phosphate levels may decline to 30% of resting stores (Bangsbo, 1994). Anaerobic performance is impaired until creatine is resynthesized. This type of fatigue that is associated with reduced muscle stores of creatine phosphate has prompted the use of creatine loading practices among athletes engaged in sports with high anaerobic components.

Synthetic creatine supplements are provided as creatine monohydrate and as various creatine salts such as creatine pyruvate or creatine citrate. The former is usually in the form of a powder, whereas creatine sources are incorporated as sports drinks or sports gels. There is experimental evidence that creatine loading enhances power output during short maximal sprints, especially when these sprints are performed in quick succession (Greenhaff et al., 1993). Supplementation with creatine also induces an increase in body mass, sometimes approaching 2 kg, and so may be counterproductive in field games where body mass must be repeatedly elevated against the resistance of gravity. Creatine supplements may be most effective in training contexts where an increase in the workload stimulates an enhanced training effect.

For best effects, users of creatine supplements should consume 20 g/day for an initial loading phase of 5 to 7 days (Terjung et al., 2000). This period is followed by a maintenance phase for the next 3 weeks. This cycle can be repeated because the effects of creatine supplementation may fade after 2 months (Derave et al., 2003). Many athletes who use creatine loading interrupt the supplementation phases with a washout period of about 4 weeks. Such a washout period is used once in every 3 months. Individuals who benefit most from creatine loading tend to have low initial stores, for example, vegetarians, who miss the intake of creatine via meat and fish.

Athletes and coaches have no ethical concerns about the use of creatine supplementation as an ergogenic aid. Early concerns about overloading kidney function seem to have been unfounded. On the contrary, creatine supplementation can accelerate the recovery from muscle atrophy caused by injury and immobilization (Hespel et al.,

2001). Objections about use of creatine in young athletes must be taken seriously, because underage participants in sport need to experience a rounded development rather than one solely oriented toward performance.

Supplements in Combination

Caffeine and creatine, separately or in combination, are used by many elite athletes. Their use does not always coincide with scientifically established loading regimens. The timing of caffeine ingestion can be delayed until 60 to 90 min precompetition whereas use of creatine requires a prolonged buildup over 1 to 4 weeks to deliver beneficial effects.

Creatine and carbohydrate have been used in combination to good effect. Ingesting 10 g of creatine together with 200 g of carbohydrate immediately after 90 min of simulated soccer activity resulted in improvements in a soccer skills task and time to fatigue 24 hr later compared with 200 g of carbohydrate alone (House of Commons Science and Technology Committee, 2007).

Co-ingestion of creatine with a carbohydrate-rich diet enhances glycogen restoration postexercise (Van Loon et al., 2004). This effect facilitates recovery after exhaustive exercise and allows reintroduction of quality training sooner.

Combining caffeine with carbohydrate loading may not be so productive. Ingestion of caffeine along with carbohydrate prior to sustained exercise may lead to no additional benefits. Jacobson and colleagues (2001) found that ingesting carbohydrate 60 min before a continuous exercise test improved performance in a time trial compared with fat ingestion, but combining caffeine (6 mg/kg) with carbohydrate had no further benefit. Where athletes are already carbohydrate loaded, there may be no further metabolic benefit of caffeine, although its effect on the nervous system may still apply.

Alkalinizers

Strenuous anaerobic exercise causes increased production of lactic acid within muscle and subsequently increased lactic acid concentration in blood. The resultant increase in acidity is associated with muscle fatigue, a transient state that lasts until the lactic acid is buffered or is oxidized upon entering the circulation. Increasing the body's buffering capacity should therefore improve performance for events limited by the athlete's anaerobic capacity, notwithstanding other physiological factors implicated in fatigue. Alkaline salts such as sodium bicarbonate and sodium citrate have buffering properties and when ingested increase the body's alkaline reserves. The increase in the bicarbonate store in blood may also facilitate coping with the efflux of hydrogen ions from muscle, thereby maintaining the pH status within the muscle (MacLaren, 1997). Alkaline salts would be expected to improve performance in all-out efforts that maximize use of the anaerobic glycolysis pathway. Alkalinizers should also improve performance in intermittent exercise where removal of lactic acid is required during recovery between bouts of exercise.

There is evidence of an ergogenic effect of sodium bicarbonate ingestion and sodium citrate on exercise performance. Although effects are typically limited to single all-out efforts between 30 and 90 s in duration, George and MacLaren (1988) reported benefits in longer exercise duration where exceeding the exercise intensity associated with the maximal lactate steady state was likely. **Alkalinizers** are attractive in helping the body tolerate exercise when lactate production is high; their use may not be suitable for all people because ingestion can lead to nausea and diarrhea.

Despite these potential problems, the positive effects of alkalinizers appear to be robust. A dose of 0.5 g/kg sodium citrate improved 5,000 m rowing performance significantly in six experienced male rowers. In a group of elite 400 to 800 m runners, a dose of 0.3 g/kg sodium citrate ingested 3 hr before five 30 s sprints on a nonmotorized treadmill resulted in higher power output in the third to fifth sprints. Furthermore, a 0.3 g/kg dose of sodium citrate ingested 3 hr before an anaerobic capacity test enhanced time to fatigue, that is, 20 m shuttles at a pace corresponding to 120% $\dot{V}O_2max$. Blood pH and bicarbonate were elevated pretest with citrate ingestion (House of Commons Science and Technology Committee, 2007).

Other Substances

The number of supplements commercially available with claims for positive effects on health or performance is legion. A supplement should be recommended to athletes only if it works in the context desired, does not cause any adverse effects on health, and is legal. The research literature is replete with reports of substances that are effective in certain exercise situations, supplements that might work in some exercise situations, substances whose effectiveness is inconclusive, and supplements that don't work.

Supplements designed to promote health include those with a targeted effect on weight control. An example is ephedra, which enhances resting energy expenditure and facilitates short-term weight loss. Other substances such as vitamin C are taken to stimulate the immune system and protect against upper respiratory tract infection. Antioxidants are advertised to prevent muscle damage by enhancing the body's defenses against formation of reactive oxygen species, known as free radicals. Glucosamine stimulates the formation of bone cartilage, with expectations of alleviating joint pain. The balance of evidence for these and other supplements in a health context was considered by Hespel and colleagues (2006).

Some athletes use supplements with health-enhancing claims, but most athletes are focused on products that improve performance. The purpose may be to enhance training effects, accelerate recovery, or boost competitive performance. Individual products include metabolites of essential amino acids (e.g., beta-hydroxy beta methylbutyrate, or HMB) to increase lean body mass and muscle strength and trace elements such as boron, which influences calcium and magnesium metabolism. Among the detailed list of products considered by Jeukendrup and Gleeson (2004) that are available in food shops are bee pollen, carnitine, ginseng, glutamine, vitamin B_{15}, fish oil, and wheat grain oil. In a majority of cases the research evidence does not support the manufacturer's claims. Some supplements are both ineffective and costly and should be replaced with advice from a sport dietitian.

Alcohol

Alcohol is used for social reasons in many cultures, and its use can affect performance both at work and in sport. Alcohol is both a drug and a fuel for providing energy. As a drug it can become addictive, and various high-profile soccer players have developed a dependence on alcohol. It can have adverse effects on health, affecting cardiac and skeletal muscle and the liver in particular. Alcohol addiction has been implicated in shortening the careers of a number of professional players.

As a source of energy, alcohol contains 7 kcal/g (29.3 kJ/g), compared with 4 kcal/g for protein and carbohydrate and 9 kcal/g for fat. Typically, wine contains about 12% alcohol and so a 1 L bottle has an energy content of 840 kcal (3,516 kJ). The alcohol concentrations in beer (~5%) and whisky (~40%) represent variations in the caloric load attributable to drinking. These energy intakes can have a large effect on weight-control programs.

The energy in alcohol cannot be used by active skeletal muscle, so exercise does not hasten the elimination of alcohol from the blood. Blood alcohol concentrations tend to peak about 45 min after ingestion, and the effects are more pronounced if the stomach is empty at the time of ingestion. Performance of sport skills is adversely affected when blood alcohol concentrations exceed 0.05% (mg/100 ml), a value below the legal driving limit in the United Kingdom (Reilly, 2005). Many road accidents are linked to impairment in driving caused by elevated blood alcohol concentrations, and driving with a blood alcohol level above the prescribed value is a criminal offense. Alcohol has some benefits in the aiming or target sports such as archery, darts, and snooker, given its reduction in spontaneous limb tremors, although the positive effects are restricted to relatively low blood alcohol concentrations (Reilly, 2005).

Acute effects of alcohol depend on the blood alcohol concentration that is induced (table 11.1). The adverse effects on performance apply to training contexts such as weight training, endurance sessions, and skill practices. The direct effects on metabolic processes likely impair endurance performance. Alcohol lowers muscle glycogen at rest and may reduce glucose output from the spleen, decrease the potential contribution of energy from liver gluconeogenesis, and lead to a decline in blood glucose. Its diuretic effect compromises thermoregulation, for example, when an athlete plays in hot conditions after drinking heavily. Alcohol inhibits glycogen resynthesis if taken after strenuous training or a match and food ingestion is delayed.

Moderate alcohol ingestion the night before competitive sport can hamper performance because of a hangover effect (O'Brien and Lyons, 2000). Alcohol has a place in social settings, suitably timed so as not to interfere with competitive or training

Table 11.1 Demonstrable Effects of Alcohol at Different Concentrations in Blood

Concentration level (mg/100 ml blood)	Effects
30	Enhanced sense of well-being; retarded simple reaction time; impaired hand–eye coordination
60	Mild loss of social inhibition; impaired judgment
90	Marked loss of social inhibition; reduced coordination; noticeably under the influence
120	Apparent clumsiness; loss of physical control; tendency toward extreme responses; definite drunkenness
150	Erratic behavior; slurred speech; staggering gait
180	Loss of control of voluntary activity; impaired vision

Adapted from T. Reilly, 2005, Alcohol, anti-anxiety drugs and alcohol. In *Drugs in sports*, 4th ed., edited by D.R. Mottram (London: Routledge), 258–287.

engagements. Nevertheless, alcohol ingestion is not essential and is shunned in many cultures. Especially in young athletes, sound dietary practices and a sensible approach to drinking alcohol are advocated.

Global Ergonomics

Ergonomics remains a fertile area of the human sciences wherever there is an interest in understanding how humans are integrated into working procedures, patterns, and processes. The focus can range from large-scale systems to routine practices, from complex multilayered processes to the design of basic consumer products. Human fallibility can never be completely eradicated but creative design features can partly compensate by accommodating fail-safe mechanisms in the event of error occurring. Commercial interest, a humanist mindset, and a quest for quality in life promote the concepts of efficiency, well-being, and stress reduction, respectively. These themes merit attention in work, sport, and leisure domains.

The nature of ergonomics interventions depends to a great extent on the degree to which industry is developed and industrial management is enlightened. As economic growth occurs in national economies, there are further shifts from physically demanding jobs to more sedentary posts, from energy processing to information processing. Ergonomics interventions in rural and agricultural settings are likely to differ from projects in urban environments and chemical plants in scale as well as in kind. Nevertheless, the principles for identifying human factors issues and the approaches to their solution may be common.

Within the field of ergonomics there is a continuing need to set out or revisit national and international standards for products and procedures. The globalization of developments necessitates standards that transfer across national boundaries. Practices and symbols for safety come into focus, especially where migrant workers are concerned. Human communication systems are also relevant, particularly where exchanges are brief and mistakes have adverse consequences. **Population stereotypes** raise problems about expectations of the relationship between displays and control devices. This notion refers to the assumption that moving a control device in one direction causes a compatible movement elsewhere. Similarly, the movement of an indicator on a dial for an increase or decrease should be in accord with expectations, for example, for a twist motion to the right or to the left. There is not universal agreement with respect to population stereotypes, but their existence must be considered in the design of human–machine systems.

Increased concern for the health of the population ensures that ergonomics has future relevance in a recreation context. Exercise programs, suitably designed and implemented, are an essential weapon in attempts to combat many morbidities that include obesity, metabolic syndrome, and cardiovascular disease. Exercise can be effective in health promotion at the work site and in preventing musculoskeletal disorders associated with poor working postures.

Reductions in infant mortality, improvements in health care systems, and elimination of notable infectious diseases have contributed to greater longevity in contemporary society. Healthy lifestyles also have increased life expectation. A demographic consequence is an increase in the proportion of elderly people in the population. This increase raises questions about their support during retirement years and their engagement in exercise and recreation programs to maintain their mobility.

The outdoor wilderness attracts visitors to enjoy their scenic qualities and to overcome their sometimes formidable challenges. The risks involved must be addressed for these encounters to be experienced with safety and satisfaction. The protection to be used embraces choice of clothing, portable assistance, correct equipment, and behavioral strategies. Human factors considerations cannot offset all aspects of danger as individuals seek challenges that stretch them to their physical and psychological limits.

The increased professionalization of competitive sport implies a continuing quest for participants who strive to achieve their potential. Record performances at national and international levels remain targets to be beaten. Tournaments and championship victories are the goals of aspiring teams and provide the motives for engaging in rigorous training programs. Individual athletes will continue to negotiate the thin divide between the benefits of training stimuli and the damage caused by harmful overload. The acquisition of a competitive edge by means of tactics, training, and nutrition is likewise likely to remain important for them.

A variety of factors determine the extent to which ergonomists continue to be engaged in the occupational, sport, and leisure domains. These factors include expansion of the existing knowledge base, further application of technology in design, and generation of creative solutions to suit novel projects. Ergonomics teams attempt to refine their models and approaches and improve their techniques for quantifying and interpreting the problems they encounter. Quality control in measurement procedures applies to measuring environmental stress, systems operations, and system output. It is important that measurement error be minimized when human capacities are assessed and the assessment tools and protocols are suited to the individual concerned. The ergonomics team can then apply the observations to match the task and the individual with confidence.

Overview and Summary

The likelihood of success in an ergonomics intervention is increased when the people concerned are involved in some way prior to its implementation. This consultative framework is an essential component of participatory ergonomics and is accepted as good practice in empowering participants on sports teams. Strategies for performance enhancement must comply with the rules for competition that apply to each sport. These rules apply not only to the regulations for equipment and behavior set by sport governing bodies but also by those of the international agencies outlining the pharmacological and nutritional agents that are accepted for use. The use of dietary practices for healthy living can be extended to engage the performance domain. Nutritional preparation for training and competing contributes to improvements in performance, although many heavily marketed supplements are ineffective. There is a need for the athlete to be fine-tuned to the culture of the sport and its working environment to continue in a safe, comfortable, and satisfying role. Ergonomics criteria do not constrain the individual into a straitjacket within a mechanistic system but hold the person to be a unique and truly valued entity. A holistic perspective considers the person, the range of individual needs, and the performance together.

THE preceding chapters emphasize the placement of the human in the center of any ergonomics evaluation. This priority has been applied irrespective of the nature of the task, the environmental conditions, and the overall objectives. Models for understanding the technological interfaces with human operators have been considered, varying both in context and complexity. Such models typically are constructed for occupational purposes, but their broad applications for sport and exercise are apparent.

A template for an ergonomics intervention is provided in figure aft.1. Implicit in the illustration are the questions the research team must formulate and the steps it must take to fulfill its investigations. The figure highlights the problem-solving nature of ergonomics and the necessity of finding solutions to the problems identified. The scheme portrayed is useful in understanding the processes involved. It illustrates that many projects do not necessarily start from scratch but constitute regenerative ergonomics.

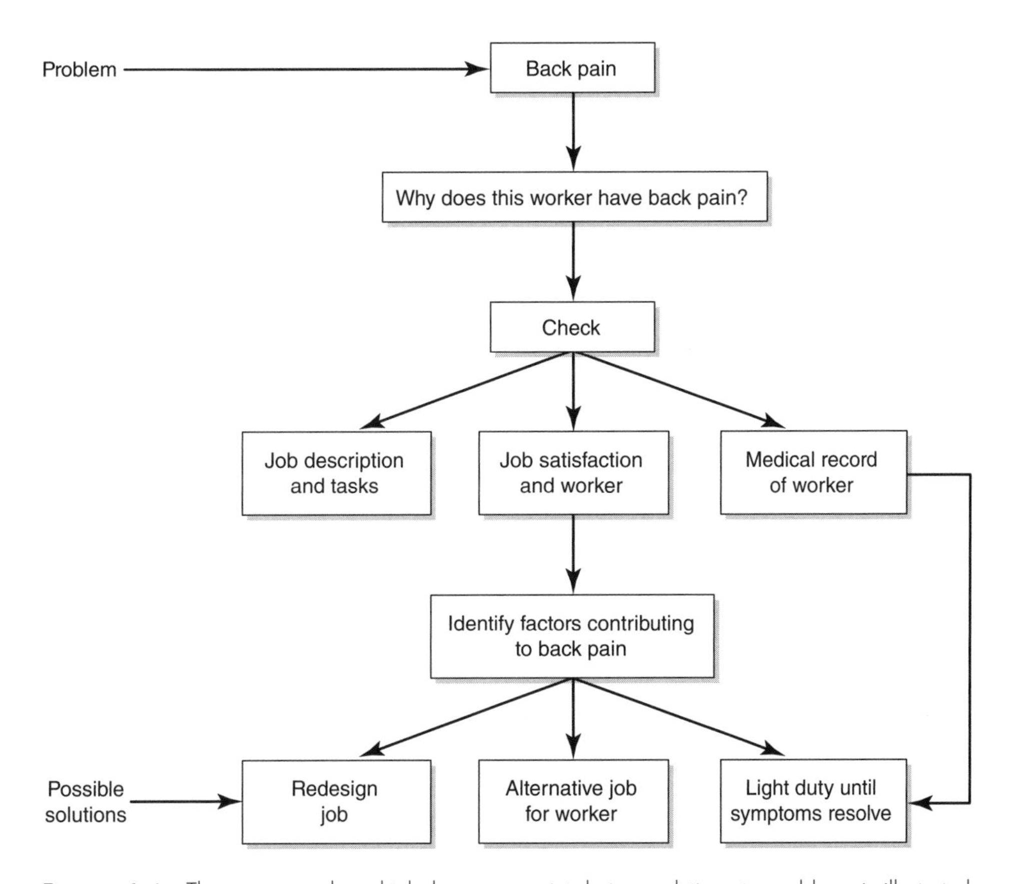

Figure aft.1 The processes by which the ergonomist derives solutions to problems is illustrated by examining likely causes of back pain.

An armory of techniques may be called on in any ergonomics investigation. Those chosen may emanate from any of the disciplines within the human sciences, but typically several analytical methods are adopted. When the sciences are integrated in this way and centered on a holistic perspective, ergonomics can be deemed to be a truly interdisciplinary approach.

A core pursuit in the field of ergonomics is finding the best means of matching the individual and the task or sport. Task analysis is a prerequisite in establishing critical features of the activity and the surroundings in which the activity takes place. An assumption is that performance is limited by physical and mental capacities, and in competitive sport these capabilities may be stretched to their limits. Training programs for sport reflect the struggles of athletes to push back these limits and reach their goals in competitions or tournaments. At the top end of the high-performance spectrum, participants benefit from the availability of sport science–support personnel, who can help the participants gain a competitive edge over opponents.

Key criteria that the ergonomist must adhere to are related to safety, efficiency, comfort, reduction of injury risk or harm, avoidance of damaging overload, prudent use of energy, fatigue reduction, and tolerance of stress. The difference in applying these principles between occupational and sport settings permeates earlier chapters of the book. It is clear that there is not complete equality between the domains, given that the athlete anticipates a level of discomfort, even pain, in competition and strenuous training that would be unacceptable in industry. Similarly, overload is accepted as an essential element of physical training so that the athlete continually treads a thin line between productive loading and harmful overload.

Stresses imposed on individuals may be attributable to environmental factors, sometimes constituting hostile and extreme conditions. Why individuals subject themselves voluntarily to environmental challenges such as climbing high mountains, sailing the seas single-handedly, or participating in endurance races across desert terrain is not completely understood. The environmental stressors encountered in these activities include heat, cold, pressure in underwater activities, hypoxia at altitude, hostile terrain, and inclement weather. Lifestyle factors that impose stress on people include disruptions of the body clock causing circadian desynchronization, lack of sleep, nocturnal shift work, and fasting among certain religious communities. The ergonomist may use various indices to evaluate heat stress or cold stress, although the effect on the individual depends on the events concerned. Altitudes of moderate heights benefit performance in short-term events because of the reduction in air density, but the prevailing hypoxic conditions impair oxygen transport when endurance activity must be sustained. There are differences also in the effectiveness of acclimatization, physiological adaptations occurring more quickly in response to heat than to cold exposure or altitude training. The quality of air is also relevant, but there is limited evidence of a beneficial adaptation to air pollution, impure air being generally harmful, especially to vulnerable individuals.

Fitness implies an ability to cope with the task at hand, a concept that applies to occupational as well as sports settings. Many jobs still retain an appreciable physical component, either large energy expenditure or generation of high forces on an intermittent basis. Work in occupations such as the armed forces may be moderate in severity but continuous and for a long duration in sometimes arduous conditions. Other occupations with appreciable physical loading are considered, as are the unique requirements of special populations associated with age, sex, and disability. In assessment of fitness for work, the establishment and validation of test batteries that are neutral to age and sex pose a challenge for ergonomics. A prerequisite is a task analysis of the operations involved in the job to provide a basis for the choice of tests.

The second part of the book concentrates on ergonomics in sport contexts. Methods of task analysis are covered in detail; some have been borrowed from classical ergonomics, whereas others have been developed more recently, particularly where complex systems are involved. Human–machine systems can be highly sophisticated or can refer to more simple combinations of athlete and craft such as occur in canoeing, cycling, motor racing, rowing, and sailing. Technology and materials science are vital to the design of contemporary sport equipment; wind tunnels are used to design helmets, and handlebars are designed to accommodate the best racing posture for cycling. Similarly, the use of water flumes for hydrodynamic experiments has contributed to design modifications in clothing for swimmers and the construction of rowing shells.

Ergonomics is relevant in clinical contexts from the viewpoint of both health care workers and patients. Many of the practices originating in physical therapy have gained acceptance in the training of athletes, notably in the rehabilitation programs used to speed recovery after injury. The principles that apply to musculoskeletal assessments are equally valid for athletes and for patients in most respects. An essential difference between the two domains involves the pharmacological agents used to improve physical capabilities. Although such manipulations are accepted in the public health domain to restore normal function when prescribed for therapeutic purposes, many of the drugs with ergogenic properties are prohibited in sport.

The final chapter of the book emphasizes the central focus on the individual. The inevitable consequence is a holistic perspective in which the individual is recognized as a unique entity rather than a cog in a mechanistic system. This view may extend to engaging the workforce or entire sports team in solving important issues. Designing or redesigning tasks with input from the participants who will ultimately perform the tasks increases the endeavor's chances of success.

Alongside the holistic perspective is the acknowledgment that the individual is influenced by what he or she eats and drinks. Athletic performance can be adversely affected by an erratic diet or inadequate hydration. Competitive performance and recovery from strenuous training can be aided by appropriate manipulation of macronutrients in the diet. Nutritional components and supplements can be used to elevate physiological capacities, although the efficacy of many nutritional supplements has not been proven. An array of pharmacological agents are used in occupational and sports settings that benefit performance, whether in strength and power events, endurance exercise, or cognitive function. Those agents deemed to constitute unethical means are listed among the substances banned by the World Anti-Doping Agency, the international institution charged with policing this area. Nevertheless, some of these substances are used for cognitive enhancement in recreational contexts to combat fatigue, maintain arousal in occupational tasks, and promote relaxation during leisure time. Drugs such as anabolic steroids are harmful to health in the doses used by bodybuilders, and alcohol—which is both a drug and a foodstuff—has damaging effects on health when used excessively and can be addictive.

Ergonomists may be criticized for their insistence on finding solutions that work rather than determining the mechanisms by which solutions work. Another critical comment is that ergonomists are prepared to accept small changes or trends that have practical significance rather than the statistical significance required in experimental designs. This difference in philosophy reflects the fact that sport contests are often determined by very small margins in a group that is homogeneous in ability and that those operating at elite level have already gone through a highly selective process.

The training of ergonomists is mainly focused on the occupational domain, although sport and leisure contexts are considered valid areas of application. Scientists with ergonomic expertise are often recruited to solve problems in sports, particularly those relating to complex technologies. Sport science support groups may be formed by specialists but an interdisciplinary mindset is essential to produce the best work.

Introduction

Atha, J. (1984). Current techniques for measuring motion. *Applied Ergonomics,* 15, 245-257.

Atkinson, G., Davison, R., Jeukendrup, A., and Passfield, L. (2003). Science and cycling: current knowledge and future directions for research. *Journal of Sports Sciences,* 21, 767-787.

Atkinson, G., and Reilly, T. (1995). *Sport, Leisure and Ergonomics*. London: Spon.

Bartlett, F.C. (1943). Fatigue following highly skilled work. *Proceedings of the Royal Society B,* 131, 247-254.

Buskirk, B.R., and Tipton, C.M. (1997). Exercise physiology. In: *The History of Exercise and Sport Science* (edited by J.D. Massengale and R.A. Swanson), pp. 367-368. Champaign, IL: Human Kinetics.

Chaffin, D. (1975). Ergonomics guide to assessment of static strength. *American Industrial Hygiene Association Journal,* 36, 505-511.

Clarys, J. (1985). Hydrodynamics and electromyography: ergonomics aspects in aquatics. *Applied Ergonomics,* 16, 11-24.

Coombes, K. (1983). *Proceedings of the Ergonomics Society: Annual Conference.* London and New York: Taylor & Francis.

Corlett, E.N., and Bishop, R.P. (1976). A technique for assessing postural discomfort. *Ergonomics,* 19, 175-182.

Costill, D.L. (1972). The physiology of marathon running. *Journal of the American Medical Association,* 22, 1024-1029.

Davison, R.C.R., Jobson, S., de Koning, J., and Balmer, J. (2008). The science of time-trial cycling. In: *Science and Sports* (edited by T. Reilly), pp. 77-93. Maastricht, The Netherlands: Shaker.

Davids, K., Smith, L., and Martin, R. (1991). Controlling systems uncertainty in sport and work. *Applied Ergonomics,* 22, 312-315.

Floyd, W.F., and Welford, A.T. (eds) (1953). *Symposium of Fatigue*. London: Taylor & Francis.

Grandjean, E. (1969). *Fitting the Task to the Man*. London: Taylor & Francis.

Hawley, J.R., Tipton, K.D., and Millard-Stafford, M.L. (2006). Promoting training adaptations through nutritional strategies. *Journal of Sports Sciences,* 24, 709-721.

Horvath, S.M., and Horvath, E.C. (1973). *The Harvard Fatigue Laboratory: Its History and Contributions*. Englewood Cliffs, NJ: Prentice Hall.

Impellizzeri, F.M., Marcora, S.M., Castagna, C., Reilly, T., Sassi, A., Iaia, F.M., and Rampinini, E. (2006). Physiological and performance effects of generic versus specific aerobic training in soccer players. *International Journal of Sports Medicine,* 27, 483-492.

Karhu, O., Harknnen, R., Sorvali, P., and Vepsalainen, P. (1981). Observing working postures in industry: examples of OWAS application. *Applied Ergonomics,* 12, 13-17.

Keller, J.B. (1976). A theory of competitive running. *Physics Today*, September, 43-67.

Klissouras, V. (1971). Heritability of adaptive variation. *Journal of Applied Physiology,* 31, 338-344.

Klissouras, V. (1976). Prediction of athletic performance: genetic considerations. *Canadian Journal of Applied Sport Sciences,* 1, 195-200.

Lees, A. (1985). Computers in sport. *Applied Ergonomics,* 16, 3-10.

Lees, A., Vanrenterghem, J., Barton, G., and Lake, M. (2007). Kinematic response characteristics of the CAREN moving platform system for use in positive and balance research. *Medical Engineering & Physics,* 29, 629-635.

Magnusson, P., and Renstrom, P. (2006). The European College of Sports Sciences Position Statement: the role of stretching exercises in sports. *European Journal of Sport Sciences,* 6, 87-91.

Mattila, M.K. (2001). OWAS: A method for analysis of working postures. In: *International Encyclopedias of Ergonomics and Human Factors,* Vol. III (edited by H. Karwowski), pp. 1880-1883. London: Taylor & Francis.

McAtamney, L., and Corlett, E.N. (1993). RULA: a survey method for the investigation of work-related upper-limb disorders. *Applied Ergonomics,* 24, 91-99.

McCormick, E.J. (1976). *Human Factors Engineering and Design.* New York: McGraw-Hill.

Meeusen, R., Duclos, M., Gleeson, M., Rietjiens, G., Steinacker, J. and Urhausen, A. (2006). Prevention, diagnosis and treatment of the overtraining syndrome. *European Journal of Sport Science,* 6, 1-14.

Morgan, W.P., and Pollock, M.L. (1977). Psychological characteristics of the elite distance runner. In: *The Marathon: Physiological, Medical, Epidemiological and Psychological Studies* (edited by P. Milvy), pp. 382-403. New York: New York Academy of Sciences.

NIOSH. (1977). *Preemployment Strength Testing.* Washington, DC: U.S. Department of Health and Human Services.

Ortega, C., and Ferrara, M.S. (2008). Athletic training and therapy. In: *Directory of Sport Science,* 5th edition (edited by J. Borms), pp. 369-381. Berlin: International Council of Sport Science and Physical Education.

Pheasant, S.T. (1986). *Bodyspace: Anthropometry, Ergonomics and Design.* London: Taylor & Francis.

Pheasant, S.T. (1991). *Ergonomics, Work and Health.* Basingstoke, UK: Macmillan.

Reilly, T. (1984). Ergonomics in sport: An overview. *Applied Ergonomics,* 15, 243-244.

Reilly, T. (1991a). Physical fitness: For whom and for what? In: *Sport for All* (edited by P. Oja and R. Telama), pp. 81-88. Amsterdam: Elsevier.

Reilly, T. (1991b). Ergonomics and sport. Applied Ergonomics, 22, 290.

Reilly, T. (1992). *Strategic Directions for Sports Science Research in the United Kingdom.* London: Sports Council.

Reilly, T. (2007). *The Science of Training: Soccer.* London: Routledge.

Reilly, T., and Atkinson, G. (2009). *Contemporary Sport, Leisure and Ergonomics.* London: Routledge.

Reilly, T., Atkinson, G., Edwards, B., Waterhouse, J., Akerstedt, T., Davenne, D., Lemmer, B., and Wirz-Justice, A. (2007). Coping with jet-lag: A position statement for the European College of Sport Science. *European Journal of Sport Science,* 7, 1-7.

Reilly, T., and Greeves, J. (2002). *Advances in Sport, Leisure and Ergonomics.* London: Taylor & Francis.

Reilly, T., and Lees, A. (2009). Sports ergonomics. In: *Encyclopaedia of Sports Medicine: The Olympic Textbook of Science in Sport* (edited by R.J. Maughan), pp. 230-247. Oxford, UK: Blackwell.

Snook, S.H., and Ciriello, B.M. (1974). Maximum weights and work loads acceptable to female workers. *Journal of Occupational Medicine,* 16, 527-534.

Snook, S.H., Irvine, C.H., and Bass, S.F. (1970). Maximum weights and work loads acceptable to male industrial workers. *American Industrial Hygiene Association Journal,* 31, 579-586.

Chapter 1

Baltzopoulas, V., and Gleeson, N.P. (2001). Skeletal muscle function. In: *Kinanthropometry and Exercise Physiology Laboratory Manual: Tests, Procedures and Data,* Vol. 2, 2nd edition (edited by R. Eston and T. Reilly), pp. 7-35. London: Routledge.

Bangsbo, J. (1994). The physiology of soccer—with special reference to intense intermittent exercise. *Acta Physiologica Scandinavica,* 151 (Suppl. 619), 1-155.

Bar-Or, O. (1987). The Wingate anaerobic test: An update on methodology, reliability and validity. *Sports Medicine,* 4, 381-394.

Boocock, M.G., Jackson, J.A., Burton, A.K., and Tillotson, K.M. (1994). Continuous measurement of lumbar posture using flexible electrogoniometers. *Ergonomics,* 37, 175-185.

Buzeck, F.L., and Cavanagh, P.R. (1990). Stance phase knee and ankle kinematics and kinetics during level and downhill running. *Medicine and Science in Sports and Exercise,* 22, 669-677.

Cavanagh, P.R., and Lafortune, M.A. (1980). Ground reaction forces in distance running. *Journal of Biomechanics,* 13, 397-406.

Chaffin, D.B. (1975). Ergonomics guide for the assessment of human static strength. *American Industrial Hygiene Association Journal,* 36, 505-511.

Christensen, E.H. (1953). Physiological valuation of work in the Nykroppa Iron Works. In: *Symposium on Fatigue* (edited by W.F. Floyd and A.T. Welford), pp. 93-108. London: H.K. Lewis.

Clarke, H.H. (1967). *Application of Measurement to Health and Physical Education.* Englewood Cliffs, NJ: Prentice Hall.

Clarys, J.P., and Cabri, J. (1993). Electromyography and the study of sports movements: A review. *Journal of Sports Sciences,* 11, 379-448.

Clarys, J.P., Martin, A.D., and Drinkwater, D.T. (1987). The skinfold: Myth and reality. *Journal of Sports Sciences,* 5, 3-33.

Coldwells, A., Atkinson, G., and Reilly, T. (1994). Sources of variation in back and leg dynamometry. *Ergonomics,* 37, 79-86.

Duquet, W., and Carter, J.E.L. (2001). Somatotyping. In: *Kinanthropometry and Exercise Physiology Laboratory Manual: Tests, Procedures and Data, Vol. 1: Anthropometry,* 2nd edition (edited by R. Eston and T. Reilly), pp. 47-64. London: Routledge.

Durnin, J.V.G.A., and Womersley, J. (1974). Body fat assessed from total body density and its estimation from skinfold thickness: Measurements on 481 men and women aged from 16 to 72 years. *British Journal of Nutrition,* 32, 77-97.

Egan, E., Reilly, T., Chantler, P., and Lawlor, J. (2006). Body composition before and after six weeks pre-season training in professional football players. In: *Kinanthropometry IX: Proceedings of the 9th International Conference of the International Society for Advancement of Kinanthropometry* (edited by M. Marfell-Jones, A. Stewart, and T. Olds), pp. 123-130. London: Routledge.

Ekstrand, J. (1982). Soccer injuries and their prevention. Medical dissertation no. 130, Linköping University, Linköping, Sweden.

Eston, R., and Reilly, T. (2001). *Kinanthropometry and Exercise Physiology Laboratory Manual: Tests, Procedures and Data, Vol. 1: Anthropometry,* 2nd edition. London: Routledge.

Eysenck, M.W., and Keane M.T. (2001). *Cognitive Psychology: A Students' Handbook,* 4th edition. Hove, UK: Psychology Press.

Geil, M.D. (2002). The role of footwear on kinematics and plantar foot pressure in fencing. *Journal of Applied Biomechanics,* 18, 155-162.

Gleeson, M. (2006). Immune system adaptation in elite athletes. *Current Opinion in Clinical Nutrition and Metabolic Care,* 9, 659-665.

Gleeson, M., Blannin, A., and Walsh, N.P. (1997) Overtraining, immunosuppression, exercise-induced muscle damage and anti-inflammatory drugs. In: The *Clinical Pharmacology of Sport and Exercise* (edited by T. Reilly and M. Orme), pp 47-57. Amsterdam, Elsevier.

Hay, J.G. (1992). The biomechanics of the triple jump: A review. *Journal of Sports Sciences,* 10, 343-378.

Heymsfield, S.B., Lohnman, T.G., Wang, Z., and Going, S.B. (2005). *Human Body Composition*, 2nd edition. Champaign, IL: Human Kinetics.

Horita, T., Komi, P.V., Nicol, C., and Kyrolainen, H. (2002). Interaction between pre-landing activities and stiffness regulation of the knee joint musculoskeletal system in the drop jump: Implications to performance. *European Journal of Applied Physiology*, 88, 76-84.

Hughes, M., Doherty, M., Jones, R., Reilly, T., Cable, N.T., and Tong, R. (2006). Reliability of repeated sprint exercise in non-motorised treadmill ergometry. *International Journal of Sports Medicine*, 27, 900-904.

Iga, J., Reilly, T., Lees, A., and George, K. (2005). Bilateral isokinetic knee strength profiles in trained junior soccer players and untrained individuals. In: *Science and Football V* (edited by T. Reilly, J. Cabri, and D. Araujo), pp. 442-447. London: Routledge.

Jacobs, I. (1986). Blood lactate: implications for training and performance. *Sports Medicine*, 3, 10-25.

Kawakami, Y., Nozaki, D., Matsuo, A., and Fukunaga, T. (1992). Reliability of measurement of oxygen uptake by a portable telemetric system. *European Journal of Applied Physiology*, 65, 409-414.

Kyrolainen, H., Avela, J., and Komi, P. (2005). Changes in muscle activity with increasing running speed. *Journal of Sports Sciences*, 23, 1101-1109.

Lake, M. (2000). Determining the protective function of sports footwear. *Ergonomics*, 43, 1610-1621.

Lees, A., Vanrenterghem, J., and de Clercq, D. (2004). Understanding how an arm swing enhances performance in the vertical jump. *Journal of Biomechanics*, 37,1929-1940.

Leger, L.A., and Lambert, J. (1982). A maximal multi-stage 20 m shuttle run test to predict $\dot{V}O_2$max. *European Journal of Applied Physiology*, 49, 1-5.

Margaria, R., Aghemo, P., and Rovelli, E. (1966). Measurement of muscular power (anaerobic) in man. *Journal of Applied Physiology*, 21, 1661-1664.

Martin, A.D., Spenst, L.F., Drinkwater, D.T., and Clarys, J.P. (1990). Anthropometric estimation of muscle mass in men. *Medicine and Science in Sports and Exercise*, 22, 729-733.

Matiegka, J. (1921). The testing of physical efficiency. *American Journal of Physical Anthropometry*, 4, 223-230.

McArdle, W.D., Katch, F.I., and Katch, V.L. (1991). Exer*cise Physiology: Energy, Nutrition and Human Performance*. Malvern, PA: Lea & Febiger.

Morris, A.D., Kemp, G.J., Lees, A., and Frostick, S.P. (1998). A study of the reproducibility of three different normalisation methods in intra-muscular dual fine wire electromyography of the shoulder. *Journal of Electromyography and Kinesiology*, 8, 317-322.

NIOSH. (1977). *Preemployment Strength Testing*. Washington, DC: U.S. Department of Health and Human Services.

Rahnama, N., Lees, A., and Reilly, T. (2006). Electromyography of selected lower-limb muscles fatigued by exercise at the intensity of soccer match-play. *Journal of Electromyography and Kinesiology*, 11, 257-263.

Rahnama, N., Reilly, T., Lees, A., and Graham-Smith, P. (2003). A comparison of musculoskeletal function in elite and sub-elite English soccer players. In: *Kinanthropometry VIII* (edited by T. Reilly and M. Marfell-Jones), pp. 155-164. London: Routledge.

Reilly (1981). *Sports Fitness and Sports Injuries*. London: Faber & Faber

Reilly, T. (1983). The energy cost and mechanical efficiency of circuit weight-training. *Journal of Human Movement Studies*, 9, 39-45.

Reilly, T. (1991). Physical fitness—for when and for what? In: *Sport for All* (edited by P. Oja and R. Telama), pp. 81-88. Amsterdam: Elsevier.

Reilly, T. (2001). Assessment of performance in team sports. In: *Kinanthropometry and Exercise Physiology Laboratory Manual: Tests, Procedures and Data* (edited by R. Eston and T. Reilly), Vol. 1, pp. 171-182. London: Routledge.

Reilly, T. (2002). Introduction to musculoskeletal diseases: The Biomed IV Project. In: *Musculoskeletal Disorders in Health-Related Occupations* (edited by T. Reilly), pp. 1-6. Amsterdam: IOS Press.

Reilly, T. (2003). Science and football: A history and an update. In: *Science and Football V* (edited by T. Reilly, J. Cabri, and D. Araújo), pp. 3-12. London: Routledge.

Reilly, T. (2007). *The Science of Training—Soccer: A Scientific Approach to Developing Strength, Speed and Endurance*. London: Routledge.

Reilly, T., and Doran, D. (2003). Fitness assessment. In: *Science and Football*, 2nd edition (edited by T. Reilly and A.M. Williams), pp. 21-46. London: Routledge.

Reilly, T., and Smith, D. (1986). Effect of work intensity on performance in a psychomotor task during exercise. *Ergonomics*, 29, 601-606.

Reilly, T., Secher, N., Snell, P., and Williams, C. (1990). *Physiology of Sports*. London: E. & F.N. Spon.

Reilly, T., Maughan, R.J., and Hardy, L. (1998). Body fat consensus statement of the Steering Groups of the British Olympic Association. *Sports, Exercise and Injury*, 2, 46-49.

Reilly, T., Williams, A.M., Nevill, A., and Franks, A. (2000). A multidisciplinary approach to talent identification in soccer. *Journal of Sports Sciences*, 18, 695-702.

Robinson, M., Lees, A., and Barton, G. (2005). An electromyographic investigation of abdominal exercises and the effects of fatigue. *Ergonomics*, 48, 1604-1612.

Sassi, R., Reilly, T., and Impellizeri, F. (2005). A comparison of small-sided games and interval training in elite professional soccer players. In: *Science and Football V* (edited by T. Reilly, J. Cabri, and D. Araujo), pp. 341-343. London: Routledge.

Togari, H., and Takahashi, K. (1977). Study of "whole-body reaction" in soccer players. *Proceedings of the Department of Physical Education* (College of General Education, University of Tokyo), 6, 33-38.

Wheeler, J.B., Gregor, R.J., and Broker, J.P. (1992). A duel piezoelectric bicycle pedal with multipleshoe/pedal interface capability. *International Journal of Sport Biomechanics*, 8, 251-258.

Williams, A.M., and Reilly, T. (2000). Talent identification and development in soccer. *Journal of Sports Sciences*, 18, 657-667.

Winter, E.M., Jones, A.M., Davison, R., Bromley, P.D., and Mercer, T. (2006). *Sport and Exercise Physiology Testing Guidelines*. London: Routledge.

Chapter 2

ACSM. (1978). Position statement on the recommended quantity and quality of exercise for developing and maintaining fitness in healthy adults. *Medicine and Science in Sports and Exercise*, 10, vii.

ACSM. (1990). The recommended quantity and quality of exercise for developing and maintaining cardiorespiratory and muscular fitness in healthy adults. *Medicine and Science in Sports and Exercise*, 22, 265-274.

ACSM. (1998). The recommended quantity and quality of exercise for developing and maintaining cardiorespiratory and muscular fitness, and flexibility in healthy adults. *Medicine and Science in Sports and Exercise*, 30, 975-991.

Corrigan, B., and Maitland, G.D. (1994). *Musculoskeletal and Sports Injuries*. Oxford, UK: Butterworth-Heinemann.

Dimitrou, L., Sharp, N.C.C., and Doherty, M. (2002). Circadian effects on the immune responses of salivary cortisol and IgA in well trained swimmers. *British Journal of Sports Medicine*, 36, 260-264.

Dunbar, M.J., Stanish, W.D., and Vincent, N.E. (1998). Chronic exertional compartment syndrome. In: *Oxford Textbook of Sports Medicine*, 2nd edition (edited by M. Harries, C. Williams, W.D. Stanish, and L.J. Micheli), pp. 670-678. Oxford, UK: Oxford University Press.

Dunning, E., Murphy, P., and Williams, J. (1988). Why "core" soccer hooligans fight: Aspects of a sociological diagnosis. In: *Science and Football* (edited by T. Reilly, A. Lees, K. Davids, and W.J. Murphy), pp. 561-571. London: Spon.

Dvorak, J., Junge, A., Chomiak, J., Graf-Baumann, T., Peterson, L., Rösch, D., and Hodgson, R. (2000). Risk factor analysis for injuries in football players: possibilities for a prevention programme. *American Journal of Sports Medicine*, 28, 69-74.

Edwards, M. (1991). Airshow disaster plans. *Aviation, Space and Environmental Medicine*, 62, 1192-1195.

Ekstrand, J. (1982). Soccer injuries and their prevention. Medical dissertation no. 130, Linköping University, Linköping, Sweden.

Ekstrand, J., and Gillqvist, J. (1982). The frequency of muscle tightness and injuries in soccer players. *American Journal of Sports Medicine*, 10, 75-78.

Fuller, C.W. (2007). Managing the risk of injury in sport. *Clinical Journal of Sports Medicine*, 17, 182-187.

Fuller, C.W., Ekstrand, J., Junge, A., Anderson, T.E., Dvorak, J., Hagglund, M., McCrory, P., and Meeuwisse, W.H. (2006). Consensus sealement on injury definitions and data collection procedures in studies of football (soccer) injuries. *British Journal of Sports Medicine*, 40, 193-201.

Garbutt, G., Boocock, M.G., and Reilly, T. (1988). Injuries and training patters in recreational marathon runners. In: *Proceedings of the Eighth Middle East Sport Science Symposium* (edited by A. Brien), pp. 56-62. Bahrain: The General Organisation for Youth and Sport.

Gleeson, M. (2006). Immune system adaptation in elite athletes. *Current Opinions in Clinical Nutrition and Metabolic Care*, 9, 659-665.

Hawkins, R.D., Hulse, M.A., Wilkinson, C., Hobson, A. and Gibson, M. (2001). The association football medical research programme: an audit of injuries in professional football. *British Journal of Sports Medicine*, 35, 43-47.

Kordich, J.A. (2004). Client consultation and health approval. In: *NSCA's Essentials of Personal Training* (edited by R.W. Earle and T.R. Baechle), pp. 161-192. Champaign, IL: Human Kinetics.

Leighton, D. and Beynon, C. (2002). The identification and measurement of risk. In: Musculoskeletal Disorders in Health Realted Occupations (edited by T. Reilly), pp. 7-24. Amsterdam: IOS Press.

Lysens, R.J., Ostyn, M.S., Vanden Auweele, Y., Lefevre, J., Vuylsteke, M., and Renson, L. (1989). The accident-prone and overuse-prone profiles of the young athlete. *American Journal of Sports Medicine*, 17, 612-619.

Mansfield, M.J., and Maeda, S. (2007). The apparent mass of the seated human exposed to single-axis and multi-axis whole-body vibration. *Journal of Biomechanics*, 40, 2543-2551

Nieman, D., and Bishop, N.C. (2006). Nutritional strategies to counter stress in the immune system in athletes, with special reference to football. *Journal of Sports Sciences*, 24, 763-772.

NIOSH. (1977). *Preemployment Strength Testing*. Washington. DC: U.S. Department of Health and Human Services.

NIOSH. (1981). *Work Practices Guide for Manual Lifting*. Washington. DC: U.S. Department of Health and Human Services.

NIOSH. (1994). Applications Manual for the Revised NIOSH Lifting Equation. Washington DC: Department of Health and Human Services.

Perrin, D.H. (1993). *Isokinetic Exercise and Assessment*. Champaign, IL: Human Kinetics.

Pheasant, S. (1991). *Ergonomics, Work and Health*. London: Macmillan.

Pollock, M.L., and Wilmore, J.H. (1990). *Exercise in Health and Disease*. Philadelphia: Saunders.

Pollock, M.L., Carroll, J.F., Graves, J.E., Leggett, S.H., Braith, R.W., Limacher, M., and Hagberg, J.M. (1991). Injuries and adherence to walking and resistance programs in the elderly. *Medicine and Science in Sports and Exercise*, 23, 1194-1200.

Rahnama, N., Reilly, T., and Lees, A. (2002). Injury risk associated with playing actions during competitive soccer. *British Journal of Sports Medicine*, 36, 354-359.

Reilly, T. (1981). *Sports Fitness and Sports Injuries*. London: Faber & Faber.

Reilly, T. (2005). Alcohol, anti-anxiety drugs and sport. In: *Drugs in Sport* (edited by D.R Mottram), pp. 258-287. London: Routledge.

Reilly, T., and Ekblom, B. (2005). The use of recovery methods post-exercise. *Journal of Sports Sciences*, 23, 617-627

Reilly, T., and Stirling, A. (1993). Flexibility, warm-up and injuries in mature games players. In: *Kinanthropometry IV* (edited by W. Duquet and J.A.P. Day), pp. 119-123. London: Spon.

Reilly, T., Lees, A., MacLaren, D., and Sanderson, F.H. (1985). Thrill and anxiety in adventure leisure parks. In: *Contemporary Ergonomics 1985* (edited by D.J. Oborne), pp. 210-214. London: Taylor & Francis.

Sanderson, F. (2003). Psychology and injury in soccer. In: *Science and Soccer* (edited by T. Reilly and A.M. Williams), pp. 148-164. London: Routledge.

Sari-Sarraf, V., Reilly, T., and Doran, D.A. (2006). Salivary IgA responses to intermittent and continuous exercise. *International Journal of Sports Medicine*, 27, 845-855.

Thomas, P. (1992). Questionnaire development: an examination of the Nordic Musculoskeletal Questionnaire. *Applied Ergonomics*, 23, 197-201.

Wichmann, S., and Martin, D.R. (1992). Exercise excess: Treating patients addicted to fitness. *Physician and Sportsmedicine*, 20, 193-200.

Chapter 3

Almond, C.S., Shin, A.Y., Fortescue, E.B., et al. (2005). Hyponatremia among runners in the Boston Marathon. *New England Journal of Medicine*, 352, 1550-1556.

Anderson, M.J., Cotter, J.D., Garnham, A.P., Casley, D.J., and Febbraio, M.A. (2001). Effect of glycerol-induced hyperhydration on thermoregulation and metabolism during exercise in heat. *International Journal of Sport Nutrition and Exercise Metabolism*, 11, 315-333.

Armstrong, L.E. (2006). Nutritional strategies for football: counteracting heat, cold, high altitude and jet lag. *Journal of Sports Sciences*, 24, 723-740.

Arngrimsson, S.A., Pettit, D.S., Stuceck, M.G., et al. (2004). Cooling vest worn during active warm-up improves 5-km run performance in the heat. *Journal of Applied Physiology*, 96, 1867-1874.

Bangsbo, J., Klausen, K., Bro-Rasmusen, T., and Larson, J., (1988). Physiological responses to acute moderate hypoxia in elite soccer players. In: *Science and Football* (edited by T. Reilly, A. Lees, K. Davids, and W.J. Murphy), pp. 257-264. London: Spon.

Bergh, U., and Ekblom, B. (1979). Effect of muscle temperature on maximal muscle strength and power in human skeletal muscles. *Acta Physiologica Scandinavica*, 107, 33-37.

Castle, P.C., Macdonald, A.L., Philip, A., Webborn, A., Watt, P.W., and Maxwell, N.S. (2006). Precooling leg muscle improves intermittent sprint exercise performance in hot, humid conditions. *Journal of Applied Physiology*, 100, 1377-1384.

Cheung, S.S., and Sleivert, G.G. (2004). Multiple triggers for hyperthermic fatigue and exhaustion. *Exercise and Sport Sciences Reviews,* 32, 100-106.

Cheuvront, S.N., Carter, R., III, Kolka, M.A., et al. (2004). Branched-chain amino acid supplementation and human performance when hypohydrated in the heat. *Journal of Applied Physiology,* 97, 1275-1282.

Coghlan, A. (2007). Dying for some peace and quiet. *New Scientist,* August 25, 6-9.

Drust, B., Cable, N.T., and Reilly, T. (2000). Investigation of the effects of precooling on the physiological responses to soccer-specific intermittent exercise. *European Journal of Applied Physiology,* 81, 11-17.

Drust, B., Rasmussen, P., Mohr, M., Nielsen, B., and Nybo, L. (2005). Elevations in core and muscle temperature impairs repeated sprint performance. *Acta Physiologica Scandinavica,* 183, 181-190.

Ekblom, B. (1986). Applied physiology of soccer. *Sports Medicine,* 3, 50-60.

Febbraio, M. (2001). Alterations in energy metabolism during exercise and heat stress. *Sports Medicine,* 31, 47-59.

Florida-James, G., Donaldson, K., and Stone, V. (2004). Athens 2004: The pollution climate and athletic performance. *Journal of Sports Sciences,* 22, 967-980.

Grahn, D.A., Cao, V.H., and Heller C. (2005). Heat extraction through the palm of one hand improves aerobic exercise endurance in a hot environment. *Journal of Applied Physiology,* 99, 972-978.

Harries, M. (1998). The lung in sport. In: *Oxford Textbook of Sports Medicine* (edited by M. Harries, C. Williams, W.D. Stanish, and L.J. Micheli), pp. 321-326. Oxford, UK: Oxford University Press.

Hasegawa, H., Takatori, T., Komura, T., and Yamasaki, M. (2006a). Combined effects of pre-cooling and water ingestion on thermoregulation and physical capacity during exercise in a hot environment. *Journal of Sports Sciences,* 24, 3-9.

Hasegawa, H., Takatori T., Komura, T., and Yamasaki, M. (2006b). Wearing a cooling jacket during exercise reduces thermal strain and improves endurance exercise performance in a warm environment. *Journal of Strength and Conditioning Research,* 19, 122-128.

Hawkins, L.H., and Barker, T. (1998). Air ions and human performance. *Ergonomics,* 21, 273-278.

Houston, C.S. (1982). Oxygen lack at high altitude: Mountaineering problem. In: *Hypoxia: Man at Altitude* (edited by J.R. Sutton, N.L. Jones, and C.S. Houston), pp. 156-159. New York: Thame Stratton.

Hsu, A.R., Hagobian, T.A., Jacobs, K.A. et al. (2005). Effects of heat removal through the hand on metabolism and performance during cycling exercise in the heat. *Canadian Journal of Applied Physiology,* 30, 87-104.

Inbar, O., Roistein, A., Dlin, R., et al. (1982). The effects of negative air ions on various physiological functions during work in hot environments. *International Journal of Biometerology,* 26, 153-156.

Ingjer, F., and Myhre, K. (1992). Physiological effects of altitude training on elite male cross-country skiers. *Journal of Sports Sciences,* 10, 37-47.

Kirk, P. (1993). Earmuff effectiveness against chainsaw noise over a 12-month period. *Applied Ergonomics,* 24, 279-283.

Krueger, A.P., and Reed, E.J. (1976). Biological impact of small air ions. *Science,* 193, 1209-1213.

Laursen, P.B., Suriano, R., Quod, M.J., et al. (2006). Core temperature and hydration status during an Ironman triathlon. *British Journal of Sports Medicine,* 40, 320-325.

Levine, B. (1997). Training and exercise at high altitudes. In: *Sport, Leisure and Ergonomics* (edited by G. Atkinson and T. Reilly), pp. 74-92. London: Spon.

Low, D., Cable, T., and Purvis, A. (2005a). Exercise thermoregulation and hyperprolactinaemia. *Ergonomics,* 48, 1547-1557.

Low, D., Purvis, A., Reilly, T., and Cable, N.T. (2005b). Prolactin responses to active and passive heating in man. *Experimental Physiology,* 90, 909-917.

MacNee, W., and Donaldson, K., (1999). Particulate air pollution: Injurious and protective mechanisms in the lungs. In: *Air Pollution and Health* (edited by T. Holgate, J. Samet, H. Koren, and R. Maynard), pp. 653-672. London: Academic Press.

Marino, F.E. (2004). Anticipatory regulation and avoidance of catastrophe during exercise-induced hyperthermia. *Comparative Biochemistry and Physiology, Part B,* 139, 561-569.

Martin, P.G., Marino, F.E., Rattey, J., et al. (2004). Reduced voluntary activation of human skeletal muscle during shortening and lengthening contractions in whole body hyperthermia. *Experimental Physiology,* 90, 225-236.

Maruyama, M., Hara, T., Hashimoto, M., et al. (2006). Alterations of calf venous and arterial compliance following acclimation to heat administered at a fixed daily time in humans. *International Journal of Biometeorology,* 50, 269-274.

McCombe, A.W., and Binnington, J. (1994). Hearing loss in Grand Prix motorists: Occupational hazard or sports injury. *British Journal of Sport Medicine,* 28, 35-37.

Mohr, M., Rasmussen, P., Drust, B., Nielsen, B., and Nybo, L. (2006). Environmental heat stress, hyperammonemia and nucleotide metabolism during intermittent exercise. *European Journal of Applied Physiology,* 97, 89-95.

Montain, S.J., Cheuvront, S.N., and Sawka, M.N. (2005). Exercise associated hyponatraemia: quantitative analysis to understand the aetiology. *British Journal of Sports Medicine,* 40, 98-106.

Morrison, S., Sleivert, G.G., and Cheung, S.S. (2004). Passive hyperthermia reduces voluntary activation and isometric force production. *European Journal of Applied Physiology,* 91, 729-736.

Mustafa, K.Y., and Mahmoud, E.D.A., (1979). Evaporative water loss in African soccer players. *Journal of Sports Medicine and Physical Fitness,* 19, 181-183.

Nielsen, B. (1994). Heat stress and acclimation. *Ergonomics,* 37, 49-58.

Nybo, L., and Secher, N.H. (2004). Cerebral perturbations provoked by prolonged exercise. *Progress in Neurobiology,* 72, 223-261.

Pollock, N.W., Godfrey, R.J., and Reilly, T. (1997). Evaluation of field measures of urine concentration. *Medicine and Science in Sports and Exercise,* 29 (Suppl. 5), S261.

Prasher, D. (2000). Noise pollution health effects reduction (NOPHER): A European Commission concerted action workplan. *Noise & Health,* 2, 79-84.

Quod, M.J., Martin, D.S., Laursen, P.B., Gardner, A.S., Halson, S.L., Marino, F.E., Tate, M.P., Mainwaring, D.E., Gore, C.J., and Hahn, A.G. (2008). Practical precooling: effect on cycling time trial performance in warm conditions. *Journal of Sports Sciences,* 26, 1477-1487.

Reilly, T. (2000). Temperature and performance: Heat. In: *ABC of Sports Medicine* (edited by M. Harries, G. McLatchie, C. Williams, and J. King), pp. 68-71. London: BMJ Brooks.

Reilly, T. (2003). Environmental stress. In: *Science and Soccer,* 2nd edition (edited by T. Reilly and A.M. Williams), pp. 165-184. London: Routledge.

Reilly, T., Drust, B., and Gregson, W. (2006). Thermoregulation in elite athletes. Curr*ent Opinion in Clinical Nutrition and Metabolic Care,* 9, 666-671.

Reilly, T., and Lewis, W. (1985). Effects of carbohydrate feeding on mental functions during sustained physical work. In: *Ergonomics International 85* (edited by I.D. Brown, R. Goldsmith, K. Coombes, and M.A. Sinclair), pp. 700-702. London: Taylor & Francis.

Reilly, T., Maughan, R.J., Budgett, R., and Davies, B. (1997). The acclimatisation of international athletes, In: *Contemporary Ergonomics 1997* (edited by S.A. Robertson), pp. 136-140. London: Taylor & Francis.

Reilly, T., and Stevenson, C. (1993). An investigation of the effects of negative air ions on responses to submaximal exercise at different times of day. *Journal of Human Ergology,* 22, 1-9.

Reilly, T., and Waterhouse, J. (2005). *Sport, Exercise and Environmental Physiology*. Edinburgh, UK: Elsevier.

Shirreffs, S.M., Aragon-Vargas, L.F., Chamorro, M., Maughan, R.J., Serratosa, L., and Zachwieja, J.J. (2005). The sweating response of elite professional soccer players to training in the heat. *International Journal of Sports Medicine,* 26, 90-95.

Sovijarvi, A.R.A., Rossel, S., Hyvarin, J., et al. (1979). Effect of air ionization on heart rate and perceived exertion during a bicycle exercise test: A double-blind cross-over study. *European Journal of Applied Physiology,* 41, 285-291.

Todd, G., Butler, J.E., Taylor, J.L., and Gandevia, S.C. (2005). Hyperthermia: A failure of the motor cortex and the muscle. *Journal of Physiology,* 563, 621-631.

Waterhouse, J., Drust, B., Weinert, D., Edwards, B., Gregson, W., Atkinson, G., Kao, S., Aizawa, S. and Reilly, T. (2005). The circadian rhythm of core temperature: Origin and some implications for exercise performance. *Chronobiology International,* 22, 207-225.

Watson, P., Hasegawa, H., Roelands, B., et al. (2005a). Acute dopamine/noradrenaline reuptake inhibition enhances human exercise performance in warm, but not temperate conditions. *Journal of Physiology,* 565, 873-883.

Watson, P., Shirreffs, S.M., and Maughan, R.J. (2005b). Blood-brain barrier integrity may be threatened by exercise in a warm environment. *American Journal of Physiology,* 288, R1689-R1694.

Webster, J., Holland, E.J., Sleivert, G., Laing, R.M., and Niven, B.E. (2005). A light-weight cooling vest enhances performance of athletes in the heat. *Ergonomics,* 48, 821-837.

Chapter 4

Akerstedt, T. (2006). Searching for the countermeasure of night-shift sleepiness. *Sleep,* 29, 19-20.

Bambaeichi, E., Reilly, T., Cable, N.T., and Giacomoni, M. (2005). The influence of time of day and partial sleep loss on muscle strength in eumenorrheic females. *Ergonomics,* 48, 1499-1511.

Bennet, G. (1973). Medical and psychological problems in the 1972 singlehanded transatlantic yacht race. *Lancet,* 2, 747-754.

Bohle, P., and Tilley, A. (1993). Predicting mood change on night shift. *Ergonomics,* 36, 125-134.

Bonnet, M.H. (2006). Acute sleep deprivation. In: *Principles and Practices of Sleep Medicine,* 4th edition (edited by M.H. Kryger, T. Roth, and W.C. Dement), pp. 51-66. New York: Elsevier.

Cajochen, C, Wyatt, J.K., Czeisler, C.A., and Dijk, D.J. (2002). Separation of circadian and wake duration-dependent modulation of EEG activation during wakefulness. *Neuroscience,* 114, 1047-1060.

Carling, C., Williams, A.M., and Reilly, T. (2005). *A Manual of Soccer Match Analysis.* London: Routledge.

Chen, H.I. (1991). Effects of 30-h sleep loss on cardiorespiratory functions at rest and in exercise. *Medicine and Science in Sports and Exercise,* 23, 193-198.

Cirelli, C. (2002). Functional genomic of sleep and circadian rhythm, invited review: How sleep deprivation affects gene expression in the brain-A review of recent findings. *Journal of Applied Physiology,* 9, 394-400.

Drust, B., Waterhouse, J., Atkinson, G., Edwards, B., and Reilly, T. (2005). Circadian rhythms in sports performance-An update. *Chronobiology International*, 22, 21-44.

Edwards, B.J., Atkinson, G., Waterhouse, J., Reilly, T., Godfrey, R., and Budgett, R. (2000). Use of melatonin in recovery from jet-lag following an eastward flight across 10 time-zones. *Ergonomics*, 43, 1501-1513.

Fisher, F.M., de Moreno, C., de Fernandez, L., Berwerth, A., dos Santos, A.M. ,and Bruni Ade, C. (1993). Day- and shiftworkers' leisure time. *Ergonomics*, 3, 43-49.

Froberg, J.C., Karlsson, C.G., Levi, L., and Lidberg, L. (1975). Circadian rhythms of catecholamine excretion, shooting range performance and self-ratings of fatigue during sleep deprivation. *Biological Psychology*, 2, 175-188.

Guezennec, C.Y., Sabatin, P., Legrand, H., and Bigard, A.X. (1994). Physical performance and metabolic changes induced by combined prolonged exercise and different energy intake in humans. *European Journal of Applied Physiology*, 68, 525-530.

Horne, J.A. (1988). *Why We Sleep*. Oxford, UK: Oxford University Press.

Horne, J.A., and Pettit, A.N. (1984). Sleep deprivation and the physiological response to exercise under steady state conditions in untrained subjects. *Sleep*, 7, 168-179.

How, J.M., Foo, S.C., Low, E., Wong, T.M., Vijayan, A., Siew, M.G., and Kanapathy, R. (1994). Effect of sleep deprivation on performance of Naval seamen: I. Total sleep deprivation on performance. *Annuals of the Academy of Medicine*, 23, 669-675.

Jehue, R., Street, D., and Huizenga, R., (1993). Effect of time zone and game time on team performance: National Football League. *Medicine and Science in Sports and Exercise*, 25, 127-131.

Koslowsky, M., and Babkoff, H. (1992). Meta-analysis of the relationship between total sleep deprivation and performance. *Chronobiology International*, 9, 132-136.

Lemmer, B. (2007). The sleep-wake cycle and sleeping pills. *Physiology & Behavior*, 90, 285-293.

Martin, B.J., and Gaddis, G.M. (1981). Exercise after sleep deprivation. *Medicine Science in Sports and Exercise*, 13, 220-223.

Meney, I., Waterhouse, J., Atkinson, G., Reilly, T., and Davenne, D. (1998). The effect of one night's sleep deprivation in temperature, mood and physical performance in subjects with different amounts of habitual activity. *Chronobiology International*, 15, 349-363.

Monk, T.H., and Folkard, S. (1992). *Making Shiftwork Tolerable*. Basingstoke, UK: Taylor & Francis

Petrilli, R.M., Jay, S.M., Dawson, D., and Lamond, N. (2005). The impact of sustained wakefulness and time of day on OSPAT performance. *Industrial Health*, 43, 86-92.

Pilcher, J.J., and Huffcutt, A.I. (1996). Effects of sleep deprivation on performance a meta-analysis. *Sleep*, 19, 318-326.

Plyley, M.J., Shephard, R.J., Davis, G.M., and Goode, R.C. (1987). Sleep deprivation and cardiorespiratory function- influence of intermittent submaximal exercise. *European Journal of Applied Physiology*, 56, 338-344.

Reilly, T. (1993). Science and football: an introduction. In: *Science and Football II* (edited by T. Reilly, J. Clarys, and A. Stibbe), pp. 3-11. London: Spon.

Reilly, T. (2003). Environmental stress. In: *Science and Soccer*, 2nd edition (edited by T. Reilly and A.M. Williams), pp. 165-184. London: Routledge.

Reilly, T. (2007). *Science of Training—Soccer: A Scientific Approach to Developing Strength, Speed and Endurance*. London: Routledge.

Reilly, T. (2009). The body clock and athletic performance. *Biological Rhythm Research*, 40,37-44.

Reilly, T., Atkinson, G., and Budgett, R. (2001). Effect of low-dose temazepam on physiological variables and performance tests following a westerly flight across five time zones. *International Journal of Sports Medicine*, 22, 166-74.

Reilly, T., Atkinson, G., Edwards, B., Waterhouse, J., Farrelly, K., and Fairhurst, E. (2007).Diurnal variation in temperature, mental and physical performance, and tasks specifically related to football (soccer). *Chronobiological International,* 24, 507-519.

Reilly, T., Atkinson, G., and Waterhouse, J. (1997). *Biological Rhythms and Exercise.* Oxford, UK: Oxford University Press.

Reilly, T., and Deykin, T. (1983). Effects of partial sleep loss on subjective states, psychomotor and physical performance tests. *Journal of Human Movement Studies,* 9, 157-170.

Reilly, T., and Edwards, B. (2007). Altered sleep-wake cycles and physical performance in athletes. *Physiology and Behavior,* 90, 274-284.

Reilly, T., and Garrett, R. (1995). Effects of time of day on self-paced performance of prolonged exercise. *Journal of Sports Medicine and Physical Fitness,* 35, 99-102.

Reilly, T., and George, A. (1983). Urinary phenylethamine levels during three days of indoor soccer play. *Journal of Sports Sciences,* 1, 70.

Reilly, T., and Hales, A. (1988). Effects of partial sleep deprivation on performance measures in females. In: *Contemporary Ergonomics* (edited by E.D. McGraw), pp. 509-513. London: Taylor & Francis.

Reilly, T., and Piercy, M. (1994). The effect of partial sleep deprivation on weight-lifting performance. *Ergonomics,* 37, 107-115.

Reilly, T.. and Walsh, T.J. (1981). Physiological, psychological and performance measures during an endurance record for 5-a-side soccer play. *British Journal of Sports Medicine,* 15, 122-128.

Reilly, T., and Waterhouse, J. (2005). *Sport, Exercise and Environmental Physiology.* Edinburgh, UK: Elsevier.

Reilly, T., and Waterhouse, J. (2007). Altered sleep-wake cycles and food intake: The Ramadan model. *Physiology and Behavior,* 90, 219-228.

Reilly, T., Waterhouse, J., and Atkinson, G. (1997b). Ageing, rhythms of physical performance, and adjustment to changes in the sleep-activity cycle. *Occupational and Environmental Medicine,* 54, 812-816.

Reilly, T., Waterhouse, J., Burke, L., and Alonso, J.M. (2007b). Nutrition for travel. *Journal of Sports Sciences,* 25, Suppl. 1, S125-S134.

Reilly, T., Waterhouse, J., and Edwards, B. (2005). Jet lag and air travel: Implications for performance. *Clinics in Sports Medicine,* 24, 367-380.

Rognum, T.O., Vartdal, F., Rodahl, K., Opstad, P.K., Knudsen-Baas, O., Kindt, E., and Withey, W.R. (1986). Physical and mental performance of soldiers on high-and low-energy diets during prolonged heavy exercise combined with sleep deprivation. *Ergonomics,* 29, 859-867.

Sinnerton, S.A., and Reilly, T. (1992). Effects of sleep loss and time of day in swimmers. In: *Biomechanics and Medicine in Swimming: Swimming Science V* (edited by D. MacLaren, T. Reilly, and A. Lees), pp. 399-405. London: Spon.

Smith, R.S., and Reilly, T. (2005). Athletic performance. In: *Sleep Deprivation: Clinical Issues. Pharmacology and Sleep Loss Effects* (edited by C. Kushida), pp. 313-334, New York: Marcel Dekker.

Smith, R.S., Walsh, J., and Dement, W.C. (1998). Sleep deprivation and the Race Across America. *Sleep,* 22, 303.

Stampi, C., Mullington, J., Rivers, M., Campos, J. and Broughton, R. (1990). Ultrashort sleep schedules: sleep architecture and recuperative value of 80-, 50- and 20-min naps. In: Why We Sleep. (edited by J. Horne), pp. 71-74. Bochum: Pontinagel.

Thomas, V., and Reilly, T. (1975). Circulatory, psychological and performance variables during 100 hours of continuous exercise under conditions of controlled energy intake and work output. *Journal of Human Movement Studies,* 1, 149-155.

Waterhouse, J., Edwards, B., Nevill, A., Carvalho, S., Atkinson, G., Buckley, P., Reilly, T., Godfrey, R., and Ramsay, R. (2002). Identifying some determinants of "jet-lag" and its symptoms: A study of athletes and other travellers. *British Journal of Sports Medicine,* 36, 54-60.

Waterhouse, J., Minors, D., Folkard, S., Owens, D., Atkinson, G., Macdonald, I., Reilly, T., Sytnik, N., and Tucker, P. (1998). Light of domestic intensity produces phase shifts of the circadian oscillator in humans. *Neuroscience Letters,* 245, 97-100.

Waterhouse, J., Reilly, T., Atkinson, G., and Edwards, B. (2007). Jet lag: Trends and coping strategies. *Lancet,* 369, 1117-1129.

Chapter 5

Atha, J. (1984). Current techniques for measuring motion. *Applied Ergonomics,* 15, 245-257.

Ball, L.T., Lucas, E.T., Miles, J.N., and Gale, A.G. (2003). Inspection times and the selection task: what do eye movements reveal about relevance effects? *Quarterly Journal of Experimental Psychology,* 56, 1053-1077.

Barton, J.G. (1999). Interpretation of gait data using Kohonen neural networks. *Gait and Posture,* 10, 85-86.

Boocock, M.G., Garbutt, G., Linge, K., Reilly, T., and Troup, J.D.G. (1990). Changes in stature following drop jumping and post exercise gravity inversion. *Medicine and Science in Sports and Exercise,* 22, 385-390.

Bosco, C. (1985). Adaptive response of human skeletal muscle to simulated hypergravity. *Acta Physiologica Scandinavica,* 124, 507-513.

Bosco, C., Cardinale, M., and Tsarpela, O. (1999). Influence of vibration on mechanical power and electromyogram activity in human arm flexor muscles. *European Journal of Applied Physiology,* 79, 306-311.

Brown, S., Haslam, R.A., and Budworth, N. (2001). Participative quality techniques for back pain management. In: *Contemporary Ergonomics 2001* (edited by M. Hanson), pp. 133-138. London: Taylor & Francis.

Cardinale, M., and Wakeling, J. (2005). Whole body vibration exercise: Are vibrations good for you? *British Journal of Sports Medicine,* 39, 585-589.

Carling, C., Williams, A.M., and Reilly, T. (2005). *Handbook of Soccer Match Analysis: A Systematic Approach to Improving Performance.* London: Routledge.

Cochrane, D.J., and Stannard, S.R. (2005). Acute whole body vibration training increases vertical jump and flexibility performance in elite female field hockey players. *British Journal of Sports Medicine,* 39, 860-865.

Davids, K., Araujo, D., and Shuttleworth, R. (2005). Application of dynamic systems theory to football. In: *Science and Football V* (edited by T. Reilly, J. Cabri, and D. Araujo), pp. 537-550. London: Routledge.

Dowzer, C.N., Reilly, T., and Cable, N.T., 1998, Effects of deep and shallow water running on spinal shrinkage. *British Journal of Sports Medicine,* 32, 44-48.

Ekblom, B. (1986). Applied physiology of soccer. *Sports Medicine,* 3, 50-60.

Fitts, P.M. (1951). *Human Engineering for an Effective Air-Navigation and Traffic-Control System.* Washington, DC: National Research Council Committee on Aviation Psychology.

Fowler, N.E., Lees, A., and Reilly, T. (1997). Changes in stature following plyometric deep jump and pendulum exercise. *Ergonomics,* 40, 1279-1286.

Graham-Smith, P., Fell, N., Gilbert, G., Burke, J., and Reilly, T. (2001). Ergonomic evaluation of a weighted vest for power training. In: *Contemporary Ergonomics 2001* (edited by M.A. Hanson), pp. 493-497. London: Taylor & Francis.

Hughes, M. (2003). Notation analysis. In: *Science and Soccer* (edited by T. Reilly and A.M. Williams), pp. 245-264. London: Routledge.

Issurin, V.B. (2005). Vibrations and their application in sport: A review. *Journal of Sports Medicine and Physical Fitness*, 45, 324-336.

Lees, A., Burton, G., and Kershaw, L. (2003). The use of Kohonen neural network analysis to qualitatively analyse technique in soccer kicking. *Journal of Sports Sciences*, 21, 243-244.

McGarry, T., Anderson, D.I., Wallace, S.A., Hughes, M.D., and Franks, I.M. (2002). Sport competition as a dynamical self-organising system. *Journal of Sports Sciences*, 20, 771, 781.

McGarry, T., and Franks, I.M. (1994). A stochastic approach to predicting competitive squash match-play. *Journal of Sports Sciences*, 12, 573-584.

Memmert, D. and Perl, J. (2009). Game creativity analysis using neural networks. *Journal of Sports Sciences*, 27, 139-149.

Mohr, M., Krustrup, P., and Bangsbo, J. (2003). Match performance of high-standard soccer players with special reference to development of fatigue. *Journal of Sports Sciences*, 21, 519-528.

Murray, A., Aitchison, T.C., Ross, G., Sutherland, K., Watt, I., McLean, D. and Grant, S. (2005). The effect of towing a range of relative resistances on sprint performance. *Journal of Sports Sciences*, 23, 927-935.

Reilly, T. (1981). *Sports Fitness and Sports Injuries*. London: Faber & Faber.

Reilly, T. (1991). Physical fitness-for whom and for what? In: *Sport for All* (edited by P. Oja and R. Telama), pp. 81-88. Amsterdam: Elsevier.

Reilly, T. (2007). *The Science of Training-Soccer: A Scientific Approach to Developing Strength, Speed and Endurance.* London: Routledge.

Reilly, T., Cable, N.T. and Dowzer C. (2001). Does deep-water running aid recovery from stretch-shortening cycle exercise? In: *Perspectives and Profiles: 8th Annual Congress of the European College of Sports Science* (edited by J. Mester, G. King, H. Struder, E. Tsolaridis and A. Osterbury) p. 752. Cologne: Sport und Buch Strauss Gmbh.

Reilly, T., Cable, N.L., and Dowzer, C.N. (2002). The effect of a 6-week land- and water-running training programme on aerobic, anaerobic and muscle strength measures. *Journal of Sports Sciences*, 21, 333-334.

Reilly, T., Dowzer, C.N., and Cable, N.T. (2003). The physiology of deep-water running. *Journal of Sports Sciences*, 21, 959-962.

Reilly, T., and Gregson, W. (2006). Special populations: the referee and assistant referee. *Journal of Sports Sciences*, 24, 795-801.

Reilly, T., and Lees, A. (2008). Sports ergonomics. In: *Encyclopaedia of Sports Science and Medicine* (edited by R. Maughan). Oxford, UK: Blackwell.

Reilly, T., and Thomas, V. (1976). A motion analysis of work-rate in different positional roles in professional match-play. *Journal of Human Movement Studies*, 2, 87-97.

Reilly, T., and Thomas, V. (1978). Multi-station equipment for physical training: design and validation of a prototype. *Applied Ergonomics*, 9, 201-206.

Rush, D.G., Edwards, P.A.M., Pountney, D.C., Shelton, J.A., and Williams, D. (1990). The design and development of an AI-assisted interactive video system for the teaching of psychomotor skills. In: *Contemporary Ergonomics 1990* (edited by E.J. Lovesy), pp. 142-148. London: Taylor & Francis.

Saltin, B. (1973). Metabolic fundamentals in exercise. *Medicine and Science in Sport*, 5, 137-146.

Stanton, N.A., Salmon, P.M., Walker, G.H., Baber, C., and Jenkins, D.P. (2005). *Human Factors Methods: A Practical Guide for Engineering and Design.* Aldershot, UK: Ashgate.

Togari, H. and Asami, T. (1972). A study of throw-in training in soccer. Proceedings of the Department of Physical Education, College of General Education, University of Tokyo, 6, 33-38.

Verhoshanski, V. (1969). Perspectives in the improvement of speed-strength for jumpers. *Review of Soviet Physical Education and Sports,* 4, 28-29.

Waterson, P.E., Older Gray, M.T., and Clegg, C.W. (2002). A sociotechnical method for designing work systems. *Human Factors,* 44, 376-391.

Wooding, D.S., Mugglestone, M.D., Purdy, K.J., and Gale, A.G. (2002). Eye movements of large populations: Implementation and performance of an autonomous public eye tracker. *Behavioural Revised Methods, Instruments and Computers,* 34, 509-517.

Chapter 6

Althoff, I., Brinkman, P., Forbin, W., Sandover, J., and Burton, K. (1992). An improved method of stature measurement for quantitative determination of spinal loading. *Spine,* 17, 682-693.

Au, G., Cook, J., and McGill, S.M. (2001). Spinal shrinkage during repetitive controlled torsion, flexion and lateral bend motion. *Ergonomics,* 44, 373-381.

Bangsbo, J. (1994). The physiology of soccer with special reference to intense intermittent exercise. *Acta Physiologica Scandinavica,* 151, Suppl.619,

Boocock, M.G., Garbutt, G., Linge, K., Reilly, T., and Troup, J.D.G. (1990). Changes in stature following drop-jumping and post exercise gravity inversion. *Medicine and Science in Sports and Exercise,* 22, 385-390.

Boocock, M.G., Garbutt, G., Reilly, T., Linge, K., and Troup, J.D.G. (1988). The effects of gravity inversion on exercise-induced spinal loading. *Ergonomics,* 31, 1631-1637.

Borg, G. (1982). Psychophysical basis of perceived exertion, *Medicine and Science in Sports and Exercise,* 14, 377-381

Bourne, N.D., and Reilly, T. (1991). Effect of a weightlifting belt on spinal shrinkage. *British Journal of Sports Medicine,* 25, 209-212.

Cannon, S.R., and James, S.E. (1984). Back pain in athletes. *British Journal of Sports Medicine,* 18, 159-164.

Clarys, J.P., Cabri, J., De Witte, B., Toussaint, H., De Groot, G., Huying, P., and Hollander, P. (1988). Electromyography applied to sport ergonomics. *Ergonomics,* 31, 1605-20

Colombini, D., Occhipinti, E., Grieco, A., and Faccini, M. (1989). Estimation of lumbar disc area by means of anthropometric parameters. *Spine,* 14, 51-55.

Cooke, C. (2001). Metabolic rate and energy balance In: *Kinanthropometry and Exercise Physiology Laboratory Manual: Tests, Procedures and Data,* 2nd edition (edited by R. Eston and T. Reilly), pp. 137-160. London: Routledge.

Corlett, E.N., Eklund, J.A.E., Reilly, T., and Troup, J.D.G. (1987). Assessment of workload from measurement of stature. *Applied Ergonomics,* 18, 65-71.

Creagh, U., Reilly, T., and Lees, A. (1998). Kinematics of running on off-road terrain, *Ergonomics,* 41, 1029-1033.

Dowzer, C.N., Reilly, T., and Cable, N.T. (1998). Effect of deep and shallow water running on spinal shrinkage. *British Journal of Sports Medicine,* 32, 44-48.

Durnin, J.V.G.A. and Passmore, R. (1970). *Energy, Work and Leisure.* London: Heinemann.

Ekblom, B. (1986). Applied physiology of soccer. *Sports Medicine,* 3, 50-60.

Florida-James, G., and Reilly, T. (1995). The physiological demands of Gaelic football. *British Journal of Sports Medicine,* 29, 41-45.

Fowler, N.E., Lees, A., and Reilly, T. (1994). Spinal shrinkage in unloaded and loaded drop-jumping. *Ergonomics,* 37, 133-139

Fowler, N.E., Trzaskoma, Z., Wit, A., Iskra, L., and Lees, A. (1995). The effectiveness of a pendulum swing for the development of leg strength and counter-movement jump performance. *Journal of Sports Sciences,* 13, 101-108.

Garbutt, G., Boocock, M.G., Reilly, T., and Troup, J.D.G. (1989). The effect of running speed on spinal shrinkage. *Journal of Sports Sciences,* 7, 77.

Garbutt, G., Boocock, M.G., Reilly, T., and Troup, J.D.G. (1990). Running speed and spinal shrinkage in runners with and without low back pain. *Medicine and Science in Sports and Exercise,* 22, 769-772.

Gerisch, G., Rutemoller, E., and Weber, K. (1988). Sports medical measurements of performance in soccer. In: *Sciences and Football* (edited by T. Reilly, A. Lees, K. Davids, and W. Murphy), pp. 60-67. London: Spon.

Hall, C.J., and Lane, A.M. (2001). Effects of rapid weight loss on mood and performance among amateur boxers. *British Journal of Sports Medicine,* 35, 350-395.

Helson, W., and Bultynck, J.B. (2004). Physical and cognitive demands on top-class refereeing in association football. *Journal of Sports Sciences,* 22, 179-189.

Hulton, A., Ford, T. and Reilly, T. (2009). The energy cost of soloing a Gaelic football. In: *Science and Football VI* (edited by T. Reilly and F. Korkusuz), pp. 307-313. London: Routledge.

Impellizzeri, F.M., Marcora, S.M., Castagna, C., Reilly, T., and Sassi, A. (2006). Physiological and performance effects of generic versus specific aerobic training in soccer players. *International Journal of Sports Medicine,* 27, 483-492.

Kawakami, Y., Nozaki, D., Matsuo, A., and Fukunaga, T. (1992). Reliability of measurement of oxygen uptake by a portable telemetric system. *European Journal of Applied Physiology,* 65, 409-414.

Kazarian, L.E. (1975). Creep characteristics of the human spinal column. *Orthopaedic Clinics of North America,* 6, 3-18.

Koeller, W., Funke, F., and Hartman, F. (1984). Biomechanical behaviour of human invertebrae discs subjected to long lasting axial loadings. *Biorheology,* 21, 175-186.

Leatt, P., Reilly, T., and Troup, J.D.G. (1985). Unloading the spine. In: *Contemporary Ergonomics 1985* (edited by D. Oborne), pp. 227-232. London: Taylor & Francis.

Leatt, P., Reilly, T., and Troup, J.D.G. (1986). Spinal loading during circuit weight-training and running. *British Journal of Sports Medicine,* 20, 119-124.

Marras, W.S., Lavender, S.A., Leurgens, S.E., Rajulu, S.L., Allread, W.G., and Fathallah, F.A. (1993). The role of dynamic three-dimensional trunk motion in occupationally-related low-back disorders: the effects of workplace factors, trunk position and trunk motion characteristics on risk of injury. *Spine,* 18, 617-628.

Martens, R., and Simon, J.A. (1976). Comparisons of three predictors of state anxiety in competitive situations. *Research Quarterly,* 47, 381-387.

McCrudden, M., and Reilly, T. (1993). A comparison of the punt and the drop-kick. In: *Science and Football* (edited by T. Reilly, J. Clarys, and A. Stibbe), pp. 362-366. London: Spon.

McNair, D., Lorr, M., and Droppleman, L.F. (1992). *Profile of Mood States Manual,* 2nd edition. San Diego, CA: Educational and Industrial Testing Services.

NIOSH. (1994). *Workplace Use of Back Belts: Review and Recommendations.* Washington, DC: National Institute for Occupational Safety and Health.

Parsons, C.A., Atkinson, G., Doggart, L., Lees, A. and Reilly, T. (1994). Evaluation of new mail delivery bag designs. In : *Contemporary Ergonomics 1994* (edited by S.R. Robertson). pp 236-240. London: Taylor and Francis.

Reilly, T. (1979). *What Research Tells the Coach About Soccer*. Reston, VA: American Alliance for Health, Physical Education, Recreation and Dance.

Reilly, T. (2007). *The Science of Training-Soccer: A Scientific Approach to Developing Strength, Speed and Endurance*. London: Routledge. Now cited in the text.

Reilly, T., and Ball, D. (1984). The net physiological cost of dribbling a soccer ball. *Research Quarterly for Exercise and Sport*, 55, 267-71.

Reilly, T., and Bowen, T. (1984). Exertional cost of changes in directional modes of running. *Perceptual and Motor Skills*, 58, 49-50.

Reilly, T., and Chana, D. (1994). Spinal shrinkage in fast bowling. *Ergonomics*, 37, 127-132.

Reilly, T., and Davies, S. (1995). Effects of a weightlifting belt on spinal loading during performance of a dead-lift. In: *Sport, Leisure and Ergonomics* (edited by G. Atkinson and T. Reilly), pp. 136-139. London: Taylor & Francis.

Reilly, T., and Freeman, K.A. (2006). Effects of loading on spinal shrinkage in males of different age groups. *Applied Ergonomics*, 37, 305-310.

Reilly, T., Grant, R., Linge, K., and Troup, J.D.G. (1988). Spinal shrinkage during treadmill running [abstract]. In: *New Horizons in Human Movement*, Vol. III, p. 142. Cheonan, South Korea: Seoul Olympic Scientific Congress Organizing Committee.

Reilly, T., Lees, A., MacLaren, D., and Sandersen, F.H. (1985). Thrill and anxiety in adventure leisure park rides. In: *Contemporary Ergonomics* (edited by D.J. Oborne), pp. 210-214. London: Taylor & Francis.

Reilly, T., and Peden, F. (1989). Investigation of external weight loading in females. *Journal of Human Movement Studies*, 17, 165-172.

Reilly, T., and Seaton, A. (1990). Physiological strain unique to field hockey. *Journal of Sports Medicine and Physical Fitness*, 30, 142-146.

Reilly, T., and Temple, J. (1993). Some ergonomic consequences of playing field hockey. In: *Contemporary Ergonomics* (edited by E.J. Lovesey), pp. 441-444. London: Taylor & Francis.

Reilly, T., Tyrrell, A., and Troup, J.D.G. (1984). Circadian variation in human stature. *Chronobiology International*, 1, 121-126.

Rhode, H.C., and Espersen, T. (1988). Work intensity during soccer training and match-play. In: *Science and Football* (edited by T. Reilly, A. Lees, K. Davids, and W. Murphy), pp. 68-75. London: Spon.

Sanderson, F.H., and Reilly, T. (1983). Trait and state anxiety in male and female cross-country runners. *British Journal of Sports Medicine*, 17, 24-26.

Sassi, R., Reilly, T., and Impellizzeri, F.M. (2005). A comparison of small-sided games and internal training in elite professional soccer players. In: *Science and Football V* (edited by T. Reilly, J. Cabri, and D. Araujo), pp. 341-343. London: Routledge.

Spielberger, C.D., Gorsuch, R.L., and Lushene, R.E. (1970). *Manual for the State-Trait Anxiety Inventory*. Palo Alto, CA: Consulting Psychological Press.

Tyrrell, A.R., Reilly, T., and Troup, J.D.G. (1985). Circadian variation in stature and the effects of spinal loading. *Spine*, 10, 161-164.

Wallace, P., and Reilly, T. (1993). Spinal and metabolic loading during simulations of golf play. *Journal of Sports Sciences*, 11, 511-515.

Wilby, J., Linge, K., Reilly, T., and Troup, J.D.G. (1987). Spinal shrinkage in females: Circadian variation and the effects of circuit weight-training. *Ergonomics*, 30, 47-54.

Chapter 7

Bowles, K-A., Steele, J.R., and Chauncharyakul, R. (2005). Do current sports brassiere designs impede respiratory function? *Medicine and Science in Sports and Exercise*, 37, 1633-1640.

Caplan, N., and Gardner, T. (2007). A mathematical model of the oar blade-water interaction in rowing. *Journal of Sports Sciences,* 25, 1025-10344.

Easterling, K.E. (1993). *Advanced Materials for Sports Equipment.* London: Chapman & Hall.

Chatard, J.C., and Wilson, B. (2008). Effect of fastskin suits on performance, drag and energy cost of swimming. *Medicine and Science in Sports and Exercise,* 40, 1148-1154.

Ekstrand, J., Timpka, T., and Hagglund, M. (2006). Risk of injury in elite football played on artificial turf versus natural grass: a prospective two-cohort study. *British Journal of Sports Medicine,* 40, 975-980.

Habberl, P., and Prokop, L. (1974). Aspects of synthetic tracks. *Biotelemetry,* 1, 171.

Hawes, M.R., Sovak, D., Miyashita, M., Kanq, S-J., Yoshihuku, Y., and Tanaka, S. (1994). Ethnic differences in forefoot shape and the determination of shoe comfort. *Ergonomics,* 37, 187-206.

Hennig, E.M. (2007). Influence of racket properties on injuries and performance in tennis. *Exercise and Sport Sciences Reviews,* 35, 62-62.

Holmer, I., and Elnas, S. (1981). Physiological evaluation of the resistance to evaporation heat transfer by clothing. *Ergonomics,* 24, 63-74.

Lees, A., and Lake, M. (2003). The biomechanics of soccer surfaces and equipment. In: *Science and Soccer* (edited by T. Reilly and A.M. Williams), pp. 120-153. London: Routledge.

Marshall, S.W., Loomis, D.P., Waller, A.E., Chalmers, D.J., Bird, Y.N., Quarrie, K.L., and Feehan, M (2005). Evaluation of protective equipment for prevention of injuries in rugby union. *International Journal of Epidemiology,* 34, 113-118.

McMahon, T.A., and Greene, P.R. (1978). Fast running tracks. *Scientific American,* 239, 112-121.

Mollendorf, J.C., Termin, A.C., Oppenheim, E., and Pendergast, D.R. (2004). Effect of swim suit design on passive drag. *Medicine and Science in Sports and Exercise,* 36, 1029-1035.

Nirschl, R.P. (1974). The etiology and treatment of tennis elbow. *Journal of Sports Medicine,* 2, 308-323

Nolte, V. (2009). Shorter oars are more effective. *Journal of Applied Biomechanics,* 25, 1-8.

O'Donoghue, P., and Liddle, D. (1998). A match analysis of elite tennis strategy for ladies' singles on clay and grass surfaces. In: *Science and Racket Sports II* (edited by A. Lees, I. Maynard, M. Hughes, and T. Reilly), pp. 247-253. London: Spon.

Reilly, T., and Gregson, W. (2006). Special populations: The referee and assistant referee. *Journal of Sports Sciences,* 24, 795-801.

Reilly, T., and Halliday, F. (1985). Influence of alcohol ingestion on tasks related to archery. *Journal of Human Ergology,* 14, 99-104.

Reilly, T., and Waterhouse, J. (2005). *Sport, Exercise and Environmental Physiology.* Amsterdam, The Netherlands: Elsevier.

Richards, C.E., Magin, P.J. and Callister, R. (2009). Is your prescription of distance running shoes evidence based? *British Journal of Sports Medicine,* 43, 159-162.

Sanderson, F. (1981). Injuries in racket sports. In: *Sports Fitness and Sports Injuries* (edited by T. Reilly), pp. 175-182. London: Faber & Faber.

Virmavirta, M., and Komi, P.V. (2001). Ski jumping boots limit effective take-off in ski jumping. *Journal of Sports Sciences,* 19, 961-968.

Wallace, E., Kingston, K., Strangwood, M., and Kenny, I. (2008). Golf science. In: *Science and Sports: Bridging the Gap* (edited by T. Reilly), pp. 94-106. Maastricht, The Netherlands: Shaker.

White, J.L., Scurr, J.C. and Smith, N.A. (2009). The effect of breast support on kinetics during overground running performance. In: *Contemporary Research in Sport, Leisure and Ergonomics* (edited by T. Reilly and G. Atkinson), pp. 201-211. London: Taylor & Francis.

Chapter 8

Aasa, U., Angquist, K.A., and Barnekow-Bergkvist, M. (2008). The effects of a 1-year physical exercise programme on development of fatigue during a simulated ambulance work task. *Ergonomics,* 51, 1179-1194.

Baker, S.J., Grice, J., Roby, L., and Matthews, C. (2000). Cardiorespiratory and thermoregulatory response of working fire-fighter protective clothing in a temperature environment. *Ergonomics,* 43, 1350-1358

Beck, T.J., Ruff, C.B., Schaffer, R.A., Betsinger, K., Trone, D.W., and Brodine, S.K. (2000). *Bone,* 27, 437-444.

Ben-Ezra, V., and Verstraete, R. (1988). Stair climbing: An alternative exercise modality for fire fighters. *Journal of Occupational Medicine,* 30, 103-105.

Bilzon, J.L.J., Scarpello, E.G., Smith, C.V., Ravenhill, N.A., and Rayson, M.R. (2001). Characterisation of the metabolic demands of simulated shipboard Royal Navy fire-fighting taks. *Ergonomics,* 44, 766-780.

Bruce-Low, S.S., Cotterrell, D., and Jones, G.E. (2007). Effect of wearing personal protective clothing and self-contained breathing apparatus on heart rate, temperature and oxygen consumption during stepping exercise and live fire training exercises. *Ergonomics,* 50, 80-98.

Carter, J.M., Rayson, M.P., Wilkinson, D.M., and Blacker, S. (2007). Strategies to combat heat stress during and after firefighting. *Journal of Thermal Biology,* 32, 109-116.

Cline, A.D., Jensen, G.R., and Melby, C.L. (1998). Stress fractures in female army recruits: Implications of bone density, calcium intake and exercise. *Journal of American College of Nutrition,* 17, 128-135.

Cooper, K.H. (1968). A means of assessing maximal oxygen uptake. *Journal of the American Medical Association,* 203, 201-204.

Davis, P., Dotson, C., and Santa Maria, D. (1982). Relationships between simulated fire fighting tasks and physical performance measures. *Medicine and Science in Sports and Exercise,* 14, 67-71.

Dreger, R.W., Jones, R.L., and Petersen, S.R. (2006). Effects of the self-contained breathing apparatus and fire protective clothing on maximal oxygen uptake. *Ergonomics,* 49, 911-920.

Durnin, J.V.G.A., and Passmore, R. (1967). *Energy, Work and Leisure.* London: Henemann.

Elsner, K.L., and Kolkhorst, F.W. (2008). Metabolic demands of simulated firefighting tasks. *Ergonomics,* 51, 1418-1425.

Fothergill, D.M., and Sims, J.R (2002). Aerobic performance of Special Operations Forces personnel after a prolonged submarine deployment. *Ergonomics,* 43, 1489-1500.

Hoffman, R., and Collingwood, T.R. (2005). *Fit for Duty: An Officer's Guide to Total Fitness,* 2nd edition. Champaign, IL: Human Kinetics.

Karvonen, M.J., and Turpeinen, O. (1954). Consumption and selection of food in competitive lumber work. *Journal of Applied Physiology,* 6, 603-612.

Knapik, J.J., Hauret, K.G., and Lange, J.L. (2003). Retention in service of recruits assigned to the army physical fitness test enhancement program in basic combat training. *Military Medicine,* 168, 490-492.

Knapik, J.J., Sharp, M.A., and Canham-Chervak, M. (2001). Risk factors for training-related injuries among men and women in basic combat training. *Medicine and Science in Sports and Exercise,* 33, 946-954.

Knapik, J.J., Sharp, M.A., Darakjy, S., Jones, S.B., Hauret, K.G., and Jones, B.J. (2006). Temporal changes in the physical fitness of US Army recruits. *Sports Medicine,* 36, 613-634.

Lappe, J.M., Stegman, M.R., and Recker, R.R. (2001). The impact of lifestyle factors on stress fractures in female army recruits. *Osteoporosis International,* 12, 35-42.

Lemon, P., and Hermiston, R. (1977a). Physiological profile of professional fire fighters. *Journal of Occupational Medicine,* 19, 337-340.

Lemon, P., and Hermiston, R. (1977b). The human energy cost of fire fighting. *Journal of Occupational Medicine,* 19, 558-562.

Love, R., Johnstone, J., Crawford, J., Tesh, K., Richie, P., Hutchinson, P., and Wetherill, G. (1996). Study of the physiological effects of wearing breathing apparatus. *FRDG Report,* 13.96.

Lundgren, N.P.V (1946). Physiological effects of time schedule on lumber-workers. *Acta Physiologica Scandinavica,* 13 (Suppl. 41),

Morris, J.N., Heady, J., Raffle, P., Roukes, G., and Parks, J. (1953). Coronary heart disease and physical activity at work. *Lancet,* 2, 1053-1057.

O'Connell, E., Thomas, P., Cady, L., and Karwasky. (1986). Energy costs of simulated stair climbing as a job related task in fire fighting. *Journal of Occupational Medicine,* 28, 282-284.

Parsons, C., Atkinson, G., Doggart, L., Lees, A., and Reilly, T. (1994). Evaluation of new mail delivery bag design. In: *Contemporary Ergonomics 1994* (edited by S.A. Robertson), pp. 236-240. London: Taylor & Francis.

Patton, J.F., Bidwell, T.E., Murphy, M.M., Mello, R.P., and Harp, M.E., (1995). Energy cost of wearing chemical protective clothing during progressive treadmill walking. *Aviation, Space and Environmental Medicine,* 66, 238-242.

Porter, J.M., and Gye, D.E. (2002). The prevalence of musculoskeletal troubles among car drivers. *Occupational Medicine,* 52, 4-12.

Puterbaugh, J.S., and Lawyer, C.H. (1983). Cardiovascular effects of an exercise program: a controlled study among firemen. *Journal of Occupational Medicine,* 25, 581-586.

Ramsbottom, R., Brewer, J., and Williams, C. (1998). A progressive shuttle run test to estimate maximal oxygen uptake. *British Journal and Sports Medicine,* 22, 141-144.

Rayson, M., Holliman, D., and Belyavin, A. (2000). Development of physical selection procedures for the British Army. Phase 2: Relationship between physical performance tests and criterion tasks. *Ergonomics,* 43, 73-105.

Reilly, T. (2005). Alcohol, anti-anxiety drugs and sport. In: *Drugs in Sport,* 4th edition (edited by D.R. Mottram), pp. 258-287. London: Routledge.

Reilly, T., Gregson, W., and Barr, D. (2007). *Evaluation of an Ergonomic Intervention to Reduce the Physiological Stress in Fire Fighters. Quarterly Report to the Merseyside Fire Service*. Liverpool, UK: Liverpool John Moores University.

Reilly, T., Igleden, C., Gennser, M., and Tipton, M. (2006b). Occupational fitness standards for beach lifeguards. Phase 2: The development of an easily administered fitness test. *Occupational Medicine,* 56, 12-17.

Reilly, T., and Waterhouse, J. (2005). *Sport, Exercise and Environmental Physiology*. Edinburgh, UK: Elsevier.

Reilly, T., Wooler, A., and Tipton, M. (2006a). Occupational fitness standards for beach lifeguards. Phase 1: The physiological demands of a beach lifeguard. *Occupational Medicine,* 56, 6-11.

Robb, M.J.M., and Mansfields, N.J. (2007). Self-reported musculoskeletal problems amongst professional truck drivers. *Ergonomics,* 50, 814-827.

Scott, A., and Hallas, K. (2006). Slips and trips in the prison service. In: *Contemporary Ergonomics 2006* (edited by P.D. Bust), pp. 531-535. London: Taylor & Francis.

Scott, G.E. (1988). *The Physical Fitness of Firemen, Vol. 2: Research Report Number 33*. London: Home Office Scientific and Development Research Branch.

Shephard, R.J. (1988). Sport, leisure and well-being- an ergonomics perspective. *Ergonomics,* 31, 1501-1517.

Sothmann, M., Saupe, K., and Jasnof, D. (1990). Advancing age and the cardio respiratory stress of fire suppression: determining the minimum standard for aerobic fitness. *Human Performance,* 3, 217-236.

Thomas, V. and Reilly, T. (1974). Effects of compression on human performance and affective states. *British Journal of Sports Medicine,* 8, 188-190.

Chapter 9

Bell, J.F., and Daniels, S. (1990). Are summer born children disadvantaged? *Oxford Review of Education,* 16, 67-80.

Bonen, A., Belcastro, A.N., Ling, W.Y., and Simpson, A.A. (1981). Profiles of selected hormones during menstrual cycles of teenage athletes. *Journal of Applied Physiology,* 50, 545-551.

Dalton, K. (1978). *Once a Month.* London: Fontana.

De Mendoza, S.G., Nuceta, H.J., Salazar, E., Zerpa, A., and Kashyap, M.L. (1979). Plasma lipids and lipoprotein lipase activator property during the menstrual cycle. *Hormone and Metabolic Research,* 11, 696-697.

Dotan, R., and Bar-Or, O. (1983). Load optimisation for the Wingate Anaerobic Test. *European Journal of Applied Physiology,* 51, 407-417.

Drinkwater, B. (1986). *Female Endurance Athletes.* Champaign, IL: Human Kinetics.

Elferink-Gemser, M.T., Visscher, C., Lemmink, K.A.P.M., and Mulder, T.W. (2004). Relation between multidimensional performance characteristics and level of performance in talented youth field hockey players. *Journal of Sports Sciences,* 22, 1053-1063.

Goldspink, D. (2005). Ageing and activity: Their effects on the functional reserve capacities of the heart and vascular smooth and skeletal muscles. *Ergonomics,* 48, 1334-1351.

Goosey, V.L., Campbell, I.G., and Fowler, N.E. (1995). Development of a treadmill test to examine the physiological responses of wheelchair athletes to submaximal exercise. In: *Sport, Leisure and Ergonomics* (edited by G. Atkinson and T. Reilly), pp. 13-18. London: Spon.

Greeves, J.P., Cable, N.T., Reilly, T., and Kingsland, C. (1999). Changes in muscle strength in women following the menopause: A longitudinal assessment of hormone replacement therapy. *Clinical Science,* 97, 79-84.

Hackney, A.C., McClacken-Compton, M.A., and Ainsworth, B. (1994). Substrate responses to submaximal evidence in the midfollicular and midluteal phases of the menstrual cycle. *International Journal of Sport Nutrition,* 4, 299-308.

Helsen, W.F., Van Winckel, J., and Williams, A.M. (2005). The relative age effect in youth soccer across Europe. *Journal of Sports Sciences,* 23, 629-636.

Jones, M.V., Paull, G.C., and Erskine, J. (2002). The impact of a team's aggressive reputation on the decisions of association football referees. *Journal of Sports Sciences,* 20, 991-1000.

Jurowski-Hall, J.E., Jones, M.L, Trews, C.J., and Sutton, J.R. (1981). Effects of menstrual cycle on blood lactate, oxygen delivery and performance during exercise. *Journal of Applied Physiology,* 51, 1493-1499.

Lees, A., and Arthur, S. (1988). An investigation into anaerobic performance of wheelchair athletes. *Ergnomics,* 31, 1529-1537.

Malina, R.M., Bouchard, C., and Bar-Or, O. (2003). *Growth, Maturation and Physical Activity.* Champaign, IL: Human Kinetics.

Matsudo, V.K.R., Rivet, R.E., and Pereira, M.H. (1987). Standard score assessment on physique and performance of Brazilian athletes in a six tiered competitive sports model. *Journal of Sports Sciences,* 5, 49-53.

Möller-Nielsen, J., and Hammar, M. (1989). Women's soccer injuries in relation to the menstrual cycle and oral contraceptive use. *Medicine and Science in Sports and Exercise,* 21, 126-129.

Musch, J., and Grondin, S. (2001). Unequal competition as an impediment to personal development: A review of the relative age effect in sport. *Developmental Reviews,* 21, 147-167.

Nolan, L., and Lees, A. (2007). The influence of lower limb amputation level on the approach in the amputee long jump. *Journal of Sports Sciences,* 25, 393-401.

O'Reilly, A., and Reilly, T. (1990). Effects of the menstrual cycle and response to exercise. In: *Contemporary Ergonomics 90* (edited by E.J. Lovesey), pp. 149-143. London: Taylor & Francis.

Phillips, S.K., Rook, K.M., Siddle, N.C., Bruce, S.A., and Woledge, R.C. (1993). Muscle weakness in women occurs at an earlier age than in men, but strength is preserved by hormonal replacement therapy. *Clinical Science,* 84, 95-98.

Pienaaer, A.E., Spamer, M.J., and Steyner, H.S. (1998). Identifying and developing rugby talent among 10-year old boys: A practical model. *Journal of Sports Sciences,* 16, 691-699.

Reilly, T., and Cartwright, S.A. (1998). Manual handling and lifting during the late stages of pregnancy. In: *Contemporary Ergonomics 1998* (edited by M.A. Hanson), pp. 96-100. London: Taylor & Francis.

Reilly, T., and Gregson, W. (2006). Special populations: The referee and assistant referee. *Journal of Sports Sciences,* 24, 795-801.

Reilly, T., and Rothwell, J. (1988). Adverse effects of overtraining in females. In: *Contemporary Ergonomics 88* (edited by E.D. Megaw), pp. 316-321. Taylor & Francis, London.

Reilly, T., Bangsbo, J., and Franks, A. (2000a). Anthropometric and physiological predispositions for elite soccer. *Journal of Sports Sciences,* 18, 669-683.

Reilly, T., Williams, A.M., Nevill, A., and Franks, A. (2000b). A multidisciplinary approach to talent identification in soccer. *Journal of Sports Sciences,* 18, 695-702.

Reilly, T., Atkinson, G., and Waterhouse, J. (1997). *Biological Rhythms and Exercise*. Oxford, UK: Oxford University Press.

Sherar, L.B., Baxter-Jones, A.D.G., Faulkner, R.A., and Russell, K.W. (2007). Do physical maturity and birth date predict talent in male youth ice hockey players? *Journal of Sports Sciences,* 25, 879-886.

Sinnerton, S., Birch, K., Reilly, T., and McFadyen, I.M (1994). Lifting tasks, perceived exertion and physical activity levels: Their relationship during pregnancy. In: *Contemporary Ergonomics* (edited by S.A. Robertson), pp. 101-105. London: Taylor & Francis.

Sutton, L., Tolfrey, V., Scott, M., Wallace, J., and Reilly, T. (2009). Body composition of female wheelchair athletes. *International Journal of Sports Medicine.*

Tanner, J.M., and Whitehouse, R.H. (1976). Clinical longitudinal standards for height, weight, height velocity, weight velocity, and stages of puberty. *Archives of Disease in Childhood,* 51, 170-181.

Van der Woude, L.H.V., Veeger, H.E.J., and Dallmeijer, A.J. (1995). The ergonomics of wheelchair sports. In: *Sport, Leisure and Ergnomics* (edited by G. Atkinson and T. Reilly), pp. 3-12. London: Spon.

Vincent, J., and Glamser, F.D. (2006). Gender differences in the relative age effect among US Olympic Development Program youth soccer players. *Journal of Sports Sciences,* 24, 423-432.

Williams, A.M., and Reilly, T. (2000). Talent identification and development in soccer. *Journal of Sports Sciences,* 18, 657-667.

Williams, A., Reilly, T., Campbell, I., and Sutherst, J. (1988). Investigation of changes in response to exercise and in mood during pregnancy. *Ergonomics,* 31, 1539-1549.

Wilson, D., Riley, P., and Reilly, T. (2005). Sports science support for the England amputee soccer team. In: *Science and Football V* (edited by T. Reilly, J. Cabri, and D. Araujo), pp. 287-291. London: Routledge.

Winner, S.J., Morgan, C.A., and Evans, J.G. (1989). Perimenopausal risk of falling and incidence of distal forearm fracture. *British Medical Journal*, 298, 1486-1488.

Chapter 10

American College of Sports Medicine. (1998). Position stand: The recommended quantity of exercise for developing and maintaining cardiorespiratory and muscular fitness, and flexibility in healthy adults. *Medicine and Science in Sports and Exercise*, 30, 975-991.

Armstrong, L.E., Herrera-Soto, J.A., Hacker, F.T., Casa, D.J., Kavouras, S.A., and Maresh, C.M. (1998). Urinary indices during dehydration, exercise and rehydration. *International Journal of Sport Nutrition*, 8, 345-355.

Bailey, O.M., Erith, S.J., Griffin, P.J., Dowson, A., Brewer, B.S., Gant, N., and Williams, C. (2007). Influence of cold-water immersion on indices of muscle damage following prolonged intermittent shuttle running. *Journal of Sports Sciences*, 25, 1163-1170.

Bajaj, P., Arendt-Nielsen, L., Madelaine, P., and Svensson, P. (2003). Prophylactic tolperisone for post-exercise muscle soreness causes reduced isometric force-A double-blind randomized crossover control study. *European Journal of Pain*, 7, 407-418.

Banffi, G., Krajewska, M., Melegati, G., and Pettachini, M. (2008). Effects of whole-body cryotherapy on haematological values in athletes. *British Journal of Sports Medicine*, 42, 558.

Barton, G., Lees, A., Lisboa, P., and Attfield, S. (2006). Visualisation of gait data with Kohonen self-organising neural maps. *Gait and Posture*, 24, 46-53.

Beynon, C., and Reilly, T. (2002). Epidemiology of musculoskeletal diseases in a sample of British nurses and physiotherapists. In: *Musculoskeletal Disorders in Health Related Occupations* (edited by T. Reilly), pp. 63-84. Amsterdam: 10S Press.

Cleak, M.J., and Eston, R.E. (1992). Delayed onset muscle soreness: mechanisms and management. *Journal of Sports Sciences*, 10, 325-341.

Coyle, E. F. (1991). Timing and method of increased carbohydrate intake to cope with heavy training, competition and recovery. *Journal of Sports Sciences*, 9, S28-52.

Dawson, B., Cow, S., Modra, S., Bishop, D,M and Stewart, G. (2005). Effect of immediate post-game recovery procedures on muscle soreness, power and flexibility levels over the next 48 hours. *Journal of Science and Medicine in Sport*, 8, 210-221.

Descatha, A., Roquelaure, Y., Chaston, J.F., Evanoff, B., Melchiot, M., Mariot, C., Ha, C., Imbernon, E., Goldberg, M., and Leclerc, A. (2007). Validity of Nordic-style questionnaires in the surveillance of upper-limb work-related musculoskeletal disorders. *Scandinavian Journal of Work and Environmental Health*, 33, 58-65.

Dohm, G.L. (2002). Regulation of skeletal muscle GLUT-4 by exercise. *Journal of Applied Physiology*, 93, 782-787.

Drust, B., Atkinson, G., Gregson, W., French, D., and Binningsley, D. (2003). The effects of massage on intra-muscular temperature in the vastus lateralis in humans. *International Journal of Sports Medicine*, 24, 395-399.

Dugdill, L., Graham, R.C., and McNair, F. (2005). Exercise referral: The public health panacea. A critical perspective of exercise referral schemes: Their development and evaluation. *Ergonomics*, 48, 1390-1410.

Dvorak, J., Junge, A., Chomiak, J., Graf-Braumann, T., Peterson, L., Rosch, D., and Hodgson, R. (2000). Risk factor analysis for injuries in soccer players: possibilities for a prevention programmes. *American Journal of Sports Medicine*, 28, 569-574.

Elliott, B., (2006). Biomechanics and tennis. *British Journal of Sports Medicine,* 40, 392-396.

Ekstrand, J. (1982). Soccer injuries and their prevention. Medical dissertation no. 130, Linköping University, Linköping, Sweden.

Ekstrand, J. (1994). Injury prevention. In: *Football (Soccer)* (edited by B. Ekblom), pp. 209-214, Oxford: Blackwell Scientific.

Eston, R., and Peters, D. (1999). Effects of cold water immersion on the symptoms of exercise-induced muscle damage. *Journal of Sports Sciences,* 17, 231-238.

Fowler, N., and Reilly, T. (1993). Assessment of muscle strength asymmetry in soccer players. In: *Contemporary Ergonomics* (edited by E.J. Lovesey), pp. 327-332. London: Taylor & Francis.

Gleeson, M., Blannen, A.K., and Walsh, W.P. (1997). Overtraining, immunosuppression, exercise-induced muscle damage and anti-inflammatory drugs. In: *The Clinical Pharmacology of Sport and Exercise* (edited by T. Reilly and M. Orme), pp. 47-57. Amsterdam: Excerpts Medicine.

Hawkins, R.D., Hulse, M.A., Wilkinson, C., Hodson, A., and Gibson, M. (2001). The association football medical research programme: An audit of injuries in professional football. *British Journal of Sports Medicine,* 35, 43-47.

Hennig, E., (2007). Influence of racket properties on injuries and performance in tennis. *Exercise and Sport Sciences Reviews,* 35, 62-66.

Hilbert, J.E., Sforzo, G.A., and Svenssen, T. (2003). The effects of massage on delayed-onset muscle soreness. *British Journal of Sports Medicine,* 37, 72-75.

Howarth, P.A., and Hodder, S.G. (2008). Characteristics of habituation to motion in a virtual environment. *Displays, 29,* 117-123.

Howartson, G., and Van Someren, K.A. (2003). Ice massage: Effects on exercise-induced muscle damage. *Journal of Sports Medicine and Physical Fitness,* 43, 500-505.

Jones, A.M., and Doust, J. (2001). Limitations to submaximal exercise performance. In: *Kinanthropometry and Exercise Physiology Laboratory Manual: Test Procedures and Data, Vol. 2: Exercise Physiology* (edited by R. Eston and T. Reilly), pp. 235-262. London: Routledge.

Kim, D.H., Gambardella, R.A., Elattrache, N.S., Yocum, L.A., and Jobe, F.W. (2006). Anthroscopic treatment of posterolateral elbow impingement from lateral synovial plicat in throwing athletes and golfers. *American Journal of Sports Medicine,* 34, 438-444.

La Fontaine, T. (2004). Clients with spinal cord injury, multiple sclerosis, epilepsy and cerebral palsy. In: *NSCA: Essentials of Personal Training* (edited by R.W. Earle and T.R. Baechle), pp. 557-578. Champaign; IL: Human Kinetics.

Lees, A., Vanrenterghem, J., Barton, G., and Lake, M. (2007). Kinematic response characteristics of the CAREN moving platform system for use in posture and balance research. *Medicine and Biology in Engineering and Physics,* 29, 629-635.

Lisboa, P.J.G. (2002). A review of evidence of health benefit from artificial neural networks in medical intervention. *Neural Networks,* 15, 11-39.

Malm, C. (2006). Susceptibility to infections in elite athletes: The S-curve. *Scandinavian Journal of Medicine and Science in Sports,* 16, 4-6.

Malm, C., Svensson, M., Ekbolm, B., and Sjodin, B. (1997). Effect of ubiquinone-10 supplementation and high intensity training on physical performance in humans. *Acta Physiologica Scandinavica,* 161, 379-384.

Maughan, R.J. (1991). Fluid and electrolyte loss and replacement in exercise. *Journal of Sports Sciences,* 9 (Special issue), 117-142.

Montgomery, P.G., Pyne, D.B., Hopkins, W.G., Dorman, J.C., Cook, K., and Minahan, C.L. (2008). The effect of recovery strategies on physical performance and cumulative fatigue in competitive basketball. *Journal of Sports Sciences,* 26, 1135-1145.

Nieman, D., and Bishop, N.C. (2006). Nutritional strategies to counter stress to the immune system in athletes, with special reference to football. *Journal of Sports Sciences,* 24, 763-772.

Passos, P., Araujo, D., Davids, K., and Serpa, S. (2006). Interpersonal dynamics in sport: The role of artificial neural networks and 3-D analysis. *Behavior Research Methods,* 38, 683-691.

Perez, M.A., and Nussbaum, M.A. (2008). A neural network model for predicting postures during non-repetitive manual materials handling. *Ergonomics,* 51, 1545-1564.

Pollock, N.W., Godfrey, R., and Reilly, T. (1997). Evaluation of field measures of urine concentration. *Medicine and Science in Sports and Exercise,* 29 (Suppl. 5), S261.

Rasmussen, B.B., Tipton, K.D., Miller, S.L., Wolfe, S.E., and Wolfe, R.R. (2000). An oral essential amino acid-carbohydrate supplement enhances muscle protein anabolism after resistance exercise. *Journal of Applied Physiology,* 88, 386-392.

Reilly, T. (2007). *The Science of Training-Soccer: A Scientific Approach to Developing Strength, Speed and Endurance.* London: Routledge.

Reilly, T., Cable, N.T., and Dowzer, C.N. (2002). The efficacy of deep-water running. In: *Contemporary Ergonomics, 2002* (edited by P.T. McCabe), pp. 162-166. London: Taylor & Francis.

Reilly, T., and Rigby, M. (2002). Effect of an active warm-down following competitive soccer. In: *Science and Football IV* (edited by W. Spinks, T. Reilly, and A. Murphy), pp. 226-229. London: Routledge.

Reilly, T., and Stirling, A. (1993). Flexibility, warm-up and injuries in mature games players. In: *Kinanthropemetry IV* (edited by W. Duquet and J.A.P. Day), pp. 119-123. London: ESpon.

Reilly, T., and Waterhouse, J. (2005). *Sport, Exercise and Environmental Physiology*. Edinburgh, UK: Elsevier.

Shirreffs, S.M. (2000). Markers of hydration status. *Journal of Sports Medicine and Physical Fitness,* 40, 80-84.

Slavotinek, J.P., Vervall, G.M., Fon, G.T., and Sage, M.R. (2005). Groin pain in footballers: The association between preseason clinical and pubic bone magnetic resonance imaging findings and athlete outcome. *American Journal of Sports Medicine,* 33, 844-899.

Thompson, D., Williams, C., Garcia-Rovers, P., McGregson, S.J., McAdle, F., and Jackson, M.J. (2003). Post-exercise vitamin C supplementation and recovery from demanding exercise. *European Journal of Applied Physiology,* 89, 393-400.

Tipton, K.D., and Wolfe, R.R. (2001). Exercise, protein metabolism and muscle growth. *International Journal of Sport Nutrition and Exercise Metabolism,* 11, 109-132.

Weldon, E.J., and Richardson, A.B. (2001). Upper extremity exercise injuries in swimming: A discussion of swimmers shoulder. *Clinics in Sports Medicine,* 20, 423-438.

Chapter 11

Bangsbo, J. (1994). The physiology of soccer-with special reference to intense intermittent exercise. *Acta Physiologica Scandinavica,* 151 (Suppl. 619), 1-155.

Brien, T., and Simon, T.H. (1987). The effects of red blood cell infusion on 10 km race time. *Journal of the American Medical Association,* 257, 2761-2765.

Costill, D.L., Dalsky, G.P., and Fink, W.J. (1978). Effect of caffeine ingestion and metabolism and exercise performance. *Medicine and Science in Sports,* 10, 155-158.

Cox, G.R., Desbrow, B., Montgomery, P.G., Anderson, M.E., Bruce, C.R., Mercedes, T.A., et al. (2002). Effect of different protocols of caffeine intake on metabolism and endurance performance. *Journal of Applied Psychology,* 93, 990-999.

Cuthbertson, D.P., and Knox, J.A.C. (1947). The effect of analeptics on the fatigued subject. *Journal of Physiology,* 106, 42-58.

Derave, W., Eijnde, B.V., and Hespel, P. (2003). Creatine supplementation in health and disease: What is the evidence for long term efficacy. *Molecular and Cellular Biochemistry*, 244, 49-55.

Dimeo, P. (2007). *A History of Drug Use in Sport 1886-1976: Beyond Good and Evil*. London: Routledge.

Doherty, M., and Smith, P.M. (2005). Effects of caffeine ingestion on ratings of perceived exertion during and after exercise: A meta-analysis. *Scandinavian Journal of Medicine and Science in Sports*, 15, 69-78.

Gallese, V., Faliga, L., Fogassi, L., and Rizzolati, G. (1996). Action recognition in the premotor cortex. *Brain*, 119, 593-609.

George, K.P., and MacLaren, D.P. (1988). The effect of induced alkalosis and acidosis on endurance running at an intensity corresponding to 4 mM blood lactate. *Ergonomics*, 31, 1639-1645.

Greenhaff, P.L., Casey, A., Short, A.H., Harris, R., and Soderland, K. (1993). Influence of oral creatine supplementation on muscle torque during repeated bouts of maximal voluntary exercise in man. *Clinical Science*, 84, 565-571.

Haslam, R.A. (2002). Targeting ergonomics interventions—learning from health promotion. *Applied Ergonomics*, 33, 241-249.

Hespel, P., Maughan, R.J., and Greenhaff, P.L. (2006). Dietary supplements for football. *Journal of Sports Sciences*, 24, 749-761.

Hespel, P., Op't Einde, B., Van Leemputte, M., Urso, B., Greenhaff, P., Labarque, V., et al. (2001). Oral creatine supplementation facilitates the rehabilitation of disuse and alters the expression of ergogenic factors in humans. *Journal of Physiology*, 92, 513-518.

Hignet, S., Wilson, J.R., and Morris, W. (2005). Finding ergonomic solutions-participatory approaches. *Occupational Medicine*, 55, 200-207.

House of Commons Science and Technology Committee. (2007). *Human Enhancement Technologies in Sport*. London: The Stationary Office.

Jacobson, T.L., Febbraio, M.A., Arkinstall, M.J., and Hawley, J.A. (2001). Effect of caffeine co-ingested with carbohydrates or fat on metabolism and performance in endurance trained men. *Experimental Physiology*, 86, 137-144.

Jensen, P.L. (1997). Can participatory ergonomics become "the way we do things in this firm": The Scandinavian approach to participatory ergonomics. *Ergonomics*, 40, 1078-1087.

Jeukendrup, A., and Gleeson, M. (2004). *Sport Nutrition: An Introduction to Energy Production and Performance*. Champaign, IL: Human Kinetics.

Junge, A., Rosch, D., Peterson, L., Graf-Bauman, T., and Dvorak, J. (2002). Prevention of soccer injuries: A prospective intervention study in youth amateur players. *American Journal of Sports Medicine*, 30, 652-659

Kogi, K. (2006). Participatory methods effective for ergonomic workplace improvement. *Applied Ergonomics*, 37, 547-554.

MacLaren, D.P.M. (1997). Influence of blood acid-base status on performance. In: *The Clinical Pharmacology of Sport and Exercise* (edited by T. Reilly and M. Orme), pp. 157-165. Amsterdam: Elsevier.

O'Brien, C.P., and Lyons, F. (2000). Alcohol and the athlete. *Sports Medicine*, 29, 295-300.

Parasuraman, R., and Rizzo, M. (2006). *Neuroergonomics: The Brain at Work*. Oxford, UK: Oxford University Press.

Reilly, T. (1988). The marathon: An ergonomics introduction. In: *Proceedings of the Eighth Middle East Sport Science Symposium* (edited by A. Brien), pp. 1-7. Bahrain: The Bahrain Sports Institute.

Reilly, T. (2005). Alcohol, anti-anxiety drugs and alcohol. In: *Drugs in Sport*, 4th edition (edited by D.R. Mottram), pp. 258-287. London: Routledge.

Rivers, W.H.R. (1908). *The Influence of Alcohol and Other Drugs on Fatigue: The Crown Lectures Delivered at the Royal College of Physicians in 1906*. London: Edward Arnold.

Rivilis, I., Cole, D.C., Frazer, M.B., Kerr, M.S., Wells, R.P., and Ibrahim, S. (2006). Evaluation of a participatory ergonomic intervention aimed at improving musculoskeletal health. *American Journal of Industrial Medicine,* 49, 801-810.

Saltin, B. (1973). Metabolic fundamentals in exercise. *Medicine and Science in Sports,* 5, 137-146.

Terjung, R.L., Clarkson, P.M., Eichnel, E.R., Greenhaff, P.L., Hespel, P., Israel, R.G., Kraemer, W.T., Meyer, R.A., Spriett, L.L., Tarnopolsky, M.A., Wagenmakers, A.J., and Williams, M.H. (2000). American College of Sports Medicine Round table. The physiological effects of oral creatine supplementation. *Medicine and Science in Sports and Exercise,* 32, 706-717.

Van Loon, L.J., Murphy, R., Oosterlaarh, A.M., Cameron-Smith, D., Hargreaves, M., Wagenmakers, A.J., Snow, R. (2004). Creatine supplementation increases glycogen storage but not GLUT-4 expression in human skeletal muscle. *Clinical Science (London),* 106, 99-106.

Vink, P., Koningsveld, E.A., and Molenbroek, J.R. (2006). Positive outcomes of participatory ergonomics in terms of greater control and higher productivity. *Applied Ergonomics,* 37, 537-546.

Whysall, Z.J., Haslam, C., and Haslam, R. (2007). Developing the stage of change approach for the reduction of work-related musculoskeletal disorders. *Journal of Health Psychology,* 12, 184-197.

Williams, C., and Serratosa, L. (2006). Nutrition on match day. *Journal of Sports Sciences,* 24, 687-697.

Young, M.S. and Stanton, N.A., (2007). What's skill got to do with it? Vehicle automation and driver mental workload. *Ergonomics,* 50, 1324-1339.

NOTE: Definitions will be found on page numbers in BOLD print.
Page numbers followed by italicized f or t indicate a figure or table on that page, respectively.

Thomas Reilly received his university education at University College Dublin and later gained postgraduate qualifications in physical education (St. Mary's College, Strawberry Hill) and ergonomics (Royal Free Hospital School of Medicine, London University). He taught for two years in the Republic of Cameroon and later for one year at Southall College of Technology before working as a research technician at the National Institute for Medical Research and then toward a PhD in exercise physiology at Liverpool Polytechnic. He was among the team that set up the first BSc (honors) degree in sport science at Liverpool Polyechnic and later the first undergraduate program in science and football. He was a visiting professor at the University of Tsukuba in Japan in 1977 and three years later was an honorary research associate at the University of California at Berkeley. As his home institution changed its name, he became the director of the School of Human Sciences in 1992 and in 1995 the director of the graduate school of Liverpool John Moores University. He became the director of the Research Institute of Sport and Exercise Sciences in 1997, a position he held until his retirement in 2009. He was awarded the degree of DSc for his work on chronobiology and exercise in 1998.

© Tom Reilly

As well as having an active academic career in research, Reilly won championship medals for track and field, cross-country running, handball, and Gaelic football. He also competed in orienteering and soccer. He coached elite athletes, including Olympic medalists, in various countries. He was chair of the Exercise Physiology Steering Committee of the British Olympic Association from 1992 to 2005 and was on the sport science support staff of team Great Britain at the 1996 and 2000 Summer Olympic Games. He acted as chair of the International Steering Group for Science and Football from 1987 to the present and was president of the World Commission for Science and Sports from 1999 to 2009. He was an inaugural member of the European College of Sport Science in 1996. He was a fellow of the European College of Sport Science and the British Association of Sport and Exercise Sciences, a fellow of the Ergonomics Society, and a fellow of the Institute of Biology.

Reilly was the first editor of the *Journal of Sports Sciences* in 1983, remaining as general editor until 1996. He lectured in sport science and ergonomics in over 40 countries and delivered the second public lecture of the Ergonomics Society in 2006. He maintained a busy consulting program, especially with professional soccer teams. He was among the research team that was awarded the President's Medal of the Ergonomics Society in 2008 for its work on chronobiology and ergonomics.

Thomas Reilly passed away on June 11, 2009.